Science and Culture in Traditional Japan

The M.I.T. East Asian Science Series
Nathan Sivin, general editor

1. Ulrich Libbrecht, *Chinese Mathematics in the Thirteenth Century: The Shu-shu chiu-chang of Ch'in Chiu-shao*
2. Shigeru Nakayama and Nathan Sivin (eds.), *Chinese Science: Explorations of an Ancient Tradition*
3. Manfred Porkert, *The Theoretical Foundations of Chinese Medicine: Systems of Correspondence*
4. Sang-woon Jeon, *Science and Technology in Korea: Traditional Instruments and Techniques*
5. Nakayama Shigeru, David L. Swain, and Yagi Eri (eds.), *Science and Society in Modern Japan: Selected Historical Sources*
6. Masayoshi Sugimoto and David L. Swain, *Science and Culture in Traditional Japan; A.D. 600–1854*

Science and Culture in Traditional Japan

A.D. 600–1854

Masayoshi Sugimoto and David L. Swain

The MIT Press
Cambridge, Massachusetts, and London, England

Copyright © 1978 by
The Massachusetts Institute of Technology

All rights reserved. No part of this book may be reproduced in any form or by any means, electronic or mechanical, including photocopying, recording, or by any information storage and retrieval system, without permission in writing from the publisher.

This book was set in Monophoto Baskerville
by Asco Trade Typesetting Limited, Hong Kong,
and printed and bound by Murray Printing Company
in the United States of America.

Library of Congress Cataloging in Publication Data

Sugimoto, Masayoshi, 1928–
  Science and culture in traditional Japan.

  (M.I.T. East Asian science series; 6)
  Bibliography: p.
  Includes index.
  1. Science—Japan—History. 2. Japan—Civilization.
I. Swain, David L., 1927–   joint author. II. Title. III. Series: Massachusetts Institute of Technology. M.I.T. East Asian science series; 6.
Q127.J3S93   509.52   77-16836
ISBN 0-262-19155-5

To Betty and Masa

Contents

Tables ix

The M.I.T. East Asian Science Series xi

Foreword xiii

Introduction xxi

Map of East Asia during Chinese Cultural Wave I xxxv

Map of Japan during Chinese Cultural Wave II xxxvi–xxxvii

1. Science in Japan's First Cultural Transformation 1
Chinese Cultural Wave I: ca. 600–894 1
Learning in Chinese Wave I 29
Science in Chinese Wave I 42

2. Five Centuries of Indigenous Development 103
The Semiseclusion Era: 894–1401 103
Learning 114
The Specific Sciences 121

3. Pressures toward Modern Society 148
Early Chinese Cultural Wave II: 1401–1639 148
Western Cultural Wave I: 1543–1639 156
Transitions in Japanese Culture 161
Techniques Strategic to Modernizing Processes 168
Learning in the Transitional Period 186
The Sciences in Transition 196

4. The Seventeenth-Century Intellectual Outburst 225
National Isolation and the Peak of Chinese Cultural Wave II: 1639–1720 225
Learning in the Seventeenth Century 234
The Sciences 250

5. The Shift from Traditional to Modern Science 291
Challenge to Isolation: 1720–1854 291
Learning and Science 295
Developments in the Sciences 347
Aftermath of Chinese Wave II and Western Wave II 395

Appendix: Chronological Charts 400

Bibliography 418

Index 441

Tables

1. Cultural Waves and Interims in Premodern Japanese History — xxv
2. Phases and Frequency of *Kentōshi* Missions — 15
3. University Curriculum and Reforms — 33–34
4. Curriculum of the Institute of Divination — 35
5. Curriculum of the Institute of Medicine — 37
6. Calendars Adopted for Use in Japan in Chinese Wave I — 51
7. The Ten Celestial Stems — 60
8. The Twelve Terrestrial Branches — 60
9. Sexagenary Cycle Coordinating the Ten Stems and Twelve Branches — 61
10. The Five Phases and Correlated Phenomena in Sets of Five — 61
11. The Twenty-four Fortnightly Periods — 68
12. *Yin-Yang* Correspondences — 92
13. Five Phases Correspondences — 93
14. Learning in the Early Phase of the Semiseclusion Era — 119
15. Early and Late Phases of Chinese Wave II and Western Waves I and II — 148
16. Steps Taken toward Isolation — 162–163
17. Key Mathematical Works Published in Japan in Early Chinese Wave II (1401–1639) — 205
18. Major Isolationist Measures of the Tokugawa Regime — 225
19. Main Events of the Seventeenth-Century Intellectual Outburst — 238
20. Proliferation of Books on Traditional Calendrics in the Seventeenth Century — 253
21. Calendrical Systems Used in Japan: From the *Senmyōreki* to the *Tenpōreki* — 254
22. Succession of Problem-Solving in Seventeenth-Century Japanese Mathematics — 265
23. Early Translations by *Rangaku* Scholars: Pioneering and Influential Works in Various Fields — 330–331

24. Leading Academies for Dutch Learning — 335
25. Prominent Figures in the Seki School of Japanese Mathematics (*Wasan*) and the Succession of Its Masters (*Sōtō*) — 369

Chronological Charts

A.1. Brief Outline of the Sociocultural History of Japan — 400
A.2. Four Major Traditions of Learning in Traditional Japan — 404
A.3. Brief Outline of Astrology and Calendrical Astronomy in Traditional Japan — 406
A.4. Brief Outline of Mathematics in Traditional Japan — 410
A.5. Brief Outline of Medicine in Traditional Japan — 414

The M.I.T. East Asian Science Series

One of the most interesting developments in historical scholarship over the past two decades has been a growing realization of the strength and importance of science and technology in ancient Asian culture. Joseph Needham's monumental exploratory survey, *Science and Civilisation in China*, has brought the Chinese tradition to the attention of educated people throughout the Occident. The level of our understanding is steadily deepening as new investigations are carried out in East Asia, Europe, and the United States.

The publication of general books and monographs in this field, because of its interdisciplinary character, presents special difficulties with which not every publisher is fully prepared to deal. The aim of the M.I.T. East Asian Science Series, under the general editorship of Nathan Sivin, is to identify and make available books which are based on original research in the Oriental sources, and which combine the high methodological standards of Asian studies with those of technical history. This series will also bring special editorial and production skills to bear on the problems which arise when scientific equations and Chinese characters must appear in close proximity, and when ideas from both worlds of discourse are interwoven. Most books in the Series will deal with science and technology before modern times in China and related Far Eastern cultures, but manuscripts concerned with contemporary scientific developments or with the survival and adaptation of traditional techniques in China, Japan, and their neighbors today will also be welcomed.

Foreword

We are just beginning to realize how complex a historical process modernization has been, and still is. The norm has been to think of "modern civilization"—a conglomerate of industrial technology, modern science, and whatever values and institutions happen to be conventional in Europe or the United States at a given time—as a force which replaces "backwardness" with the ineluctability of a gas rushing in to fill a vacuum. More recently, we have noted that societies which did not entirely volunteer to compete on our terms for survival resent being reminded in such an abrupt manner that we are still ahead of them at our own game. Consequently we have at least tried to make the bitter pill more palatable by referring to their condition as "underdevelopment" instead of "backwardness." Still the image in most of our minds remains that of development irresistibly supplanting underdevelopment.

The issues of modernization become clearer as we place ourselves temporarily in the past. In the seventeenth and eighteenth centuries European science and technology had to compete on their own merits with their traditional Asian analogues. Before the mid-nineteenth century this Western learning did not have the enterprising resources of the West, seconded by gunboats, to back it up on the spot. Its propagators were often unprepared to present it adequately fleshed out or in its most current state (in fact we find that the first accurate descriptions of the Copernican heliostatic system were not published in China and Japan until the end of the eighteenth century). Asians who took the new ideas seriously overtly measured them against—and, so far as they could, reconciled them with—their own traditions. In fact one of the unanticipated applications of European science was to bring about the strengthening of ancient sciences, or even the revitalization of some which for some time had not been practiced at their highest levels (for instance, Chinese mathematics).

By 1900 most people being trained in modern science were hardly aware that their own cultures had ever evolved sophisticated sciences. Even the very few who took an interest in both ancient and modern were motivated most generally by a desire to enrich the latter. The possibility of reviving the former could no longer be considered seriously because between 1543 and 1900 European science had become modern science and had joined forces with technology to spawn the Industrial Revolution. But we would do well to keep in mind too that Chinese and Japanese who were becoming scientists and engineers in 1900 no longer lived their professional lives according to the old pattern or took their values from it. In China there were now large numbers of talented people who no longer had to think of their futures as tied to the millennial but failing institutions of their homeland. They no longer had to depend for advancement on the civil service examination system, which had rigidly determined for so long the shape of young men's ambitions toward socially useful work. In Japan a Westernizing class had been created by government policy. The very existence of this alienated sector of the population was a symptom that the tradition had succumbed to the cataclysmic solvent power of modernization—or to be more precise, of rationalized commerce and industry—drawing expansive vitality from engineering, which in its turn was a marriage of science and technology. Once that transition to a scientific community insulated from the traditional society had taken place, we find it all too tempting to consider the interaction between traditional and modern as just one more of history's countless conflicts of political interest, for it had become an interplay between individuals or subcultures who represented old and new rather than between ideas competing in one man's mind. We also find it easy to dismiss traditional responses as xenophobia, superstition, or blind reaction because the outcome was no longer in doubt. Why devote serious study to the loser when the winner is so much easier to understand? Because unless we comprehend both, we will never be able to understand the variety of processes by which modern science finds a foothold within societies and transforms them.

If what we want ultimately is a general theory of the introduction of modern science into traditional societies, we will save a great deal of undirected effort by realizing that what we are looking for is only a special case of the general theory of scientific revolutions. Historians of science have been studying for some time the process by which one theory, model, or general conception replaces another, and we are beginning to think of the great Scientific Revolution of the sixteenth and seventeenth centuries in Europe as only a particularly obvious instance due to its comprehensiveness, traumatic implications, and fertility.[1] In that revolution the "traditional science" was the comprehensive theoretical vision of late Aristotelian philosophy (which incorporated loosely such appendages as the rational medicine of Galen and the computational astronomy of Ptolemy and—when Ptolemy would no longer be supported—of Tycho Brahe). Learned men held to this ungainly body of belief in the face of Copernican cosmology and Galilean mechanics not out of some abstract recalcitrance but because it is only reasonable that a whole vision of the cosmos and man's place in it, however flawed, be replaced by another, equally whole, vision; for after all, making coherent sense of the natural world is the deepest concern of science. Galileo knew that he could not convince the schoolmen of his own time. The basis of their educations, and thus the precondition of their own intellectual activity, was a system which held that cosmology was the concern of metaphysicians and not of astronomers, and that number and measure could not be applicable to physical situations on the earth in the same way as to those in the sky. Galileo chose to write in Italian rather than Latin in the hope of appealing to a new scientific community which was just beginning to form.

The missionaries and the imperial government in China, and the latter in Japan, opened schools in the late nineteenth century for very much the same reason: to bring about a Scientific Revolution by replacing the scientific sector of

[1] The most important book of this kind is Thomas Kuhn, *The Structure of Scientific Revolutions* (*International Encyclopedia of Unified Science*, vol. 2, no. 2; Chicago: Phoenix Books, 1964).

society. In the Far East the old science was not that of the European scholastics. Aristotle had evoked infinitely less response from East Asian scientific thinkers in the period of early Western contacts than had Tycho in astronomy or Galen in medicine. Although the Jesuits had begun by trying to introduce an integral system of philosophy and science, they abandoned this aim very quickly. Neither the dialectical orderliness nor the conceptual flaccidity of late Aristotelianism turned out to be very appealing; Chinese tended to be bored with long, patient arguments that four elements are better than five. New technical ideas evoked the true engagement of minds. In the early period, the conceptual pattern to which these new ideas were referred (and reconciled so far as possible) was that of traditional China. The old scientific pattern had grown up in China out of that society's inner necessities over two millennia. In Japan, as this book demonstrates, despite great early dependence upon China for cultural and institutional forms, Chinese science itself was seriously taken up and adapted to Japanese realities only in the seventeenth century. It too, in other words, was a latecomer; but its compatibility greatly accelerated its appropriation and gave it an initial competitive advantage.

In the preliminary Japanese exploration of Western science before the nineteenth century, the relationship between it and the Chinese complex of thought about nature was parasitic rather than symbiotic. The former was sustained by the latter, but they were bound to compete with increasing explicitness for survival. Natural selection is not, however, an appropriate model for the outcome, for the competition was short-circuited by political decisions to sponsor modern education. These decisions in turn cannot be explained simply as the outcome of intellectual encounters. Rather more crucial in precipitating the decisive phase of the East Asian Scientific Revolution, which replaced the whole intellectual matrix of science, were the initiative of British narcotics merchants in China around 1840 and the implied threat of the American "Black Ships" in Japan in 1853.

There is no need to pursue these complex issues further, for the point is plain. Models which envisage a succession of inevitable steps in the introduction of European science into non-European cultures cannot account for the complexity of history and thus have nothing to contribute to social science. There is no alternative to focusing on the dynamic interaction of two integral systems; this one can do only after studying both seriously.

It is hardly remarkable that the most useful and solid work on the introduction of modern science into the Far East has been done by Asians, who have studied their own ancient traditions in depth and who are less prone to scientific provincialism than Americans and Western Europeans. As the volume of important publication continues to mount, we will see their discoveries increasingly distilled into Western languages. Shigeru Nakayama has recently provided a most distinguished study of the Chinese astronomical paradigm and its displacement in Japan.[2] Wang P'ing's survey of the reception of Western mathematics and astronomy into China, although written in Chinese, is organized and documented in accordance with very contemporary standards of accessibility.[3] It makes up for some of the many serious defects of Fr. D'Elia's *Galileo in China*, which largely ignores nonmissionary sources.[4]

Masayoshi Sugimoto and David Swain have made a remarkable contribution to understanding the role of science in the formation and transformation of culture. I have stressed the issue of modernization above because most readers will have been drawn to this book by curiosity about it, and be-

---

[2] Shigeru Nakayama, *A History of Japanese Astronomy: Chinese Background and Western Impact* (Harvard-Yenching Institute Monograph Series, vol. 18; Cambridge, Mass.: Harvard University Press, 1969).
[3] Wang P'ing, *Hsi-fang li-suan-hsüeh chih shu-ju* (The introduction of Western astronomical and mathematical sciences into China; Institute of Modern History, Monograph No. 17; Taipei: Academia Sinica, 1966).
[4] Pasquale M. D'Elia, S. J., *Galileo in China, Relations through the Roman College between Galileo and the Jesuit Scientist-Missionaries (1610–1640)* (Cambridge, Mass.: Harvard University Press, 1960).

cause the authors have presented for the first time in a Western language an adequate synthetic study of its inception in Japan. (They are now completing a second volume to examine the character and success of modernization in the past century.) But the theme of the book is much larger, among the largest of all historical themes—the movement of knowledge and ideas into a society and the rhythm of their assimilation. What we now have is in fact a new history of Japan seen from the standpoint of the development and formative role of science, and a new periodization based on the intermittent flow of cultural influences from China (via Korea at first) and from the West. Thus emphasis falls on the social and institutional matrix of science rather than on its technical content. The interplay of politics, social structure, scientific institutions, and the orientations of individual scientists and popularizers of science are demonstrated clearly throughout. The general content of the book has been carefully composed for its bearing on the reception of science; but the ensemble is a kind of cultural history in which for the first time science is, as it should be, integrated. We have long since passed the point at which a man might still delude himself that culture consists simply of religion, arts, and letters. The book brims with hypotheses which specialists on Japan as well as historians of science will find suggestive. In addition, it will be of exceptional interest to people who are trying to understand world history in terms of cultural interactions—who in order to reconstruct the whole feel a need to consider not only the parts but their boundary conditions.

Finally a few words about the authors. I met Masayoshi Sugimoto, then Associate Professor of Physics at Kanagawa University, Yokohama, when during a sojourn in Cambridge several years ago he visited my Chinese Science seminar at M.I.T. I found him remarkably knowledgeable about traditional Chinese and Japanese mathematics and have come to think of him as a man whom that great pioneer Yoshio Mikami would gladly have accepted as a disciple and successor. Subsequently in Japan I met and also came to know his collaborator, David Swain, a permanent resident of Tokyo in such capacities as Secretary for Research and Publications of the World Student Christian Federation. Both authors are at ease in both Japanese and English; David Swain, among his other accomplishments, has published translations of important Japanese scholarly works on urban sociology. This book is the fruit of more than a decade's vehemently shared pursuit of a grand intellectual goal.

N. Sivin

Introduction

Japan alone in the non-Western world has in the past century fully adopted and applied modern science and technology in industry, economy, education, and public welfare. It has thereby moved rapidly into the circle of "developed" countries and today shares with them various uncertainties as to the future role of science and technology in society. Japan's mastery of modern science and technology for its "modernization" was, nonetheless, a remarkable achievement, one viewed generally with admiration, not least by those nations still engrossed in the developmental struggle.

When first moved, more than a decade ago, to investigate the background of Japanese science, we had in mind a continuous survey from ancient beginnings to the present. We soon found that, because of vast differences in content and applicable methodologies, our work would have to be divided into two parts: premodern developments, up to 1854, and those since. For both parts, however, we have been guided by the conviction that neither premodern nor modern science can be properly understood apart from the broader intellectual and social milieu.

The place of natural science in the cultural history of premodern Japan, with which this volume deals, is interesting by itself; at the same time, its analysis is indispensable to an understanding of Japan's modernization. Among reasons why it is inherently interesting is that, compared with Japan's religious, literary, and aesthetic traditions, the tradition of premodern science—and of Confucian learning as well—is relatively new. Science and Confucianism became established in Japan as comprehensive systems only in the seventeenth century.

"Science" in the above context means science borrowed from China, and this suggests another point of interest. All science in premodern Japan was initially imported, first from China and then from the West. Nearly a millennium transpired between the time when full-fledged borrowing from China began in the seventh century and the seventeenth-century rooting of Chinese-style science in Japanese soil.

Meantime, Western science made its first appearance in the middle of the sixteenth century, only to be ruled out in mid-seventeenth century by the Tokugawa government's isolation policy. Remaining Western influences were meager compared with the subsequently flourishing tradition of Chinese-style science, although from the eighteenth century this tradition came to be criticized by a growing number of professional scholars thoroughly trained in its concepts and methods. Indeed, it was precisely professionals aware of the defects of their own tradition who first grasped the merits of modern Western science and in the nineteenth century urged its adoption—as the necessary basis for mastery of Western military technology—if Japan were to remain independent in an age of colonization by Western powers.

The importance of appraising Japan's premodern scientific enterprise becomes clear, then, if one asks how Japan managed to meet, not merely see, the need for acquiring modern Western science and technology. Although the specific forms and most of the content of traditional science were discarded in the acquisition process, it was not simply a matter of jettisoning an outworn tradition in order to import a newer, better one. While we argue, in Chapter 3, that, but for isolation, Japan could have made a much earlier entry into the process of comparing its traditional science with that of the West, adapting and adopting where feasible, the actual process would have been exceedingly difficult without most of the elements of the Tokugawa legacy of learning available by the middle of the nineteenth century. That legacy consisted, at its best, of a deep love for learning, a disciplined pursuit of detailed knowledge, an appreciation of rational criticism, and a respect—often despite poverty—for the scholarly profession. This legacy was manifested in an ever-growing body of professionals who, because of their thorough grounding in traditional learning and science, knew from experience something of what the role of science in society should be.

There were, of course, many other factors favoring modernization. To mention only a few directly related to the function of professional scholars, there were an administratively competent samurai class that formed the core of the new Meiji bureaucracy and created a fairly stable and efficient government; a relatively high rate of literacy among upper-class merchants and farmers who, with proper leadership, were able to participate in the nation's goals as local agents for marshaling the masses; a relatively uniform social structure and language, and thus easy communications between the national capital of Edo and the various domain capitals ("little Edos" with similar administrative and academic institutions); a nationally integrated economy and transportation network; and, not least, a feudal system that had exhausted virtually all its possibilities and was ripe for change.

Mid-nineteenth-century Japan did not lack advocates of radical change—those spokesmen of the need for drastic reform who tend to be well remembered and reported at the expense of the professionals without whom change could not have been premised upon the uses of modern science. Advocates and also administrators to plan and carry out changes are necessary, but not sufficient. Professional skills and knowledge are needed to produce factories, railroads, communications systems, and the like; and professional scholars are needed to prepare a society for these things. The work of introducing modern science into Japan was in the hands of such professionals who, rooted in traditional learning and science, had late in the eighteenth century already begun to acquire a considerable understanding of Western science. By the mid-nineteenth century their grasp of Western learning and science had expanded sufficiently for them to serve as the transitional figures needed to undergird the work of planners and politicians.

Purpose
This book and its projected companion volume on science in modern Japan are intended neither for specialists in Japanese history nor for specialists in the history of science, but for the general reader attracted to the theme of "science in Japan

within a wider social and intellectual context." To offer a reliable perspective on science in Japan, data on each phase of Japanese history have been selected and organized to indicate the social and intellectual situations relative to the scientific scene; for further details the reader should consult the widely available literature on Japanese history and society. Those with specific interests in scientific matters will, we hope, find it easier to consult specialized monographs and articles after reading this more general survey. A few basic features of this book do, however, warrant further clarification.

Wave Periods

The most striking departure from custom, at least for readers familiar with standard texts on Japanese history, is our scheme of periodization. The history of any country is a combination of domestic development and foreign influence, and for island countries like Japan these two factors often stand out in sharp relief. As far as the history of learning and science is concerned, the undulations of foreign influx and its retardations were of crucial importance. Hence, our periodization is based on successive waves of cultural influx—particularly learning and science—into Japan from China and the West, including interim periods when influx subsided or was purposely thwarted. Each chapter shows the kind and mode of influx that affected the fortunes of science in that period.

The influx from China is seen in two major waves—Chinese Cultural Waves I and II—the former a period when Chinese learning and science were formally introduced but only partially retained, the latter a stage in which the most sophisticated forms of Chinese culture—Confucianism and science—were finally rooted in Japan. A long period of influx of material culture into Japan preceded the first wave, a period of about eight centuries during which Japan absorbed the agricultural, metallurgical, and domestic skills of China and was thereby transformed from a primitive tribal society to a settled agrarian one. Between the two major waves of Chinese cultural influx there was an interim period of semiseclusion from the continent when influx—and with it, Japanese scientific activity—subsided noticeably. The dates of the waves are suggestive, not definitive, and run as follows:

Chinese Wave I: seventh century to the end of the ninth century

Semiseclusion Era: tenth century to the end of the fourteenth century

Chinese Wave II: beginning of the fifteenth century to mid-nineteenth century.

The dominant, Chinese-style tradition was challenged twice by smaller waves of cultural influx from the West. Western Cultural Wave I (1543–1639), when Portuguese traders and Jesuit missionaries brought Scholasticism, Western science, and various techniques to Japan, had its challenge cut short by political decision—the Tokugawa isolation policy. Only after the seventeenth-century interim of remarkable activity in traditional learning and science was the isolation policy relaxed, in 1720, to allow Western Cultural Wave II to bring a second challenge that gradually intensified until 1854 when Japan was forced to open its doors to international intercourse. This second challenge was ultimately successful, but again

Table 1. Cultural Waves and Interims in Premodern Japanese History

	A.D. 500	1000	1500	1900
Waves and Interims	ca. 600————894 Chinese Wave I		1401————————1854 Chinese Wave II	
		894————————1401 Semiseclusion Era	1639————1854 Isolation[a]	
			1543–1639 Western Wave I	1720–1854 Western Wave II
Conventional Periods	552————710–794————	1192————1336————	1467————1603————	1868
	Asuka Nara Heian	Kamakura Muromachi	Sengoku	Tokugawa (Edo)

[a] Throughout isolation, limited trade was permitted with the Chinese and Dutch.

because of political decision—the Tokugawa shogunate's moves to acquire Western military technology, followed by the Meiji program for all-out importation of Western science and technology.

The chapter divisions are adjusted to these successive Chinese and Western waves and the interim periods of semi-seclusion and isolation. Charts are also provided in the appendix to give the reader a bird's-eye view of premodern Japanese science and culture. A general chart indicates the overall sociocultural framework of the book, correlating domestic developments with foreign influences. One chart shows the four main traditions of learning in premodern Japan— Chinese, Buddhist, Japanese, and Western. There is one chart each for the three specific scientific fields on which this book focuses: astrology and astronomy, mathematics, and medicine. These charts are extremely simplified so that the reader may not get lost in details; we hope that the text will be read with frequent reference to these aids. Other tabular material accompanies the text. As this book is for general use, technicalities of the specific sciences are not treated in detail; various examples in the text, illustrations, tables, and notes will perhaps provide some flavor of the sciences.

Special Themes

To place premodern science properly within the larger social and intellectual context, some rather detailed explanations of particularly important events, processes, and institutions are presented, themes which might not appear in a more straightforward history of science. The non-Japanese reader may have less difficulty with general social, political, and economic factors that affected Japanese learning and science—especially those related to Western influence—than with certain special agencies of influx and with institutions of learning directly related to the development of premodern Japanese science. For example, the efforts of the *ritsuryō* government in the Nara and early Heian eras to transform Japanese society according to the contemporary Chinese model involved a number of academic institutions which must be explained if the beginnings of Chinese-style science in Japan are to be understood. Likewise, an explanation of the system of *kentōshi* missions to China is essential to clarify the mode of influx undergirding the *ritsuryō* program. Or, subsequent trade patterns demand elaboration in order to account for the radical changes in science and learning in the Semiseclusion Era as well as the unprogrammed renewal of Chinese-style learning and science in Chinese Wave II. Again, overseas trade and certain features of Jesuit activity in Japan were crucial to the course of Western Wave I, particularly the novel possibilities related to techniques in that wave. The necessity for developing such themes will, it is hoped, make understandable the allocations of space to social and cultural data in proportions different from those commonly encountered in more general political or cultural histories of Japan.

The Specific Sciences
Throughout this book we concentrate on only three fields: astrology and calendrical astronomy, mathematics, and medicine. These fields have not been arbitrarily selected; they were the three specific scientific fields that developed earliest in China and there, as in all East Asia, including Japan, occupied the central place in the pursuit of natural knowledge and its social uses. All East Asian astronomy was done for calendrical purposes; there was no observational astronomy independent of the astrological and calendrical arts. While astrology is pertinent to understanding the intellectual ethos of a given era, we take a special interest in calendars because a "calendrical reform" meant the working out of a complete and integral set of computational techniques by which celestial phenomena could be annually predicted—that is, such a reform produced a new astronomical system on which the annual civil calendar was based. Thus, this activity is here termed "calendrical astronomy." While mathematics is not, strictly speaking, a natural science but a branch of logic, in premodern China and thus Japan the logical aspect of mathe-

matics was quite limited; it was, rather, practically oriented to calendrical science, land-surveying, taxation, commerce, and other social uses—though a special kind of Japanese mathematics (*wasan*), developed in the seventeenth century, was not performed so much for logical or social purposes as for enjoyment as an intellectual pastime. (Like other sophisticated pastimes in other times and places, *wasan* attracted professionals who devoted their lives to it.) Medicine in the Chinese tradition had an increasingly philosophical body of theory but was, throughout the premodern era, focused mainly on drug therapy and therefore active in the search for materia medica, which later developed into natural history. In its Japanese adaptations, medicine was often the forerunner in scholarly criticism and innovation. By systematically pursuing the development of these three specific fields we gain a general framework with which to coordinate data on related fields. (Physics, chemistry, and biology, in their modern forms, appeared in Japan only at the end of the premodern era and thus receive only passing mention in this volume.)

Multiple Methodology
Our methodology for pursuing these three scientific fields is a multiple one that isolates, yet treats in an integral way, four key aspects of each field. From concrete data we seek to describe *historical* trends related to sustained central traditions in each field. The content and methods internal to these fields we call the *logical* aspects. *Intellectual* aspects include relations with various thought systems and the concepts, values, attitudes, and academic facilities that aided or inhibited scientific development. Nonacademic processes, agents, and institutions that affected or resulted from scientific activity are discussed under the heading *social* aspects.

Only in Chapter 1, however, are these four aspects consistently separated for treatment. This is done partly to introduce our approach, but more especially because of the need simultaneously to introduce the unfamiliar forms of Chinese-style science. In subsequent chapters one or more of the key aspects is singled out from time to time as the data warrant it. We do not, of course, provide extensive discussion of Chinese science as such, only the minimum needed to grasp the course of science in Japan.

Caveat on Technology
Science as more or less systematized knowledge of nature (and its sometime instrument, mathematics) is the focus of this book—not technology as more or less conscious applications of that knowledge to human goals, or even techniques which make little use of science as such. The whole spectrum of techniques, from agriculture and fishing to shipbuilding and printing, is introduced only insofar as there is some intrinsic relationship with the three basic scientific fields. Otherwise, we sometimes refer briefly to techniques to help describe the general social or intellectual milieu of a given historical period.

Sometimes, even when the potential for science implicit in new techniques was great, the outcome was meager—as in the interrelations in Western Wave I between, for example, navigation and astronomy, gunnery and dynamics, mining and mensuration, and the place of mathematics in all these samples. At other times the force of intrinsic potential was greatly magnified—as in Western Wave II when the desire for Western military technology and ordnance directly affected early decisions to introduce modern Western science.

Historical Perspectives
Science in the popular mind has become so deified that even the history of science is sometimes expected to reveal a smooth and straightforward, if not assured, course toward success. Certainly the introduction of Chinese and Western sciences into Japan did not constitute a simple success story. Therefore, we also present regressions or unrealized potential as an essential part of that story. Later Japanese developments of Chinese or Western sciences were not direct outcomes of ideas and techniques introduced in Chinese Wave I or Western Wave I but had to wait for new influences in quite different social

settings—which can be fully appreciated only after having first comprehended the earlier waves.

Moreover, despite the many interesting facets of its historical development, the specific achievements of premodern Japanese science were meager compared with those of China and the West. Japanese scientists made virtually no direct contributions to the mainstream of world science up to the middle of the nineteenth century. This does not obviate, however, the need for a comprehensive evaluation of science in premodern Japan substantiated by facts and placed in a perspective constructed from clear historical, logical, intellectual, and social concepts. It is not too early for a balanced assessment of achievements combined with a clear account of limitations to replace mere guesswork on the character and development of science in Japan. We hope our book contributes to that end.

In assessing Japan's scientific background we are not insensitive to the questions being raised today as to the value of science. Nor are we naively disposed to overestimate its contribution to the welfare of either developing or advanced societies. On the other hand, we see little virtue in ignoring accumulated knowledge based on proven theories and tested methods, particularly if such neglect results in the wasting of resources, natural or human. Social goals cannot, after all, be realized beyond the capabilities of the persons in a given society or irrespective of the pressures—internal and external—that aid or hinder the nurture and use of their abilities. If the Japanese experience of science in modernization is to be of any use whatsoever to other developing countries or yield any lessons for Japan's own facing of the future, the continued and careful assessment of Japanese science is hardly optional. From this conviction, and on the basis of what is established in this volume, its sequel on science and culture in modern Japan is projected.

Notes on Style

Kenkyusha's *New Japanese-English Dictionary* (4th ed., 1974) has been followed in the romanization of Japanese words.

The romanization of Chinese words follows the Wade-Giles system; of Korean, the McCune-Reischauer system.

Macrons appear over long vowels in all Japanese proper names and all common terms except well-known, and thus unitalicized, terms such as shogun and daimyo. In proper names of agencies, temples, and the like, our usual practice is to retain the suffix—such as *-shi* for envoy, and *-ji* for temple —and follow with the English equivalent without a capital letter, e.g., *kentōshi* envoys and Hōryūji temple. In exceptions to this practice, the English part is capitalized and comes first, as in Mt. Hiei (instead of Hiei-zan mountain).

Chinese, Korean, and Japanese personal names are given in the normal East Asian order: surname first, followed by the given name (for example, Manase Dōsan, where Manase is the surname). In most cases, birth and death dates follow the initial appearance of personal names; otherwise, the time of significant activity of a person should be clear from the context. Chinese characters for all personal names, as well as for East Asian place-names, book titles, and all terms in italics, are given in the Index and Bibliography; all characters are in the older classical form, as modern simplified forms may vary in Japan and China.

Parentheses and brackets inside parentheses in the text include translations or other explanatory material; brackets alone mark the authors' or a translator's addition to a quotation. English equivalents are given on initial appearances of all italicized terms and, as this book is for the general reader, an attempt was made to provide English equivalents for all East Asian book or manuscript titles on first appearance, though not with complete confidence (renderings of particularly difficult titles are marked with our own question mark).

Approximations of Japanese measures are provided, with the warning that, because of variations in different historical periods, they are not necessarily precise. Dates for historical eras are given on initial occurrences and sometimes when the era name appears after a long interval.

Finally, the use of "scroll" for *kan* 卷 (of a manuscript) and "volume" for *satsu* 冊 (of a printed text) could not be systematically followed. The same division of a text sometimes appears in the literature as *kan* and *satsu*. As a general rule, prior to the expansion of printing on a commercial scale in the seventeenth century, most texts were in manuscript form (printed books before the seventeenth century were usually Buddhist or Confucian classics, and these comprised only a minority). From the seventeenth century on, texts were increasingly printed (but not always); an effort has been made to indicate the difference where known.

Acknowledgments

It would be entirely misleading to offer this general survey as purely our own product. We have relied on a great number of books and articles and have tried to make clear the sources of data and evaluations. Although it is impossible to acknowledge all indebtedness, we offer an apology for instances where we failed to do so.

Among our daily acquaintances over the past decade not many have been greatly interested in what to them may well have seemed an esoteric exercise. Consequently, the steadfast support of a very few and the timely assistance of a number of persons and agencies are especially appreciated. To Professor Nathan Sivin, historian of Chinese science at the University of Pennsylvania, we are particularly indebted. He was the first to encourage us and for nearly a decade has unstintingly given his personal support; repeatedly we have had the benefit of his scholarly criticism and editorial advice, as he has read and criticized the manuscript in detail at various stages of its development. His generous attention to the preparation of this book has not only done much to improve its quality but also heightened our appreciation of our own material. In Japan, Professor Nakayama Shigeru of the University of Tokyo, a prominent historian of Japanese astronomy and a recognized leader in Japanese studies in the history of science, has throughout most of the process of writing been a valued friend and counselor, providing support and advice and frequently access to materials not easily attainable.

During the academic year 1969–1970 the East Asian Research Center (EARC) of Harvard University kindly provided financial assistance and full access to its research facilities. There, for the first time since we began this project, we were able to depart from our pattern of weekly meetings to read, discuss, write, and revise together (while holding full-time jobs in other fields) and give undivided attention to finishing a complete draft of the manuscript. Then and since we have had the helpful counsel of Professors Edwin O. Reischauer and Albert M. Craig, former and present directors respectively of Japanese studies at the EARC. Professor George Elison of Colby College made some helpful suggestions on Jesuit activity in Chapter 3 during this time; and Professor James K. Ash of Fort Lewis College (Colorado) supplied some pertinent data on matters related to Korea.

Generous grants from the EARC permitted Sugimoto to concentrate on this volume during the year 1969–1970 and to stay on at the EARC in 1970–1971 to engage in research on science in modern Japan (while Swain returned to Japan to make further revisions of this volume). Further grants from the Danforth Foundation and the Hazen Foundation enabled Sugimoto to extend his period of work on Japan's modern century at the EARC through 1971–1972. We deeply appreciate all this assistance, as well as the long-time interest in this endeavor of our mutual friend, Dr. J. Edward Dirks, a former vice-president of the Danforth Foundation and now Vice Chancellor, Division of Humanities, of the University of California at Santa Cruz.

Throughout the period of research and writing Swain has been supported by the Board of Global Ministries of the United Methodist Church for work in Japan related to university education and publishing. We are grateful for this continued support, but especially during the year spent at the EARC together (1969–1970); and we wish to express particular appreciation to the Board's former East Asia secretaries, Dr. Charles H. Germany and Pharis Harvey for their sustained interest and encouragement. Harvard University kindly supplemented the Board's support for Swain's year at the EARC.

In recent years, as we made final revisions, Professors Peter Duus, formerly of Harvard University and presently of Stanford University, and David Reid, professor of the study of religion at Japan Biblical Seminary and director of research at the International Institute for the Study of Religions (Tokyo), have kindly read the manuscript and made a number of helpful criticisms and suggestions. Likewise, Michael Cooper, S.J., editor of *Monumenta Nipponica*, made a number of helpful contributions to the refinement of Chapter 3. Not all advice or criticisms could be accommodated in a work so long in the making, and in any case we alone are responsible for any remaining errors or omissions.

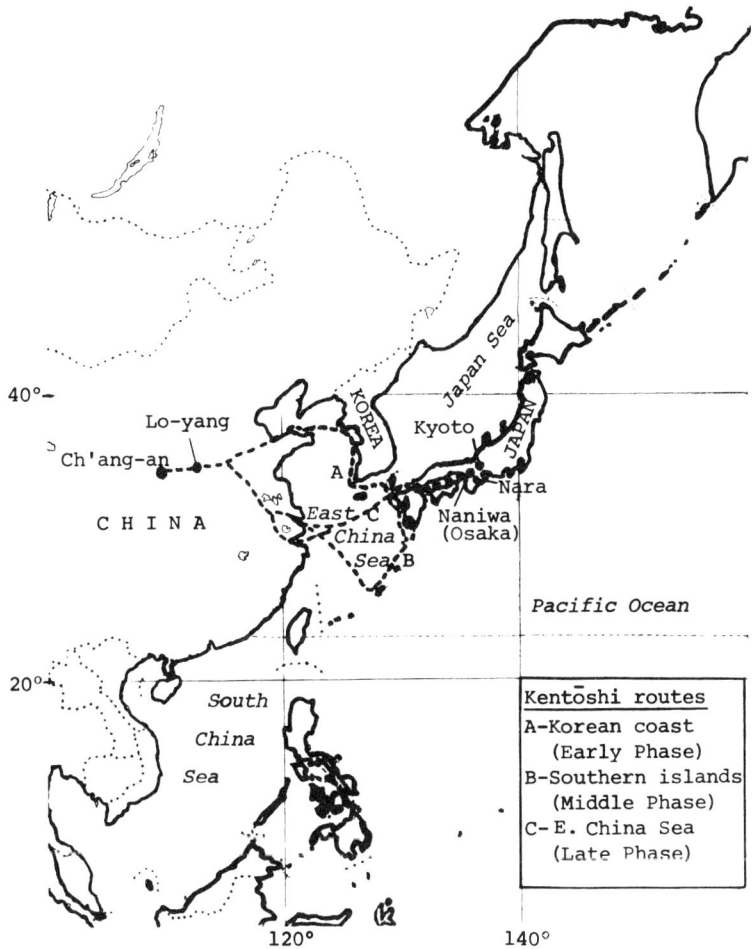

East Asia during Chinese Cultural Wave I.

Overleaf:
Japan during Chinese Cultural Wave II.

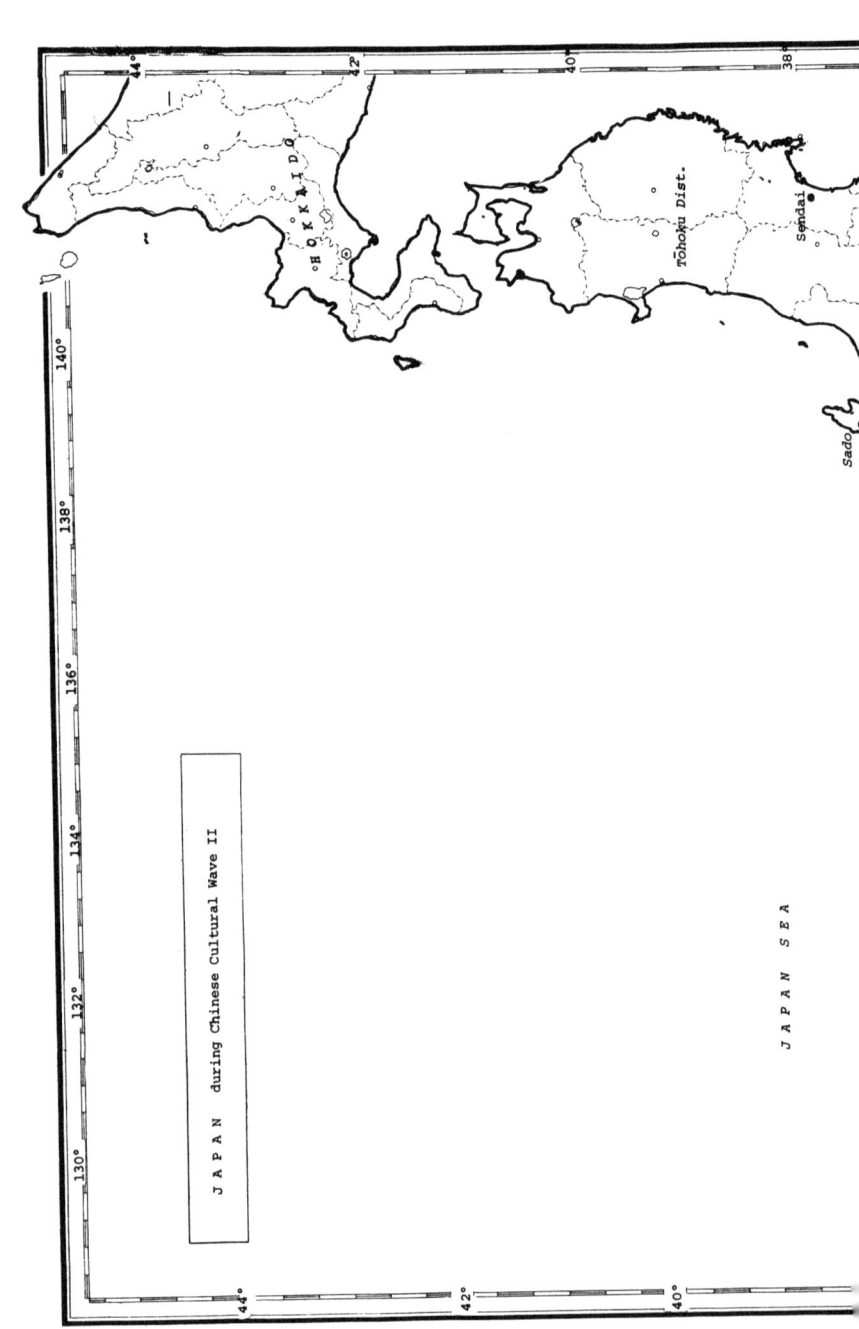

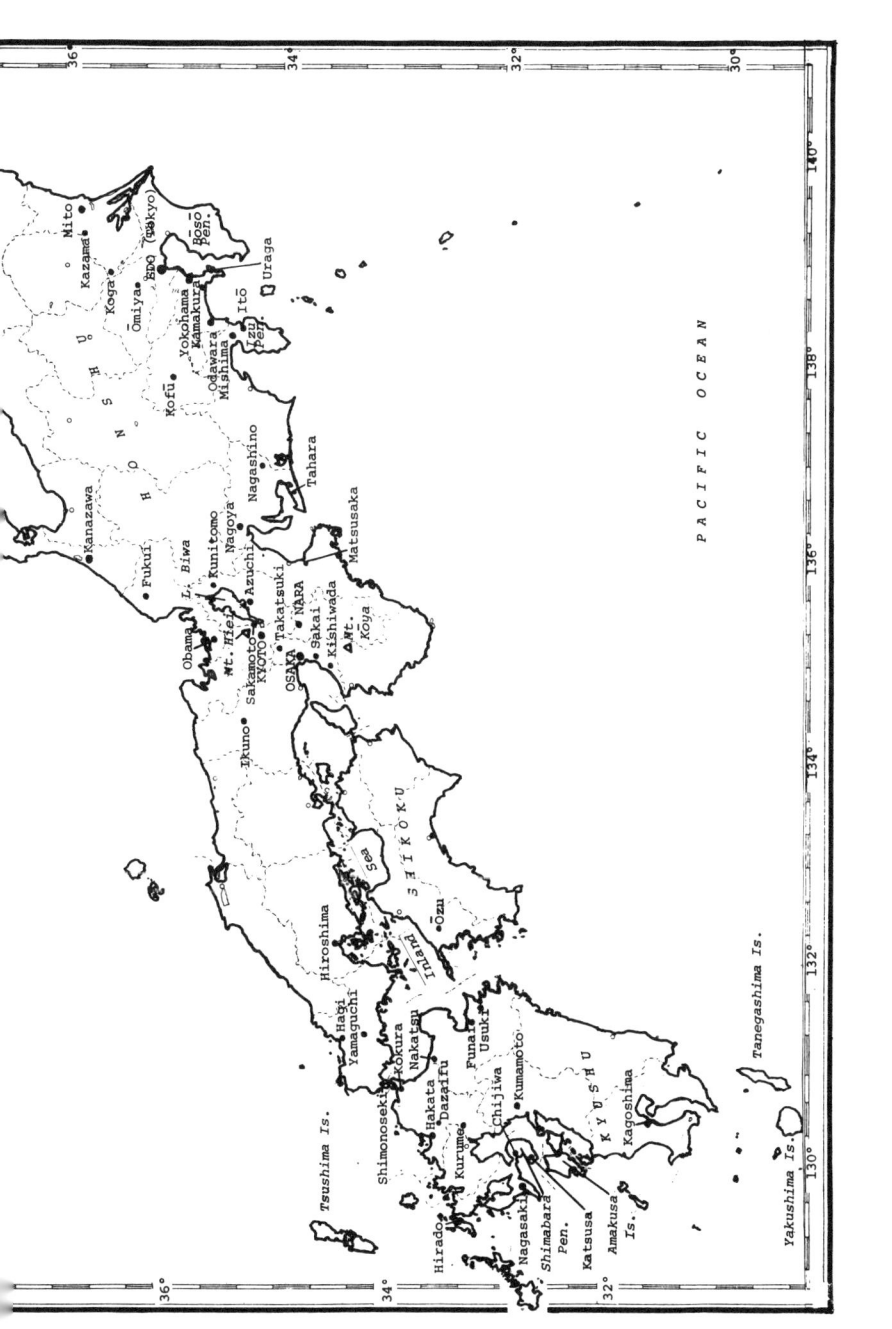

1

Science in Japan's First Cultural Transformation

Chinese Cultural Wave I: ca. 600–894
Only twice in Japanese history has it been national policy to undertake an overall transformation—or "modernization"—of the entire social system according to an imported foreign model. In both cases the importation of new knowledge and the creation of new systems for its advancement and use in society were at the heart of the process. The current, and in many respects highly successful, pursuit of the Western model, associated with pressures leading to the Meiji Restoration of 1868 and the sweeping reforms following it, is the more recent of these two times. The first began much earlier with the Taika Reforms of A.D. 646 and continued through the Nara and early Heian eras (seventh–ninth centuries inclusive) when Japan sought to adopt the Chinese model of T'ang society. Japan had been assimilating more elementary material forms of continental culture for several centuries, but the Taika transformation was the first effort consciously based on systematic, large-scale importation of high culture directly from China, and thus we designate it Chinese Cultural Wave I.

The overall character of Chinese learning and science as introduced in Chinese Wave I—the standards and style of scholarship, and the image and role of the scholar and scientist in society—was so deeply implanted in Japanese culture in this period as to dominate the intellectual life of Japan throughout the premodern period. Such a determinative impact came only after a long period of growing exposure to continental influences that prompted in Japan a gradual evolution from a primitive to an agrarian society with village states eventually unified by an imperial court.

JAPAN'S EARLY INVOLVEMENT IN EAST ASIAN CULTURE

Even as late as China's Warring States era (403–221 B.C.) when the Confucianists, Taoists, Legalists, and other schools of Chinese thought were already in the process of formation, the peoples living in the islands now called Japan were still in the late Stone Age. Dependent upon hunting, fishing, and plant-gathering for their livelihood, they were unaware of movements on the continent that would shortly extend to the virtually isolated islands and revolutionize their technical, social, and political patterns.

The initial movement was the penetration into northern Korea by the Yen state of northeastern China during the Warring States era. Large numbers of Chinese people subsequently settled in northern Korea, forming a distinct political unit loosely subject to Chinese sovereignty. During the Han dynasty (Former Han 206 B.C.–A.D. 8; Later Han, A.D. 25–220) cultural influences spread from these settlements down the Korean peninsula and across the Korea Strait into northern Kyushu, bringing wetland rice cultivation, metalworking, weaving, and other crafts that precipitated a gradual shift in Japan from migratory tribal life to a more sedentary agrarian mode of existence.[1] Along with ideas and ways of doing things, a considerable number of immigrants from Korea settled in western Japan. They seem to have been assimilated along with the new techniques.[2]

The force of cultural influx into Japan was strengthened when in 109 B.C. the Chinese settlements in northern Korea were brought under direct Han control after rebellions there in the previous year had been put down. Concurrently, village states arose in many places in Japan, notably in Kyushu and

[1] Attempts have been made to verify the existence of "incipient agriculture" such as cultivated millet, deccan grass, etc., in the late part of the Jōmon period (about 8000 B.C. to third century B.C.), but decisive proof is still lacking. Cf. Ishida and Izumi, pp. 200–201.

[2] Ibid., pp. 154–155. As we discern no subsequent cause to resort to "racial" explanations of the course science took in Japan, the thorny problem of the various racial types that eventually combined to form the Japanese race is not treated here. For some recent studies in English of the various possibilities, e.g., Ainu, Manchu-Korean, and Mongoloid types, thought to have entered the Japanese islands, see Levin.

the central part of Honshu, the main island.³ These small states maintained some relations with similar village states in southern Korea. Separately or in groups, the Japanese states are known to have sent tribute to the Han-controlled Chinese settlements in northern Korea or directly to the Han court.⁴ Continued influx was assured as some of the small states in Japan became minor members of the loose East Asian political order dominated by Han China, though the greater part of the cultural influences came by way of the Korean peninsula and its peoples, not directly from China proper or through face-to-face Sino-Japanese contacts. Moreover, cultural flow was primarily confined to rudimentary material techniques at this stage; the more sophisticated elements of Chinese culture —its administrative system, literature, religion, philosophy and science—were not yet included.

From the beginning of the third century A.D., and especially as iron came into use for tools and weapons, one of the village states, the Yamato, built up a strong power base in central Japan and by the middle of the fourth century brought the other small states into loose subordination to form Japan's first unified state.⁵ This shift in domestic politics came at a

---

³ The existence of such simple political units in northern Kyushu is known as early as the middle of the first century A.D.

⁴ The first known example of this intercourse, according to official Chinese records, was a certain head of a northern Kyushu state called Nu (Na in Japanese) who paid tribute to the Han court in A.D. 57 and received from the Han court a gold seal for his use as "king" of the Nu region of Wo (Wa in Japanese; i.e., Japan). Cf. the "Tung-i ch'uan" (Record of the eastern barbarians [Japanese]) of the *Hou Han shu* (History of the Later Han) compiled in the fifth century; also Inoue Mitsusada (2), pp. 174–182.

⁵ This date for the original unification of Japan contends with an earlier one, about the middle of the third century, that traditionally has received strong support among Japanese scholars. The discrepancy is due to an unsettled debate over the location of the state of "Yamatai". According to the official Chinese record "Wo-jen ch'uan" (Record of the Japanese) in the *Wei-chih* (History of Wei), a Queen Himiko of Yamatai loosely unified thirty states in Japan, and in 239 sent her envoy to Lo-yang, the capital of Wei, by way of Tai-fang in northern Korea. If "Yamatai" is "Yamato" in central Japan,

time when Chinese controls over the Korean peninsula were declining due to political instability in China proper during the Six Dynasties period (A.D. 222–589), following the collapse of Han power. When direct Chinese control over Korea ceased in A.D. 313, the Koguryŏ, Paekche, and Silla states emerged as the ruling powers of northern, southwestern, and southeastern Korea respectively.

From late in the fourth century and well into the fifth, the newly unified Japanese state at various times dispatched troops to Korea to secure its foothold on the peninsula's southern tip—a district called Karak by the Koreans and Mimana by the Japanese.[6] To press its claims in Korea the Yamato court reinstituted its tributes to the Chinese court, after a lapse of about 150 years. Between A.D. 413 and A.D. 502 tribute-bearing envoys were sent thirteen times to the Chin and later the Southern courts.[7] Greater involvement on the far more culturally advanced Korean peninsula—thus far the main gateway for cultural contact with the continent—and promotion of direct contacts with China inevitably led to the initial penetration of Chinese-style[8] high culture into Japan.

---

the earlier date is likely; if "Yamatai" was in northern Kyushu, the fourth-century date is more appropriate. Cf. Inoue Mitsusada (2), pp. 197–247. This dispute is complicated by another over *who* first unified Japan (see n. 6). While awaiting settlement of these issues, the later date is acceptable since, for our purposes, it is sufficient that unification occurred far in advance of the seventh-century initiation of aggressive importation of China's high culture.

[6] "Mimana" is traditionally viewed as a colony or military outpost maintained by the Yamato state to establish its control over Korea's southern tip. In 1948 Egami Namio proposed the theory that it was not Japan that invaded southern Korea in the second half of the fourth century but, rather, that a "horse-riding people" (*kiba minzoku*) from Korea invaded Japan, conquered, and unified it. Hence, Mimana was, in this theory, a point of return to the peninsula from which the imperial clan had originated and still had much to gain—by force if necessary. The "internal unification" theory is most widely accepted among Japanese scholars; the "invasion" theory has recently found support among some Japanese, most Korean, and a growing number of Western scholars. See especially Kamstra, also Kidder, and Levin.

[7] Inoue Mitsusada (2), p. 369.

[8] The term "Chinese-style" is used to allow for Korean modifications of cultural forms basically Chinese in origin prior to their influx into Japan, and for further Japanese adaptations after influx. Subsequent uses of this term are often made in a more general sense.

In this preparatory phase the role of Koreans was clearly paramount. In the sixth century the kingdom of Paekche dispatched men schooled in Chinese learning to the Yamato court, both on its own initiative and in response to Japanese requests. Many of these learned men became residents of Japan and assumed important roles in the diplomatic and financial affairs of the Yamato court and in the compilation of its official records. Emissaries to the Yamato court also presented gifts of Buddhist and Confucian literature, as well as Buddhist statues and other liturgical equipment. From around A.D. 578 various powerful groups in Japan brought temple experts from Paekche for a temple-building campaign that produced about one temple per year for the next forty-six years. Thus, through these contacts the more sophisticated, and not simply the material, elements of Chinese-style culture prior to the impressive Sui and T'ang periods began to seep into Japan.

Meanwhile, the wealth, power, and technical capabilities evident in the large burial mounds constructed for the Yamato emperors in the fourth and fifth centuries were also acquired by many provincial clan leaders. By the sixth century the strength of some of these leaders rivaled that of the imperial court and posed for it an increasingly serious problem. In the same century Mimana sought independence from Japan, while Paekche and Silla began to press their claims on that territory. Japan repeatedly sent troops to the peninsula but, forced to retreat again and again, in A.D. 562 relinquished Mimana to Silla's control.

Confronted with deepening crises at home and abroad, Japan began moving, by the end of the sixth century, toward a revamping of her own internal structures along the lines of the centralized government developed by the Sui dynasty, which in A.D. 589 once again unified all of China. Of short duration, the Sui reign was succeeded by the T'ang court (618–907), which took up the mission for national unity and continued it for three centuries. The Sui and T'ang dynasties adopted aggressive policies toward surrounding countries, which responded by carrying out internal reforms according to the superior Chinese model. As a result, the cultural levels of the satellite societies rose swiftly, producing a distinctly T'ang-colored cultural sphere that extended north and east across Manchuria and Korea into Japan, south into Indo-China, and westward across central Asia, opening up intercourse with the Indian and Arab worlds. Throughout Chinese Wave I, then, Japan was seeking to adjust to what is regarded as the most cosmopolitan of all premodern eras in Chinese history.[9]

CHINESE CULTURAL BACKGROUND

The Sui-T'ang system was in its own way projected as the answer to internal and external troubles. Since the downfall of the Later Han dynasty in A.D. 220, China had suffered several centuries of internal disunion (in the Six Dynasties period, 222–589), as well as repeated invasions by the less civilized clans from the steppes north of China. In particular, though, the Sui and T'ang governments sought to establish the unchallenged supremacy of the emperor over powerful clan leaders (though former aristocrats could not be fully deprived of hereditary powers until later in the Sung era, 960–1279, when this goal was finally realized).

One measure used to reduce provincial powers was to redistribute privately-held lands equitably among peasants as central government grants. This policy was a crucial one, as agriculture constituted the economic basis of society. But the administrative bureaucracy is a matter of more immediate concern to us, as positions in it were made accessible through civil service examinations (though most bureaucrats were recruited by other means, and the highest positions were divided between examination graduates and those eligible by birth). Already in the Han era it had become accepted belief that the right to political office belonged to the literati trained in classical Chinese learning, of which Confucian thought formed the core. This assumption was vigorously revived by the Sui and T'ang courts, and it is this emphasis upon learning and, where

[9] Reischauer and Fairbank, pp. 176–177.

relevant, science as functions of government that was of special importance in the impact of Chinese Wave I upon Japan.

While learning as a function of government was largely a Confucian monopoly, the total Chinese intellectual enterprise was much more diverse and complex. The fountainhead of much of Chinese culture can be traced as far back as the Shang dynasty—the oldest major political unit of the Han people that is historically confirmed. This dynasty displayed a marked rise in cultural level about the sixteenth century B.C. Inscriptions on oracle bones (records on tortoise shells and animal bones of divinations on behalf of the king) dating from this early period reveal the existence of a written language that was clearly an early stage of the Chinese script. In the deciphered content of these inscriptions are discernible a few definitive thought forms, including a concept of heaven, a belief in efficacious divination, use of a fairly accurate lunisolar calendar, and careful recording of solar and lunar eclipses.

After iron came into use about the sixth century B.C., agriculture made remarkable progress, serf labor yielded to independent farmers with small landholdings, and commerce flourished in the many towns that arose in various regions. It was thus in a much changed social milieu that there appeared later, in the Warring States era, many scholars and their disciples from whose labors came the first major differentiation into specific schools of thought, such as the Confucian, Taoist, Legalist, and other schools. Intimately related to these schools was the pervasive influence of the *yin-yang* active and passive principles to dichotomize qualities, and the notion that a set of Five Phases (*wu-hsing*) characterized the cyclic changes of all nature. Although the *wu-hsing*—Wood, Fire, Earth, Metal, and Water—may have originated as constituents of things, by the time of the earliest records they were being used to characterize functions, not compositions. Material constituents were included among their many associations but were relatively unimportant in general philosophy and quite negligible in scientific discourse (see pp. 61 f.). The early Chinese sciences of astrology, calendrical astronomy, mathematics, and medicine also evolved in this period, and various treatises were produced that later, in the Han era, were incorporated into a number of standard texts.

The Han era was decisive in at least two respects. On the one hand, a synthesis took place in which Confucianism emerged as the ideal prototype of learning, and into it were fused a number of thought strands (including cosmological ideas) derived from non-Confucian sources. The other decisive move was to make learning and science, under the aegis of Confucianism, matters of government support and sponsorship; this included specification of the ranks and responsibilities of scholars and scientists in government employ. The Rituals of Chou (*Chou li*), an idealized depiction of government in Chou times, for example, describes not only the place of astrology, calendrical astronomy, and medicine in government, but also the administrative roles, grades, and duties of official scientists. The Sui and T'ang courts further refined and fixed these patterns of learning and science as functions of government.

Confucianism in its public forms was centered on a variety of rites, and especially on its literary corpus. By the Han period a special position was assured a group of Five Classics, of which the Book of Changes (*I-ching*), a manual of divination, represented the most significant assimilation of the functional theories of *yin-yang* and the Five Phases—at least as interpreted in early Confucian commentaries. The interdependent *yin-yang* principles referred to such natural complements as day and night, summer and winter, light and dark, hot and cold, and male and female; while the Five Phases—Wood, Fire, Earth, Metal, and Water—could be used for a more detailed analysis of the aspects of configurations in space or time. The application of these binary and fivefold systems in science led to considerable further abstraction within elaborate metaphysical schemes. Only loosely organized until well into the Han era, they were elaborated throughout the T'ang era and then arranged in an extremely complicated system after a breakthrough in the later Sung period, at the same time as and in contact with the development of Neo-Confucianism. As all these schemes were applied as systematically as possible to a wide range of phenomena, they were easily amenable to use in the sciences as well as in the pseudosciences, such as numerology, geomancy, and alchemy.

Although Confucianists took for granted many notions of the *yin-yang* and Five Phases theories, it was popular thought (recorded mainly in the canons of religious Taoism, as distinguished from the philosophical tradition identified with Lao Tzu) that tended to exploit the pseudoscientific possibilities. Rather vulgar in the Han period, natural knowledge among upper-class "eccentrics" and the lower orders was gradually refined from the era of the Three Kingdoms (half a century after the Later Han era) by absorbing elements of the more orthodox stream of Chinese science. Indeed, research into natural phenomena within circles closely or distantly associated with Taoism was quite remarkable, producing results notably in alchemy and pharmacology.[10] These achievements were in time fused with the main body of Chinese science, so that by T'ang times no distinct and unified system of heterodox science existed—only a miscellany of autonomous popular sciences and pseudosciences remain discernible and to varying extents identified with Taoism.

Buddhism started its influx into China in the Later Han era, after both Confucianism and Taoism were firmly established in society. Initially regarded as another Taoist sect, it gradually won acceptance and flourished from the fourth to the eighth century, reaching its peak around A.D. 700 in the reign of, and under the patronage of, the Empress Wu. Though promoted by various rulers, Buddhism's ascendancy was implemented mainly through the efforts of many lay believers—Indian, Chinese, and Central Asian—who translated voluminous sutras and erected numerous temples and monasteries. In the early half of the T'ang era Buddhism enjoyed a special place as the state religion of China.

---

[10] On Chinese alchemy, see Sivin (1), Yabuuchi (2), and Needham, Vol. 5, Part 2 (published separately from Part 1).

Buddhism shared Taoism's appeal to the masses, partly because of its emphasis on mysticism and estatic experiences, and partly because of its interests in divination and magical cures. It had, in fact, a considerable influence on the systematization of Taoist thought and the development of an extensive Taoist literature. Buddhist influence on Chinese learning as such was much more limited, consisting principally of the new concepts and terminology introduced by energetic translation projects.[11] It made a limited contribution to the content, but could not alter the basic framework of Chinese learning, centered firmly on Confucianism—and in the process became thoroughly sinicized.

It was rather Buddhism's economic strength, based on large landholdings, as well as its tendency to move great numbers of people out of the mainstream of productive life into religious communities, that posed the major challenge to Chinese ways. This challenge was met largely by accommodating Buddhism to government policies as a guardian of peace and order. Because the time of its zenith in China coincided with the period of greatest Japanese interest in Chinese culture, this role as public guardian left a strong imprint on Japanese Buddhism. In this context of Japanese enthusiasm for things Chinese, two other factors merit mention. One is that the esoteric elements of Indian Buddhism can be seen in the early stages of Chinese Buddhism but came into their own in China only in the T'ang era. In Japan, as we shall see, it was esoteric Buddhism, far more than Confucian learning or Chinese science, that outlasted Chinese Wave I.[12] Secondly, in the

---

[11] It seems that many of the translations distorted original meanings, and later Chinese Buddhist scholars, unable to refer to or understand the original Indian texts, could not easily check the earlier translations. Moreover, the logic of exegesis and interpretation was largely overlooked, as very few logical works, and even then only the simplest, were translated into Chinese. Cf. Nakamura, pp. 191–193.

[12] Zen Buddhism did not appear in Chinese Wave I as an independent sect. Saichō (Dengyō Daishi; 767–822) incorporated Zen elements into the Tendai sect, but Zen developed as an independent sect in Japan only after Chinese Wave I. Cf. Kitagawa, p. 173.

seventh and eighth centuries of heightened Chinese cultural influx into Japan, Buddhism eclipsed Taoism as the religion of China; and in any case, Taoism was the most ethnic and thus the least exportable element of Chinese culture. Its part in Chinese Wave I is almost negligible.

In T'ang China it was, after all, the this-worldly, sobermindedly rationalistic, and humanistically ethical norms of Confucianism that occupied an unassailable centrality in official society. It was for this school of thought, concerned with the role of cultured men and with the proper relations of intellectual and political power, that government supported institutes were maintained. It was Confucian doctrine that informed the larger political framework—based on a strong central government with its hierarchical bureaucracy, its administrative codes, and its civil examination system—which was firmly fixed in Chinese society during the T'ang dynastic reign. It was preeminently Confucianism that was at the heart of the culture Japan sought energetically to adopt during Chinese Wave I.

THE RITSURYŌ EXPERIMENT IN SOCIAL TRANSFORMATION

While no single date marks the beginning of Chinese Wave I, the Asuka period (552–645) served as the time of transition that saw the coming of Buddhism, commencement of diplomatic contacts with the Sui and T'ang courts, and random institutional innovations. But the initiation of sustained, systematic, and large-scale importation of Chinese culture accompanied by sweeping institutional adjustments dates from the unprecedented reforms promulgated in the famous Taika Reforms edict of 646 (second year of the Taika era). Our consideration of Japanese culture, learning, and science in Chinese Wave I necessarily focuses on that event and its consequences.

THE ADOPTED SOCIOPOLITICAL SYSTEM

Japan's all-inclusive remodeling according to the Sui-T'ang model involved a redefinition of political authority, a reorganization of all administrative apparatus to support and implement that authority, and the introduction of the social, economic, educational, and religious institutions necessary to the new system. The new directions first projected in the Taika Reforms of 646 were further refined in the Taihō Code of 701 and a revised form of the latter, the Yōrō Code of 718. These codes provided the formal basis for the overall system known as the *ritsuryō seido*. *Ritsu* refers to the penal code, and *ryō* refers to the administrative directives for the new order. *Ritsuryō seido*, however, connotes the social system as an integrated political unit under central controls, of which the legal codes were merely the rationale. The political system had its roots in China's Han dynasty, and similar legal and social institutions had been formed in Korea by Koguryŏ and Paekche in the fourth century and by Silla in the sixth century. After the T'ang court perfected the system early in the eighth century, the Korean versions were updated, and this more advanced form was soon thereafter introduced into Japan.

In the older Yamato hegemony, each of the ruling groups— the imperial household, the clan chiefs allied with the court, and the provincial chiefs subordinate to the court alliance— had its own private realm of lands and peoples. There was no single tax system or administrative bureaucracy. As the *ritsuryō* reforms sought to establish the singular and supreme authority of the emperor over all the semi-independent realms and their rulers, it instituted a new system of political subdivisions whose aministrative officers were subject to eight ministries. These ministries were supervised by a Grand Council of State (Dajōkan) directly responsible to the emperor.

The entire populace was reorganized into a new status system. Most people belonged to either of two main groups: one of ordinary peasants; the other embracing both servants and slaves (including some specialized technicians), the latter being regarded as the private property of temples, shrines, and ranking families. The ruling stratum was dominated by the imperial family and those with court rankings that accompanied all high official posts. The established clan heads who had put the emperor on the throne actually filled most official positions, though the rankings implied a merit system coordinated with the new educational agencies founded to train those who would hold offices and ranks.

Most important of the *ritsuryō* policies was the land reform aimed at doing away with independent ownership by forcibly distributing lands according to T'ang-inspired principles of equity. Henceforth, all lands were government-owned and allocated for cultivation, with appropriate tax measures designed to free people from clan controls by counting individuals as taxable units. It is not surprising that the reform government met with considerable resistance and was frequently forced to compromise with pre-*ritsuryō* customs. Since central and provincial posts were filled mostly by former clan heads, the lands granted them under the new system did not usually revert to the government upon death, as the new policy required; the lands remained private property in all save title. As in China, farmers faced with unbearable tax burdens and food shortages abandoned fields to seek work in the cities or became tenants on gradually expanded landholdings of powerful aristocratic families. This not only reduced tax income but also increased the estates of those able to control their own land, in time seriously threatening the *ritsuryō* government. Lands available for equitable distribution became increasingly scarce, and when private ownership and rights to reclaim new lands were eventually legalized, the *ritsuryō* system was destined to collapse.

Although the *ritsuryō* experiment could not be permanently maintained, during the greater part of Chinese Wave I, while the attempt was sustained, it did produce several major activities crucial to this study. Most important of these was the cultivation of a corps of trained personnel to staff the administrative organs through an official network of academic institutions for teaching and utilizing Chinese-style sciences and learning (discussed later in this chapter). The major supportive activity for this effort was a system of envoys (*kentōshi*) dispatched with varying frequency to the T'ang court, providing an important pipeline for continued importation of Chinese culture as required by the new government. An important supplementary activity was the introduction of the Sinicized Buddhism that served the T'ang regime; it served Japan as the primary instrument for the spiritual promotion of the *ritsuryō* order, providing a further source of new intellectual perspectives. Not only was each of these activities without precedent, but as a whole they remained unsurpassed in subsequent Japanese history until a full, frontal engagement with Western culture commenced in the middle of the nineteenth century.

KENTŌSHI: PIPELINE FOR CULTURAL INFLUX

Japan sent four missions, called *kenzuishi*, to the Sui court,[13] and after it was replaced by the T'ang court, resumed them under the name *kentōshi*. The four *kenzuishi* and the earliest *kentōshi* had primarily diplomatic goals: to restore Japan's fading control over southern Korea, or, after the combined armies of Silla and T'ang in 663 decisively expelled Japanese troops from the Korean peninsula (and annihilated Paekche), to placate these continental powers. The diplomatic flavor of the *kentōshi* missions soon faded, however, as Japanese energies turned entirely to the absorption of T'ang culture. The diplomatic missions from the outset included teams of officially selected students and student-priests, and their role became increasingly important in making the missions a means for the introduction of Chinese learning, science, and religion.

While Japan intended these envoys, it seems, as cultural emissaries between two equal countries, China viewed them as tribute bearers from one of the many subordinate states on her periphery. From antiquity the Chinese court had regarded itself as the "central kingdom" and treated all other states as "barbarian" subjects in its foreign relations. Activity that was essentially trade was formally designated "tribute presented to the imperial court," which in turn proffered its own "gifts," giving rise later to the term "tribute-trade" (discussed in Chapter 3).

Eighteen missions were appointed during the 265-year

[13] Mori (2), pp. 9–10.

period from the first *kentōshi* mission of 630 to the final one appointed in 894. The twelfth mission was shipwrecked, the thirteenth called off "because of unfavorable winds," and the final one canceled soon after appointment. Only fifteen actually made the full journey to China. In their early phase *kentōshi* appointments occurred with greater frequency following the Taika Reforms edict of 646, as the need for new knowledge

Table 2. Phases and Frequency of *Kentōshi* Missions

Mission Phase/Number	Year of Appointment	Number of Years Since Last Appointment	Remarks
Early			Generally proceeded by way of Korea's western coast.
I	630		(646: Taika Reforms)
II	653	23	
III	654	1	
IV	659	5	
V	665	6	
VI	669	4	
Middle			Usually passed by islands off the southern tip of Kyushu.
VII	702	33	(701: Taihō Code)
VIII	717	15	(710: Capital moved to Nara)
IX	733	16	(718: Yōrō Code)
X	752	19	
XI	759	7	
XII	761	2	Shipwrecked
XIII	762	1	Canceled: Unfavorable winds
XIV	777	15	
XV	779	2	(794: Capital moved to Kyoto)
Late			Usually sailed directly across the East China Sea.
XVI	804	25	
XVII	838	34	Last actual mission
XVIII	894	56	Canceled soon after appointment

Source: Mori (1).

became more urgent. Though the *ritsuryō* system was at least formally established with the refinements of the Taihō and Yōrō codes of 701 and 718 respectively, fulfillment of the legal provisions was another matter. The famed Nara era (710–794) marked the peak of enthusiasm for T'ang culture, and the frequency of appointed missions during this middle phase reflects a sustained sense of their importance. After the capital city was moved to Kyoto (Heiankyō), enthusiasm for the imitation of T'ang culture declined considerably as Japanese society displayed increasing independence on all fronts. That the sending of *kentōshi* missions became less and less an urgent matter is seen in the progressively long intervals between appointments in the late phase of this activity.

The *kenzuishi* and *kentōshi* missions were far more extensive in scale and organization than those sent by the Yamato court to the Chin and Southern courts in the fifth century. In the early phase a single mission consisted of only one or two ships. As the new *ritsuryō* government increased its power and wealth, missions in the middle and late phases usually had four ships per mission. A single vessel carried from 120 to 160 persons. The largest complement of personnel was that of the seventeenth mission (838), with more than 600 persons aboard four ships.[14]

A *kentōshi* mission included (apart from servants) four major groups:[15] a *diplomatic* team consisting of the chief envoy and his staff; a *technical* team made up of specialists in various fields (physicians, pharmacists, practitioners of divination, artists, jewelers, smiths, foundrymen, and other craftsmen); a *study* team with two subgroups, one of students of Chinese law, philosophy, literature, science, etc., and another of student-priests trained in Buddhist doctrine and discipline; and the *ship's crew*, which constituted about forty percent of each mission (ship's outfitting, and especially the rigging, was yet in its infancy, and the crews resorted frequently to the oars, thus accounting for the large numbers needed).

[14] Ibid., p. 118.
[15] Ibid., p. 38.

Normally one or two months were required for the full journey from Naniwa (present-day Osaka) to the T'ang capital of Ch'ang-an. Only those receiving permission from the T'ang court went inland to Ch'ang-an; the rest remained in one of the port cities. The missions generally returned to Japan in the following year, leaving only the study team in the care of the Chinese government, which provided living and study expenses for the Japanese students (student-priests received room and board at some sponsoring temple). No system of maintaining an official embassy or resident envoy was established.

Fig. 1. Stylized depiction of a *kentōshi* ship entering a Chinese port; from a scroll relating the story of the Chinese Buddhist priest Ganjin's journey to Japan.

Besides enormous quantities of gifts (e.g., silk and cotton thread and cloth, silver, quartz, and agates) sent to the Chinese from the Japanese court and considerable amounts of goods (e.g., robes, brocades, spices, and perfumes) received in return, the envoys requested and generally were granted great quantities of literature.[16] Personal budgets granted the *kentōshi* personnel by the Japanese government were quite generous,

[16] Ibid., p. 109.

and thus they were able to acquire many representative Chinese cultural items, especially books.[17] Student-priests brought home large collections of religious literature and plentiful supplies of Buddhist statues, paintings, and liturgical furnishings. According to personal tastes, copies of Chinese poetry and other writings were also acquired.

Though the diplomatic and technical teams rarely stayed in China more than one full year, they were able to learn much about T'ang institutions and techniques through direct contacts. Since members of distinguished families who had received some academic training were chosen for the diplomatic staff, they were able to assume high-level positions back home and, in many cases, realize social and political ideals nurtured while in China.[18] Technicians were chosen from families of relatively low standing, presumably on the basis of demonstrated ability. Their contributions to the remarkable technical progress made during Chinese Wave I in Japan presumably were substantial, though their names for the most part have not been preserved. The number of students selected for a given mission appears to have been usually more than ten but rarely more than twenty. The secular students were chosen from among sons of middle-ranked government officials, on the basis of some previous performance. Of those mentioned in existing historical records, twenty-six are of the secular class and ninety-two of the student-priest class.[19] Students accompanying earlier *kenzuishi* and the early- and middle-phase *kentōshi* missions remained in China for long periods, returning home only after first becoming specialists abroad. Ten or more years was common, while some stayed two or three decades. Not all returned, as some died abroad or at sea, and a few married and settled in China.[20] Later, as the academic level in the *ritsuryō* educational institutions rose considerably, it became usual for students first to gain recognition as specialists at home before studying abroad.

[17] Ibid., pp. 112–113.
[18] Ibid., p. 102.
[19] Ibid., p. 121.
[20] Ibid., p. 115.

They went to China for specialized research—in, for example, medicine or Buddhist doctrine—and stayed for comparatively shorter periods.[21]

In the same year it was appointed, the eighteenth mission was canceled at the request of the head envoy himself. No further missions were sent to the T'ang court. Official reasons for discontinuance were two: (1) T'ang society was in confusion due to civil wars, and it was increasingly difficult to guarantee the safety of the *kentōshi* (the T'ang dynasty gave way in 907 to a period of conflict under the Five Dynasties); and (2) worsened relations with Silla made the direct route across the East China Sea imperative, with greater risks of disasters at sea.[22] Other substantial reasons may also be surmised: Japanese culture had progressed sufficiently to render cultural importation less urgent; saturation had lessened the thirst for things Chinese; expansion of tax-exempt landholdings cut down government funds needed to support the costly *kentōshi* operations; and from 842 on, Chinese trading ships began visiting Japanese ports almost annually, and thus cultural importation no longer needed to await the appointment of widely separated missions or decades of study abroad by Japanese students.[23]

After cessation of the *kentōshi* missions, overseas travel by Japanese citizens was prohibited—except in the case of a few Buddhist priests given special permission. Restrictions were also placed on the entry of Chinese vessels into Japanese harbors and on the kinds of trade permitted. This policy resulted in semiseclusion from the continent, a condition that continued for some five centuries until Ashikaga Yoshimitsu (1358–1408) entered into the Tally Trade relations with the Ming court in 1401 and official relations with China were partially resumed.

KOREAN IMMIGRANTS: AN AUXILIARY PIPELINE

The peak of refugee influx from Korea to Japan was reached shortly after the allied forces of T'ang and Silla crushed

[21] Ibid., pp. 116–117.
[22] Ibid., p. 196.
[23] Ibid., pp. 187–214.

Paekche (663) and Koguryŏ (668), and Silla unified the whole of Korea. The refugees' escape to the offshore nation of Japan marked the last sizable migration of Koreans to Japan in premodern times.[24] The timing of this large influx, at the height of Japanese enthusiasm for continental culture, was most propitious. The refugees constituted an auxiliary pipeline for cultural imports, since many served as top-level government officials, envoys of the *kentōshi* missions, students to accompany the T'ang missions, and as farmers, weavers, and other technicians. They and their descendants played especially important roles in the adoption of Chinese culture while the political center was still in Nara. After the capital was moved to Kyoto in 794, however, these new citizens were largely fused with the general populace of Japan, and no special service was expected of anyone merely because he was a descendant of an immigrant family. Nonetheless, a document produced by court order in 814 to make clear which households possessed legitimate claims to official appointments and other privileges shows that about thirty percent of 1,059 households listed were descended from Korean immigrants.[25] Known as the *Shinsen shōjiroku* (New compilation of the register of families), this document is still extant.

The entire *ritsuryō* program, including the introduction of Buddhism, learning, and science, would have been impossible in so short a time without the aid of these expatriates. The same is true, of course, of the incomparably higher level of technical development achieved during Chinese Wave I. The most important immigrants for the processes of Chinese Wave I were those newly arrived after the fall of Paekche and Koguryŏ, and not the descendants of earlier arrivals of the fourth and fifth centuries following the fall of the Han dynasty, though the preparatory role of the earlier immigrants should not be underrated.

BUDDHISM AND SHINTO

The dissemination of Buddhism in East Asia began with China proper in the first century A.D. From there it spread to the

[24] Seki, pp. 138–139.
[25] Ueda Masaaki, p. 178.

Korean peninsula, first to Koguryŏ about A.D. 372 and to Paekche about A.D. 384. It was opposed by the Silla court until A.D. 527 when an imperial edict encouraging Buddhism was issued. Buddhism reached Japan first through Korean immigrants prior to Chinese Wave I, official recognition being first given in A.D. 538 when some sutras and other items were presented to the Yamato court by envoys from Paekche.[26] It continued to enter Japan from Korea, especially Paekche, until it was imported wholesale and directly as an integral part of Chinese Wave I.

In the T'ang cultural format, official emphasis was on the role of Buddhism in the stabilization and integration of national life—though this did not preclude, of course, the individual's concern for his own spiritual welfare. With the extension of T'ang power over all East Asia, Buddhism's sociopolitical role was likewise extended throughout the region. It was in this context that it was transplanted and nurtured in Japanese soil, though the Taika Reforms of 646 marked a shift from vague to systematic promotion of Buddhism as a state religion.[27] One result was the proliferation of temples throughout the land, funded mainly by the imperial court or powerful clans.

To consider the social role of Buddhism alone, however, would be to obscure the key role it also played in the diffusion of the whole complex of sinicized Buddhist learning and art in the capital region and provincial centers. Leading temples in this process were such great temples as Hōryūji, Tōdaiji, and Kōfukuji, and to a lesser degree the many official provincial temples (*kokubunji*). It may be presumed that teams of temple workmen became agents of considerable technical diffusion and that copious copying of sutras and recitation of prayers had their part in the Japanese mastery of the Chinese language and the eventual evolution of Japanese writing and literature (see p. 113).

Buddhism's role in the *ritsuryō* program did not go unchallenged. While the capital was in Nara the temples gained such vast landholdings and exercised such great economic and

---

[26] Inoue Mitsusada (2), p. 475. This date is disputed.
[27] *Nihon shoki*, 2: 332, 330, 342.

Fig. 2. Miniature three-storied wooden pagodas (approximate height, 13 cm.), each containing a printed charm (*darani*). One million were produced and distributed among ten leading temples. (Tokyo National Museum)

political influence that an attempt was made, when the capital was moved to Kyoto, to eliminate their undue influence by restricting temple construction to as few as possible in order to strengthen the emperor's hand in government.[28] That many temples were in due time erected in Kyoto indicates the degree of patronage the religion enjoyed among the nobility. This patronage, along with further extended land ownership, helped Buddhism survive beyond Chinese Wave I into the feudal age of Japan; but its vast landholdings also contributed much to the breakdown of the *ritsuryō* system.

During the historical phase preparatory to Chinese Wave I, and certainly by the Kofun period (Ancient Tomb period; ca. A.D. 250–552), there had emerged an indigenous Japanese culture centered in its spoken language, and religious, aesthe-

[28] Yazaki, p. 48.

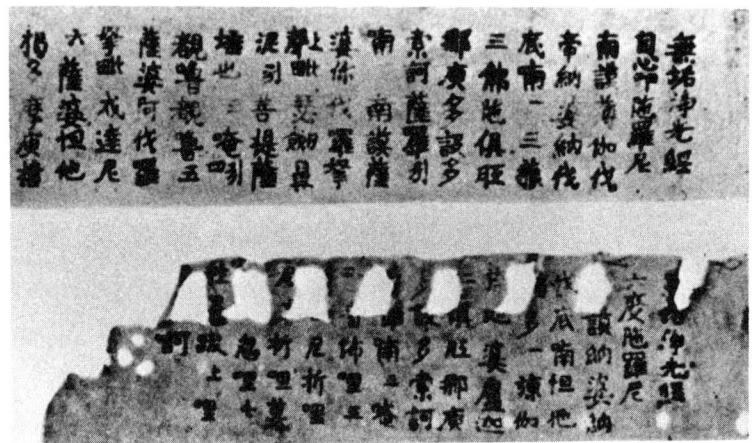

Fig. 3. Printed charms (*darani*) placed in the miniature pagodas of Fig. 2; regarded as the oldest extant printed matter in Japan (see note 36, for chapter 3). (Tokyo National Museum)

tic, and other ethnic sentiments related to the accepted forms of behavior and social organization. This early cultural formation, which included primitive Shinto as one of its key elements, constituted the substratum upon which the full impact of Chinese Wave I was received. Despite almost fanatic enthusiasm for Buddhsim in Chinese Wave I, the indigenous substratum managed to coexist with it, without any seriously disabling doctrinal or social conflict. While Buddhism enjoyed the more favored position, when the wave of cultural influx subsided the larger Shinto shrines, such as the Grand Shrine of Ise, Kasuga Shrine in Nara, and the Kamigamo and Shimogamo Shrines in Kyoto, boasted ample personnel, large landholdings, and, in a word, a very secure place in society. Shrines of more modest scale were erected far and wide in the provinces, and these too won a sure foothold among the people.

The ceremonies performed at the Yamato court in Chinese Wave I were almost exact imitations of the Chinese court rites. In time, though, these ceremonies became inextricably fused with Shinto rituals and were preserved in this form throughout succeeding centuries (some being retained in the routine rites of the imperial court today). Shinto failed, however, to develop any relations to the learning and science introduced, or to the academic system instituted. After Chinese Wave I subsided, relations between Shinto and Buddhism became an overt problem that was intellectually resolved by regarding Shinto deities as secondary manifestations of the Buddhas or Bodhisattvas, though in later generations the order of priority was sometimes reversed.[29] For all its failure to enter the *ritsuryō* mainstream and its subsequent accomodation to Buddhism, Shinto remained a unique expression of the ethnic religious and aesthetic sensibilities of the Japanese people—and remains so to the present day.

TECHNICAL INFLUX AND DIFFUSION

Transmission into Japan of agricultural techniques centered on irrigated rice cultivation marked the very beginning of the eight-century preparatory period leading to Chinese Wave I. When metalworking techniques were added, better tools stimulated farm productivity which, along with better weapons, led to the early village states. Better houses, clothing, and household implements contributed to the improved living conditions that emerged in the new agrarian folk society of Japan.

Following the collapse of Han power, the influx of immigrants, especially refugees, produced a rise in the volume of cultural influx and an acceleration of its pace. Hence, from the fourth century there appeared the first large-scale civil engineering projects in Japan: huge burial mounds and extensive irrigation works. The size of the sepulchral mounds reached a zenith with the construction of the tombs of the Emperors Ōjin (died *ca.* A.D. 310) and Nintoku (died *ca.* A.D. 399).[30] The largest known tomb is Nintoku's (located on a wide plain near Osaka), estimated to have cost 1,800,000 labor days for its construction.[31] Enthusiasm for these costly symbols of wealth

---

[29] Kamstra, p. 469.
[30] Both dates are traditional. Recent studies indicate that Ōjin died much later in the fourth century, and Nintoku's death falls proportionately later. Inoue Mitsusada (1), p. 124.
[31] Ibid., p. 130.

and power waned by the sixth century, after which construction of Buddhist temples captured the interest of emperors and clan chiefs. It is worth noting in passing that from about A.D. 400 the relatively simple Shinto ceremonial items placed in the tombs gave way to large quantities of elaborate continental artifacts, such as personal ornaments of gold and silver, and swords and armor,[32] reflecting intensified cultural inflow, presumably due to Japanese contacts with the Korean peninsula.

The large-scale irrigation projects carried out in the central district and surrounding regions, coincident with the tomb fad, often included riparian and reservoir construction.[33] In contrast to the wholly unproductive burial mounds, these projects boosted agricultural productivity and strengthened the economic and political power of the court and of the clan leaders able to undertake them. Like large tombs, however, the irrigation works required more than ordinary skills in measurement and calculation, and it is thought that Korean immigrants were relied upon for such tasks.

All skills of the preparatory stage were rather unsophisticated compared to the broad range of more advanced techniques introduced in Chinese Wave I. The immigrant families, and especially those newly arrived in the seventh century, were the central agents of transmission of the newer techniques; though after the opening of the Nara period influx came directly from China, the immigrants were the most skilled technicians available for the rapid assimilation of continental skills into Japanese society. The Japanese experience was, after all, but one phase of the expansion of Sui and T'ang culture into the surrounding East Asian realm, and contemporary Paekche and Silla far excelled Japan in the mastery of continental techniques.

Korean Buddhism provided the technical as well as the spiritual background and leadership for the buildings, statues, and ornamentation of the famous Hōryūji temple (completed in

---

[32] Kobayashi, pp. 153–154.
[33] Asahi, pp. 631–633.

607) and many others built very early in Chinese Wave I,[34] though direct Chinese influences are more evident from the Nara period, as in Tōdaiji (752) and Tōshōdaiji (759). The richly aesthetic forms of Buddhist temples, statues, ritual implements, and decorative art are better preserved in Japan than in China, where they eventually deteriorated, or in Korea, where they were ruined by successive Chinese, Mongol, and Japanese invasions.[35] Thus, a sizable legacy of Buddhist art and architecture from this period still exists in Japan and for many centuries has exerted a strong influence on the fine arts, architecture, and various crafts.

The major medium for technical diffusion, however, was capital city construction. During most of the Asuka period, and even before, the imperial residence was relocated rather frequently, but these moves were dwarfed by the construction of a new capital city called Fujiwarakyō in 694. Copied after the great T'ang capital of Ch'ang-an, it boasted large city gates, a grid street-pattern, and palace and temple structures comparable to some of the later Nara and Kyoto buildings. Unprecedented as it was, the scale of operations was escalated each time the capital was moved—in 710 to Nara (Heijōkyō) and again in 794 to Kyoto (Heiankyō). Heiankyō was more magnificent than anything seen before in Japan, though still measurably smaller than its model, Ch'ang-an, which had attained an estimated population of two million. These major moves necessitated huge concentrations of manpower, wealth,

---

[34] Cf. Kamstra, pp. 312–315, for a discussion of the original Hōryūji ground plan and other temples inspired by Paekche temple specialists.

[35] It is one of the great tragedies of East Asian history that so much devastation was wrought on Korea and Korean Buddhism by invading armies that usually sacked and burned the various capitals of Korea: Paekche's treasures were obliterated by the allied armies of T'ang and Silla; many tombs of Koguryŏ were defaced and their treasures carried off first by the Mu-yung clan of the Hsien-pi people, and then in the late seventh century by the armies of T'ang and Silla; many treasures of the Koryŏ dynasty were destroyed by Mongol invaders and great libraries were burned to the ground; Yi dynasty Korea was sacked by the armies of Hideyoshi, who carried off the finest treasures and left cities burning, while Korea's own allies in that struggle, the armies of Ming China, also looted freely. Cf. Hatada (1) and (2).

and materials, and they afforded repeated occasions to unprecedented numbers of workmen for assimilation and mastery of the architectural and decorative skills of the continent. There emerged groups of specialized technicians, some of whom were institutionalized in the *ritsuryō* system. Provincial governors were responsible for providing corps of laborers annually, facilitating diffusion of the imported techniques into outlying areas. Technical diffusion was, of course, unevenly in favor of the upper classes and the capital city over commoners and provincial centers.

The pervasive influence of T'ang culture in Chinese Wave I is uniquely exemplified in the collection of handicraft specimens preserved in the Shōsōin Repository located in the precincts of Tōdaiji temple in Nara. The nucleus of the collection is a group of objects used by the Emperor Shōmu (reigned 724–749), founder of the temple. The items cover a wide range: furniture, stationery, games, liturgical implements, musical instruments, armor and weapons, ceramics, wood and metal work, examples of weaving, dyeing and embroidery, and so on—valuable properties from the age when importation from the continent and promotion of the Buddhist complex of religion, arts, and crafts were at their peak. Yet nothing in particular remains from which to guess imperial interest in learning or the sciences.

Most of the techniques adopted were, by the end of Chinese Wave I, successfully assimilated into Japanese society, helping to open up a whole new way of life among the people. As we shall see below, the assimilation of Chinese sciences was much less successful, and the sciences had much less impact on the lives of ordinary citizens.

THE EBBING OF CHINESE WAVE I

The *ritsuryō seido* reached its peak in the Nara era, only to enter into a process of dissolution with the advent of the Heian period (794–1192). Gradual accumulation of vast, tax-exempt landholdings by the imperial family, the court nobles, and the large temples and shrines undercut the equitable land plan, and monopolization of top government posts by the powerful nobility emasculated the merit system and the academic institutions designed for leadership selection. Among the court nobles the simple dedication to Chinese and Buddhist culture so prominent in the Nara era gradually faded away, as what had promised to be a single East Asian political community was fractured into smaller regional and national units. Chinese culture was no longer a central concern of political leaders in Japan. The Heian nobility continued to regard certain Chinese cultural forms as symbols of sophistication and prestige, but no longer as the necessary substance of the new social order.

With the abandonment of the *kentōshi* in 894, official Sino-Japanese relations ceased, and contact with the continent was thereafter limited to the slim pipeline of visiting Chinese merchant vessels and more particularly to the few Japanese Buddhists who, with court permission, occasionally caught rides to the mainland on these vessels. Japan gradually broke away from direct imitation of Chinese culture and exhibited a distinct cultural character of its own. Over the indigenous cultural substratum that had formed by the beginning of Chinese Wave I was poured a new layer of Chinese-style culture which, during that wave, completely dominated Japanese culture at its more sophisticated levels. As the wave ebbed, though, there emerged, from the bottom as it were, distinctly Japanese traits that were able to maintain themselves within the overall framework of Chinese cultural influence yet develop along new lines. This suggests a reassertion of primitive Shinto—though not in the simple forms prior to Chinese Wave I, but the product of an encounter with and a partial reaction to it. The new thrust is seen positively in the reassertion of indigenous aesthetic, literary, and, to a degree, religious sentiments; and negatively in the decline of interest in Chinese science and learning even to the point of virtual disappearance of certain forms.

Confucian thought and Chinese rationalism and science proved incapable of penetrating Japanese life and thought sufficiently to survive the ebbing of Chinese Wave I. They were to win a secure, self-sustaining place in Japan only from the seventeenth century on. By way of contrast, the literary, aesthetic, technical, and even magical elements of Chinese culture were absorbed and diffused within Japanese society with relative ease, and from a fairly early stage precipitated various Japanese adaptations that held a firm and remarkably durable place in Japanese cultural traditions.

For more than three centuries the Japanese had followed the Chinese star with almost fanatic enthusiasm. Their cultural perspectives were widened to embrace most of East Asia—it was Japan's most cosmopolitan premodern era, if still naive by the standards of Ch'ang-an. Japan's first centralized state was born, in a time of tremendous upsurge of sentiment for nation-building. The great technical impact of Chinese Wave I was visibly represented in the splendid capital cities and in the Buddhist temples that graced them. But the foundations of Chinese-style learning and science that came with the whole *ritsuryō* package were left in such a shambles that it would take over half a millennium to rebuild them, and then only after a long, arduous process of reintroduction.

Learning in Chinese Wave I
The cultural gap that had to be bridged to introduce new knowledge from China for the reshaping of Japan was overwhelming. The Japanese had not even developed their own system of writing, yet they wanted to import the highest levels of Chinese learning and science, along with sinicized Buddhist teaching. From the fifth century a limited number of Japanese had, it is true, acquired some facility in reading and writing Chinese characters. Even so, documentary functions of government remained in the hands of educated immigrants and their descendants until early Chinese Wave I. Despite such slim cultural resources, the formal structure of Chinese learning as it existed in the T'ang era was, in fact, adopted and built into the *ritsuryō* system, without significant alteration of its intellectual and institutional characteristics.

Through the *kentōshi* it was the officially endorsed system of learning that was sought and introduced from China. Private scholarship, of course, had an important place in China; but in the T'ang model, state resources supported the institutions that incorporated the essentials of Confucianism into the state cult and those that nurtured and utilized the sciences for state purposes. As the intermittent state religion in T'ang times, Buddhism afforded a secondary source of ideas and values. Direct Taoist influence was scarce, if not totally nonexistent. It was encountered on the Japanese side of Chinese Wave I only as suggestions of earlier incorporations into either Confucianism or Buddhism, or as elements in some of the sciences.[36]

The Japanese attempt to adopt Chinese learning gave a great boost to the general intellectual level of Japan, broadening its scope and enriching its content—an epoch-making event in Japanese cultural history. Yet, though the intellectual imprint of Chinese Wave I endured for the rest of Japan's premodern period, the Japanese net for the reception of Chinese learning undoubtedly had some gaping holes. When the *ritsuryō* system began to crumble late in Chinese Wave I, and particularly after official relations with China came to an end in 894, it soon became evident that the core, not the marginal, elements were the most poorly received. The penetration of the basic Confucian values integral to the *ritsuryō* institutions, and especially to the academic system, as well as that of the spirit of Chinese rationalism and the norms of Chinese science, turned out to have been quite shallow. While the general cultural level rose throughout and beyond Chinese Wave I, the more purely rational interests of the Japanese gradually declined from late in the wave period. These trends can be seen in both the curricular changes within the *ritsuryō* academic institutes and the fate of these institutes in society at large.

[36] Philosophical Taoism was excluded from the Daigakuryō, and Taoism as an organized religion did not enter Japan. Exorcistic incantations recorded in Taoist documents were part of the course on ritual healing (*jugon*) in the Ten'yakuryō. Buddhist priests were formally prohibited to engage in Taoist practices, and though such practices did make their way into Japan, efforts were made to suppress them. On how the immortality concept, the most essential element in Taoist thought, developed in Japan, see Shimode.

THE RITSURYŌ ACADEMIC SYSTEM

Four different kinds of institutes were founded in the *ritsuryō* system to propagate and utilize learning and science. Three of these were located in the capital, and there was only one of each: a University (Daigakuryō), an Institute of Divination (Onmyōryō), and an Institute of Medicine (Ten'yakuryō).

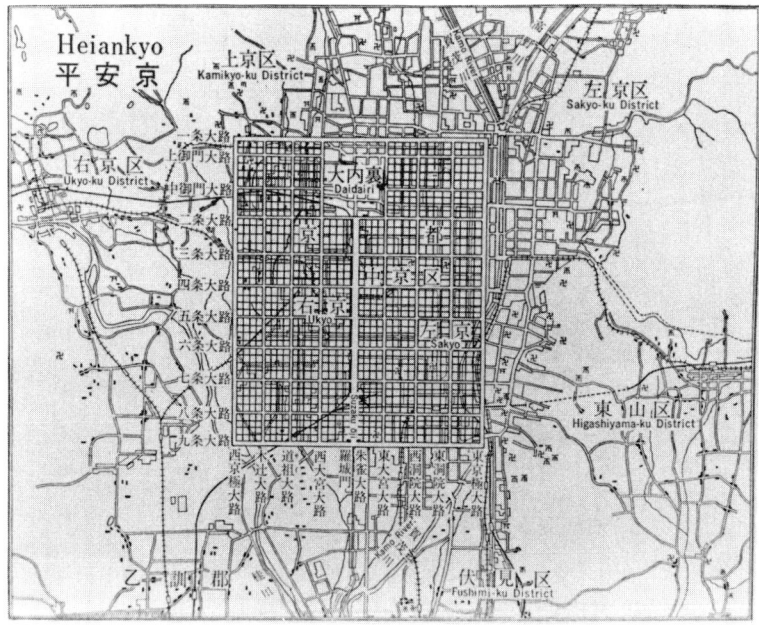

Fig. 4. City plan of Heiankyō superimposed on a map of present-day Kyoto. The Institute of Medicine (Ten'yakuryō) and the Institute of Divination (Onmyōryō) were located inside the square marked "Daidairi"; the University (Daigakuryō) was outside, to the right of the main gate (facing Daidairi).

The University was established for the express purpose of training higher-level government officials and was the central educational organ of the new government. The Institute of Divination both trained specialists and performed official tasks for the state in astrology, calendrical astronomy, and divination. The Institute of Medicine likewise undertook both the training of personnel and the actual performance of medical services. The fourth kind consisted of provincial colleges (Kokugaku) which were small-scale versions of the University. Their principal purpose was to train personnel for provincial governments, but they also trained medical doctors for service in provincial localities. Inasmuch as all four kinds of institutes were for educating public officials, all expenses were borne by the state.

These institutes were all patterned after T'ang models, though on a much smaller scale. Starting from social and intellectual bases far inferior to those of contemporary China, and in so different a cultural milieu, the *ritsuryō* academic program, though promulgated as the law of the land, was not in the beginning immediately implemented in its entirety, and some aspects of the program were never realized. With the passage of time, revisions in the program and improvements in its content were made. When the *ritsuryō* system began to crumble, however, the program of learning also fell into disrepair. Most of the institutes had ceased to perform meaningful services other than routine administrative work by the end of the Heian period (late twelfth century), though nominal staff appointments were still made.

THE *RITSURYŌ* UNIVERSITY (DAIGAKURYŌ)

Founded most probably in the year 670,[37] and administered by the ministry for ceremonials (Shikibushō), the University was an elite school for training top-echelon officials.[38] Its basic course was a doctrinal course in the Confucian classics. Preparatory courses in the proper pronunciation of the Chinese language and in writing its characters were also provided. Initially the University's curriculum consisted of only two courses: Confucian studies (*myōgyōdō*) and mathematics (*sandō*). In a 728 reform, a course in Chinese literature (*monjōdō*) and one in law (*myōbōdō*) were added. Another in history (*kidendō*) was added in 808, though this was incorporated into the

---

[37] Hisaki, p. 19.
[38] Ibid., pp. 23–24, for a comparison of the Daigakuryō with its Chinese counterpart.

literature course in 834. Mathematics was a special course in the University curriculum for training bureaucrats to perform administrative tasks such as public finance, land measurement, and construction projects. It was completely independent of the rest of the University curriculum, with no relation to students in other courses.

In addition to teaching staff, the University had a rector, five administrative staff members, and a number of clerical workers. Entrance was limited to sons and grandsons of the upper nobility (fifth rank and above), and to sons of the Yamato no Fumi and Kawachi no Fumi households of immigrant families employed by the imperial court for documentary work. Sons of the lower nobility (sixth to eighth ranks) could make special application for admission.[39] The ages of students at the time of admission fell between thirteen and sixteen. The number of years required for completion of a course of study was not definitely set. In Confucian studies, for example, a student could, upon completion of two standard doctrinal texts, take the civil service examination for a government post. The texts, content of courses, and methods of examination were prescribed in detail by law.

THE INSTITUTE OF DIVINATION (ONMYŌRYŌ)

This institute was under the central administrative ministry (Nakatsukasashō). Because its tasks were to train specialists and to perform official services, intellectual efforts in astrology

Table 3. University Curriculum and Reforms

I. Initial Curriculum (presumably 670)

Courses	Teachers	Students
Confucian studies (*myōgyōdō*)	1 professor 2 assistant professors 2 phonetics professors 2 calligraphy professors	400
Mathematics (*sandō*)	2 professors	30

Table continues, next page.

[39] Ogata Hiroyasu, p. 27.

Table 3 (continued)

II. Reform of A.D. 728[a]

Courses	Teachers	Students
Confucian studies	1 professor 2 assistant professors 3 lecturers 2 phonetics professors 2 calligraphy professors	400
Chinese literature (monjōdō)	1 professor	20
Law (myōbōdō)[b]	2 professors	10[c]
Mathematics	2 professors	30

III. Reform of A.D. 808

Courses	Teachers	Students
Confucian studies	1 professor 2 assistant professors 2 lecturers 2 phonetics professors 2 calligraphy professors	400
Chinese literature	1 professor	20
History (kidendō)	1 professor	—[d]
Law	2 professors	20
Mathematics	2 professors	20

IV. Reform of A.D. 834

Courses	Teachers	Students
Confucian studies	1 professor 2 assistant professors 2 lecturers 2 phonetics professors 2 calligraphy professors	400
Chinese literature[e]	2 professors	20
Law	2 professors	20
Mathematics	2 professors	20

Sources: Ogata Hiroyasu (1) and Hisaki (1).
[a] In A.D. 730 a new system of graduate students was inaugurated: 4 in Confucian studies, 2 each in literature, law, and mathematics, a total of 10.
[b] Legalized in the A.D. 730 reform.
[c] Increased to 20 in A.D. 802.
[d] Not specified.
[e] Kidendō incorporated here in this reform.

and calendrical astronomy were largely oriented to social, i.e., state purposes. It was not at all strange—except by modern standards—that divination was included in the curriculum and even in the institute's name.

The administrative staff of the Institute of Divination, with one director, two vice-directors, and three lesser officials, was hardly comparable to the T'ang operation, which maintained a separate Bureau of Divination (T'ai-pu shu) besides the main Institute of Astrology (T'ai-shih chu, sometimes rendered Office of the Grand Astrologer). In the Japanese facility there were only 5 on the entire teaching staff and 30 students, as against 9 professors and assistant professors for 551 students in the T'ang Institute of Astrology and 4 teachers for 45 students in the T'ang Bureau of Divination. As in China,

Table 4. Curriculum of the Institute of Divination[a]

Course	Teachers	Students	Remarks
Divination[b] (*onmyōdō*)	1	10	Plus 6 practitioners of divination
Calendrical astronomy[c] (*rekidō*)	1	10	Doctors (*hakase*) of calendrics engaged in both teaching and annual calendar production
Astrology[d] (*tenmondō*)	1	10	Professor responsible for both teaching and astronomical observation
Timekeeping[e] (*rōkoku*)	2	0	Professor in charge of 20 minor officials who announced time publicly with bells and drums

Source: Nakayama (6).

[a] This single *ritsuryō* institute combined the functions of two separate T'ang institutes: the Institute of Astrology, which handled both services and education in astrology, calendrical astronomy, and reporting time; and the Bureau of Divination, an organ for divination.

[b] Business of the T'ang Bureau of Divination; 1 director, 2 vice-directors, 37 practitioners, 2 professors, 2 assistant professors, and 45 students.

[c] Business of the T'ang Institute of Astrology: 1 professor, 2 technicians, and 41 students.

[d] Business of the T'ang Institute of Astrology; 2 professors, 5 observers, and 150 students.

[e] Business of the T'ang Institute of Astrology; 6 professors, 37 technicians, 360 students, 280 bell clerks, and 160 drum clerks (the latter two categories for public announcing of time).

though, the director of the Japanese Institute of Divination was obliged to submit to the emperor a sealed report of astronomical or geophysical abnormalities. Legal prescriptions for admission, terms of study, curriculum, and examinations, as well as for appointment to government posts, were similar to those of the University.

THE INSTITUTE OF MEDICINE (TEN'YAKURYŌ)[40]

Administered by the imperial household ministry (Kunaishō), the program of this institute was divided into five courses, with the medical course at the center of the curriculum. For medicinal needs, there was a course in the cultivation of herbs. In line with ancient Chinese medical practice, courses in acupuncture (including moxibustion) and massage (including bonesetting) therapies were offered. Equally typical of Chinese practice was inclusion of a course in exorcistic incantations (what Joseph Needham calls apotropaic medicine).

The head of this institute was aided by four staff members, and, as in other institutes, texts, curriculum, and examinations were set by law. Sons of medical families had priority for admission, though others could apply when places were available. Admitted students ranged from thirteen to sixteen years of age.[41] Periods of study varied according to the specialty.

PROVINCIAL COLLEGE (KOKUGAKU)

The provincial colleges were most probably authorized by the Taihō Code of 701, as the first record of their existence is dated 703.[42] With one college for each province (*kuni*), they were smaller versions of the capital schools, with curricula combining the courses of Confucian studies and medicine. There was one teacher for doctrinal studies and one medical doctor who both taught and practiced. The number of students ranged from twenty to fifty, depending upon the grade (based on wealth and population) of each province.[43] Each college

---

[40] The following is based mainly on Hattori (1), pp. 76–79, Yamazaki Tasuku (2), pp. 41–42.
[41] Hattori (1), pp. 78–79.
[42] Hisaki, p. 155.
[43] Hattori (1), pp. 109–111.

Table 5. Curriculum of the Institute of Medicine

Course	Teachers[a]	Students		Years required
Medicine (*i*)	11	40 {	24 Internal medicine (*tairyō*)	7
			6 Surgery (*sōshu*)	5
			6 Pediatrics (*shōshō*)	5
			4 Eye, ear, mouth, teeth (*moku, ji, kō, shi*)	4
Acupuncture (including moxibustion) (*hari*)	6	20		7
Massage (including bonesetting) (*anma*)	3	10		3
Exorcistic incantations (*jugon*)	3	6		3
Herb cultivation (*yakuen*)	2	6		—[b]

Source: Fujikawa Yū (1).
[a] A "doctor" (*hakase*) was designated head of each section except herb cultivation; the rest of the instructional staff in all sections were designated "practitioners' (*shi*).
[b] Not specified.

was required to enroll one-fifth of its students in the medical course. In the case of Saikaidō region (Kyushu), one regional college called Fugaku was established in Dazaifu, capital of Saikaidō, for all the provinces in the region. This college also included the law course in its curriculum. The doctrinal course in the provincial colleges was restricted to sons of high provincial officials, though sons of lower-ranking officials were admitted when space was available.[44] Texts, curriculum, and examinations were patterned after the capital schools.

ACTUAL CONDITIONS IN THE *RITSURYŌ* ACADEMIC PROGRAM

Given the difficulties in obtaining adequate teaching personnel and in providing necessary facilities such as libraries, it is hard to imagine the provincial colleges (if, indeed, they were actually built in all provinces) as having achieved much more than, at best, modest and, at worst, nominal academic results. Only the regional college in Kyushu appears to have been the exception to this generalization, and its real achievements are hard to ascertain.

The capital institutes of Divination and Medicine used texts representing the highest levels of learning and science of contemporary China. Whether all provisions of the curriculum were fully carried out, or if so, how fully they were understood or assimilated, remains in question. Had the materials for the course in calendrical astronomy, for example, been clearly understood and adequately utilized, it is not to be expected that Japanese calendars would have sunk to the low level discernible after the last one adopted in the year 862. The Japanese specialists in the calendrical section (*rekidō*) of the Institute of Divination were able, with the methods adopted, to turn out annual calendars; but with the lapse of time discrepancies between their calendars and solar and lunar phenomena (especially summer and winter solstices) increased. Their solar and lunar eclipse predictions increasingly failed. They were unable, however, to revise their methods or devise new ones for eight more centuries (see p. 254).

---

[44] Hisaki, p. 155.

The major changes in academic patterns were related to the University. In its curriculum Confucianism occupied the central position; it was the core course of the new system of learning. The ideal of that system was to imbue government officials with a profound knowledge of Confucianism so that they could conduct the duties of public office according to its precepts. Given the intellectual and social circumstances of Japan at that time, this was too idealistic a goal and one too far removed from specific needs. The abilities required by top-level bureaucrats were not in the realm of doctrinal comprehension, but of more practical skills, such as literary and legal competence, for use in remolding national life. The Confucian course became overshadowed by literary interests and needs. It was in response to such needs and interests that the curriculum reform of 728 was carried out to include literature and law. The supposed "core course" in Confucian studies fell gradually into decline, as most young aristocrats thronged to the literature course. For a while sons of the upper nobility monopolized this course.

By 810 a lecture hall and dormitory for students in the literature course came into existence.[45] Called Monjōin, it had two sections, East and West. After Chinese Wave I ended, these acquired a more or less independent existence under the hereditary management of certain noble households. This contributed to the undermining of the prestige and quality of the University itself. During the ninth century, when the University was still at its peak, a few powerful families among the nobility founded their own private academies. These had two functions: one, to prepare younger members of the households for the University's entrance examinations, and secondly, to supplement the education of their sons after they were enrolled in the University. The oldest of the private academies was the Kangakuin, founded by the Fujiwara clan in 821.[46] In 872 this academy was given official status as an affiliate

---

[45] Ibid., pp. 84–87.
[46] Ibid., pp. 133–134.

institute of the University (Daigakuryō bessō). Other private academies founded in Chinese Wave I were also to receive such status after the wave came to an end, and the University lost its preeminent position.

The T'ang ideal was that a student could be placed in the bureaucracy on the basis of examination results alone—a principle designed, of course, to deprive the divisive nobility of former powers and to consolidate the power of the central government. But such centralization was hardly feasible. Access to the examinations was largely restricted to powerful families, and the *yin* privilege awarded posts to the offspring of nobles and high officials outside the examination system. Even so, the examination system gave scope to talent, and mobility was considerable. Japan's *ritsuryō* government was to an even greater degree a compromise with and among the most powerful clans. Imperial rule could not easily have been maintained without their collaboration, and it was only natural that the leading nobility should gain hereditary monopolies of the top positions in the central and provincial governments. From the very beginning the principle of placement according to merit had little possibility of realization; and a "privileged-rank system" (*on'isei*) appeared on an even greater scale than in China. By this system, sons and grandsons of the imperial family and upper nobility of third rank and above, as well as sons of nobility of fourth and fifth ranks, received court ranks and administrative appointments at age twenty, irrespective of their academic records. Thus the principle of access to talent was largely compromised.

BUDDHIST LEARNING

Japanese studies in Buddhism during Chinese Wave I were limited mainly to sinicized forms, carried on mainly in the Chinese language, and the Japanese Buddhists were often quick to insist on their faithfulness to the parent Chinese traditions. While the sincerity of canonical studies for religious purposes is beyond question, circumstances under which Buddhism was promoted in Chinese Wave I inevitably colored

Buddhist learning. In keeping with the strong emphasis upon stabilizing national life, sutras were searched for doctrinal confirmation of imperial authority and of the traditional Japanese family system. The original virtue of compassion was directly linked to filial piety and hierarchical loyalty.[47]

Buddhist learning in Chinese Wave I was completely separated from the academic agencies of the *ritsuryō* system. Though studies were promoted in some of the larger temples, they were not as extensive or systematic as those offered in the government's academic program. More important, Buddhist institutions did not take over the educational curricula or tasks of those agencies after the *ritsuryō* experiment collapsed. Patronage from the ruling class provided Buddhist scholars with considerable opportunity to contribute to the general uplifting of cultural life, and yet that patronage was not an unmixed blessing. From late in Chinese Wave I and after it ended, the mystical, intuitive side of Buddhism came to the fore and on the whole contributed to lowering the rational level of the times. This kind of faith, supported by priests and nobles, was inhospitable to science.

The major challenge of Buddhist learning to the government-sponsored academic program was an academy founded about 828 by the Buddhist priest Kūkai (774–835), founder of the Shingon sect in Japan after going to China with the *kentōshi* mission of 804. The academy, called Shugei shuchiin, was for the education of the general public, regardless of social status, and especially for those not eligible for admission to the government's University or provincial colleges. Its curriculum differed from that of the University or of the private academies of the nobility in that it included both Buddhism and Confucianism. A potentially different model for Japanese education, this academy closed in 845, a decade after Kūkai died and only seventeen years after it began.[48]

Also in the 804 mission to China was the priest Saichō

---

[47] Nakamura, p. 326.
[48] Hisaki, pp. 171–173.

(767–822), founder of the Tendai sect in Japan. Saichō established his base on Mt. Hiei, near Kyoto. Scholarly traditions flourished here, as they did also at the Shingon base on Mt. Kōya, near Nara, and there was considerable rivalry between the two. Saichō hoped that the ablest of his student-monks would remain in religious service but that others would enter service to the state as lower servants in government offices or as workers in agricultural and construction projects.[49] The early influence of the Mt. Hiei monastic center, called Enryakuji, was due in large part to the third abbot of the Tendai sect, Ennin (794–864), who accompanied the last *kentōshi* mission in 838.[50] He returned in 847 to intensify the emphasis upon esoteric doctrines and magical rituals that contributed to the lowering of rational interests among Buddhists and their followers. The longer-range influence of Enryakuji was more positive, in that it served as a breeding ground for many priests later involved in the introduction of the newer forms of Buddhism that flourished late in the Semiseclusion Era. They cultivated scholarly interests which in time helped revive rational concerns in Japan.

Science in Chinese Wave I

As Japan entered Chinese Wave I without her own written language and the few documents presumably produced by immigrant scholars in the sixth century are now lost,[51] it is virtually impossible to confirm, for the preparatory period, the existence in Japan of formulated concepts of natural knowledge, much less their arrangement into any kind of system. The many social and technical advances made prior to Chinese Wave I do imply a considerable accumulation and transmission of an oral tradition of practical knowledge related

---

[49] Tsunoda, p. 133.
[50] Cf. Reischauer (1) and (2).
[51] Such include the *Teiki* (Genealogy of the imperial family) and the *Kyūji* (Collected tales of the Yamato court), said to have been written in the sixth century. Neither is extant, though both are considered to have been used as source materials for compiling the *Kojiki* and *Nihon shoki*.

to weather, farming, fishing, simple architecture, and other domestic crafts. It is impossible, however, to distinguish between indigenous and continental elements.

While the prehistory of science in Japan cannot be easily assessed, a very general picture can be drawn from various sources. Mythology abounded in the early folklore of Japan, but there was no explicit cosmology defining the structure or extent of the universe, and no astrology or astronomy are known to have been practiced prior to Chinese Wave I.[52] Social and technical developments before the wave began clearly indicate use of rudimentary skills in numerical calculations, as well as simple measurements of area and volume. Lacking written symbols, conceptualization of these operations must have been limited to a very rudimentary level. The evolution of a systematic mathematics was hardly feasible in the preparatory period, and lack of mathematics must surely have been a major block to early appreciation of Chinese calendrical astronomy.

Slightly more data are available from which to guess the situation in medicine, but they suggest no more than accumulated practical wisdom about certain symptoms, cures and medicinal materials. The famous folk tale of the "White Rabbit of Inaba"[53] implies treatment of skin irritation using a kind of bulrush named *gama*; other sources permit an assumption that medicinal effects of a wider range of roots, barks, grasses, and other substances were recognized, and a few simple surgical techniques were evidently known. A folkloric pathology often related the causes of sickness to religious be-

---

[52] Attempts were made as early as the seventeenth century to infer an independent native calendar system that predates the adoption of Chinese calendar systems. The inference rests on the fact that the dating system of the *Nihon shoki* is not entirely consistent with any Chinese system. Though opposed even then, this theory was kept alive until after World War II when it was demonstrated that the apparent uniqueness of a supposed Japanese system was merely the result of careless omissions made in copying Chinese calendrical indexes and that, when these errors are corrected, the system is revealed as purely Chinese. For details, see Nakayama (6), pp. 7–9, 44, 52.
[53] *Kojiki*, p. 94, n. 8.

liefs in the displeasure of the gods (*kami*) and personal impurity (*kegare*), and exorcistic incantations and purification rituals were widely employed as prophylaxis and therapy. These beliefs may be taken as integrally related to primitive Shinto.

Learned men among the immigrants of the fourth and fifth centuries probably knew how to use, if not produce, the Chinese-style calendar, used at least simple Chinese arithmetic in fiscal and construction work, and had an everyday knowledge of ills and cures according to Chinese medicine (or at least its Korean adaptations). Initiatives to seek the services of continental scholars cannot be firmly ascertained until the fifth century, as records of dates and events in the semihistorical *Kojiki* (Record of ancient matters; compiled in A.D. 712) and in the *Nihon shoki* (Chronicles of Japan, also known as the *Nihongi*; compiled A.D. 720) are not very reliable before A.D. 400. In the year 414 "An envoy was sent to Silla to procure a good physician," who arrived that same year, treated the Emperor Ingyō successfully, and received a generous reward before being sent home.[54] Another such request brought a physician named Te Lai (Tokurai in Japanese) from Koguryŏ in 459; it is said that he settled permanently in Japan and that his descendants continued his practice.[55] The Yamato court in 553 made requests to Paekche to send, and replace on a rotation basis, specialists in medicine, divination, and calendrics, asking also for "books of divination, calendars, and drugs of various kinds."[56] The following year there appeared ". . . a man learned in the calendar, . . . a physician, . . . and herbalists," as well as teachers of Confucianism, Buddhism, and music.[57] From the Wu state of China in the Northern-Southern Dynasties era (A.D. 420–589) a physician named Chih Ts'ung came in 562 and settled permanently in Japan. It is reported that he brought with him 164 books,

---

[54] *Nihon shoki*, 1: 315–316.
[55] Medical practitioners descended from him were known as *Naniwa no kusushi* (Medical practitioners of Naniwa). See Nihon Gakushiin (1), 5: 326.
[56] *Nihon shoki*, 2: 68.
[57] Ibid., p. 72.

including some on materia medica and a chart of the human body showing the application points for acupuncture and moxibustion. This is the earliest known record of direct access to China proper for medicine.

Although Korean influence decreased considerably in Chinese Wave I, the most important example of continued Korean contacts in the transitional Asuka era appears as an entry for the year 602 in the *Nihon shoki*:[58]

A Paekche priest named Kwal-lŭk arrived and presented as tribute books on calendar-making, astrology, and geography, and also books on the arts of prognostication and magic. At this time three or four pupils were selected and put to study under him.... Each of them mastered [his subject of study] and made it his profession.

Of this entry Nakayama Shigeru says, "This statement is the first record of a serious attempt by the Japanese to study the astronomical arts."[59]

While serious efforts by the Japanese to study Chinese sciences can first be discerned fairly early in Chinese Wave I, a systematic program did not get under way until the *ritsuryō* academic institutes were founded following the 646 Taika Reforms. Of course, the sciences introduced had already undergone a millennium and a half of complex development in China. From the Spring and Autumn era through the Warring States era and up to the two Han eras (altogether, eighth century B.C.–A.D. second century), Chinese scientific enterprise rose to a high level of creative activity. By the T'ang period this creative thrust waned. Institutionalization into a state-supported system from Han times on had assured public utilization of Chinese science, but state control seems to have thwarted its creative development.

A ready-made system of institutionalized science was available for adoption, but it proved too difficult to maintain indefinitely in Japan at this stage. The difficulty arose, however, not from any weakness of diminished creativity, but

---

[58] Ibid., p. 126. This particular translation is taken from Nakayama (6), p. 9.
[59] Nakayama (6), pp. 9–10.

rather from the main strength of the Chinese scientific enterprise—its tradition of rationalism, nurtured by Confucianism but encountered also as a strong factor in the *yin-yang* and Five Phases theories, or as an aspect of Taoist thought and practice. The Japanese were less attracted to rationalistic norms than to the more mystical Buddhist ways of thinking which, by the T'ang era, had become very peripherally related to Chinese science proper in areas such as natural philosophy and astrological, calendrical, and medical arts.

ASTROLOGY AND CALENDRICAL ASTRONOMY

HISTORICAL TRENDS

During the millennium from Shang to Han times, Chinese astrology and calendrical astronomy moved from infancy to early maturity, from emergence of early forms to the formulation of a cumulative literature and assured public status in government agencies. The objectives of Chinese astrology and calendrical science during this long history were shared in great part, of course, with those of their counterparts elsewhere. But as there was virtually no concourse with other cultural realms during this millennium, the Chinese versions evolved and maintained some distinct characteristics in internal methods and public roles. In the half-millennium from Han to T'ang times some Indian and Western influences entered to give them a certain cosmopolitan color, notably in astrological practice, but the major technical advances were purely Chinese. The fundamental forms and criteria of Chinese astrological and calendrical science remained essentially unaltered to shape all Chinese endeavors from T'ang to modern times.

The original form of the calendar used consistently until early in the twentieth century is traceable to the Shang lunisolar calendar consisting of long months of 30 days and short months of 29 days. By the fourth century B.C. (and possibly as early as the sixth century B.C.) a method was devised to compensate for discrepancies between the calendar, on the one hand, and the average (synodic) month and (tropical) year, on the other, by inserting an intercalary (leap) month seven times during every 19-year period (*chang*). (This method was used in the Babylonian calendar as early as the fourth century B.C., and the same cycle is also credited to Meton of Athens, fl. 432 B.C.) This method was used in the *Ssu-fen li* calendar; its name means "one-quarter calendar," derived from the adoption of a $365\frac{1}{4}$–day year, the same year length as that of the Julian calendar.[60] In the Later Han era, a revised form of this calendar combined four of the nineteen-year cycles into a seventy-six-year *pu* cycle (same as that of Callippus, fl. 370–332 B.C.), for greater precision of intercalation (see later in this chapter).

By Han times there appeared an important book, the *Chou-pi suan-ching* (Arithmetical classic of the Chou gnomon), considered the product of an original compilation based on observational data of the preceding two centuries. In addition to some cosmological and mathematical content (see pp. 53, 75), it also included knowledge of the solstices (available from measurement of the gnomon shadow) and determination of the equinoxes by interpolation.

Chinese astrological and calendrical work became increasingly organized during the Han period. Administrative routines of the dynastic system were firmly established according to the Confucian doctrine that the emperor rules under heaven's mandate. On this premise the imperial court undertook two specific functions: one, to compute and promulgate the official calendar; and two, to interpret unusual phenomena as the will or warning of heaven. Hence, both the teaching and practice of astrology and calendrics were incorporated into the machinery of government. Official histories of the Han and succeeding dynasties usually included special records such as

---

[60] The Roman calendar before Julius Caesar ordered a reform in 46 B.C. was a lunisolar one (but with an intercalary scheme and month lengths somewhat different from the Chinese system). It was subject to political abuse, e.g., months were added to increase taxes or permit officials to remain longer in office. In the Julian reform, lunar periods were disregarded.

the *T'ien-wen chih* (Astrological treatise) and the *Lu-li chih* (Treatise of harmonics and calendrical astronomy; later the harmonics section was separated, and it became simply *Li chih*, Calendrical astronomy).

These records give considerable evidence of progress made in various calendrical reforms from Han to T'ang times. By the beginning of the first century A.D. the *T'ai-chu li* calendar was reformed to produce the *San-t'ung li* calendar which, like all other compilations from this time on, was organized as a calendrical system complete with astronomical tables. In it and in the Later Han revision of the *Ssu-fen li*, techniques were introduced for calculating planetary positions and lunar and solar eclipses. Computation of the moon's motion was greatly improved in the second century.[61] In the fourth and fifth centuries A.D. experiments were made with various alternate intercalary cycles, such as one that intercalated 221 months in a 600-year cycle, or another that inserted 114 months in a 391-year cycle.

The Chinese calendars sought to have the fifteenth of each month coincide with the appearance of a full moon, by use of the mean synodic month of approximately 29.53 days obtained by determining the mean conjunction of the sun and moon. A method for determining true conjunction was first employed in an official calendar in 627;[62] it derived its advanced character from a calendrical system compiled about 600 by Liu Ch'o, who introduced for the first time both anomalistic solar motion and an interpolation formula for equal intervals (see later in this chapter).[63] The two most

---

[61] These improvements were made in the *Ch'ien-hsiang li* (compiled in the Kung-ho era, 178–183, Later Han) and in the *Ching-ch'u li* (during the reign of Emperor Ming Ti, 226–239, of the Wei dynasty of the Three Kingdoms period) respectively. Cf. Yabuuchi (2), pp. 452, 453.
[62] Ibid., pp. 455, 456.
[63] In addition to twenty-five official calendars listed by Yabuuchi as in use from Han to T'ang times, twenty-two that were not approved by governments for official use are also named. Though most have been lost, the *Huang-chi li* has been fully preserved in the official history of the Sui dynasty (*Sui shu*). Ibid., p. 458.

representative calendars of the T'ang era, and the primary models for succeeding generations, came one and two centuries later respectively. The *Ta-yen li*, compiled in 727 by the famous Buddhist monk, I-hsing (683–727), revised computations on anomalistic solar motion and introduced an interpolation formula for unequal intervals (discussed in this chapter). The *Hsüan-ming li*, compiled by Hsu Ang and used in China from 822 to 892, showed great improvement in the computation of solar eclipses.[64]

I-hsing was the only Buddhist monk to produce an official calendar in China. But the Buddhist, or more precisely, the Indian influence dates much earlier. Indian astrology began to make its way into China in the wake of Buddhist influence, in the third century A.D., particularly in the A.D. 230 translation of an Indian astrological work, the *Mo-teng-ch'ieh ching* (Canon of astrology based on lunar mansions).[65] An Indian astronomer named Chu-t'an Lo compiled two unofficial calendars; and Chu-t'an Hsi-ta, an Indian who served as court astronomer in the T'ang dynasty, in 718 translated into Chinese an Indian astronomical book, the *Chiu-chih li*. The title means "Calendrical system of the nine upholders," referring to the sun and moon, the five planets, and two "invisible" planets, *rahu* and *ketu*, thought to be located at the moon's nodes.[66] Only one text in a secret 120-volume set devoted almost entirely to astrology and divination, the *Chiu-chih li* belonged to a class of texts abridged (omitting theoretical sections) for use by compilers of ephemerides. Distinctly Indian features were present, such as a description of Indian numerals, including a dot symbol for zero, and a table of sine functions (of Greek origin); but these were not put to use in this book. The Chinese seem to have been more interested in the results than in the methods of Indian astronomy and astrology. On the other hand, in the middle of the eighth century A.D., a

---

[64] Ibid., p. 456.
[65] Ibid., p. 159f. The *Mo-teng-ch'ieh ching* lists the lunar mansions and the number of stars in each.
[66] For a translation in English of the *Chiu-chih li*, see Ibid., pp. 493–538.

Chinese translation, *Hsiu-yao ching* (Canon of lunar mansions and planets; A.D. 759), of an Indian astrological treatise did begin to exercise an influence on Chinese astrology. The Indian method was horoscopic and described solar, lunar, and planetary positions according to the twenty-eight (or twenty-seven) lunar mansions and the week (rather than the twelve-sector zodiac), thus indicating largely Indian, not Greek, influence.[67]

During the era of division into Northern and Southern dynasties, astronomers related to Taoism were quite active, and one Taoist, Ts'ui Hao of the Northern Wei (fourth–fifth centuries A.D.), compiled one of the unofficial calendars. By the T'ang era, however, the independent character of Taoist activity had decreased considerably, though many Taoists remained active in astrological or calendrical work.

Initial Japanese knowledge of astrology and calendrical astronomy was gained from Korea late in the pre-Chinese Wave I preparatory phase. Early in Chinese Wave I the two related branches of Chinese science were brought over directly from China and made the official business of the *ritsuryō* Institute of Divination. In 628 the Chinese system of timekeeping was adopted, and a water clock was constructed. An astronomical observatory began to function in 675. Water clocks were probably used only for timekeeping in Japan, in contrast to extensive Chinese use, along with the gnomon, for astronomical research. The Japanese acquired the gnomon at least as early as 736, but little use of it prior to the Tokugawa era (seventeenth to mid-nineteenth centuries) is recorded.[68]

The Japanese did not, of course, import the entire Chinese astrological and calendrical corpus, but the texts assigned in 757 to students of the calendrical astronomy section of the Institute of Divination indicate that some of the most advanced materials were used. Treatises on harmonics and calendrical astronomy (*Lu-li chih*) of the Han and Ch'in dynastic histories were included, as were the *Chou-pi* astronomical classic and

---

[67] Nakayama (6), p. 62.
[68] Ibid., p. 71.

the *Chiu-chang suan-shu* (Nine chapters on mathematical art, discussed later in this chapter), the most important classics in their respective fields.

Likewise, the Japanese did not adopt (nor need to) all of the many Chinese calendrical systems representing so many revisions and reforms. Only four (possibly five[69]) of the official calendars were adopted during Chinese Wave I, and as shown in Table 6, there was a considerable time lapse between Chinese compilation and Japanese adoption and use. Although the circumstances relative to adoption and use of the first of these calendrical systems are not clear, it is known that the *Ta-yen li* was brought to Japan in 735, three decades before it was adopted there for official use. An imperial edict announced adoption of the *Wu-chi li* in the year following its importation in 780, though three-quarters of a century elapsed before it was put into official use; it is said that failure to study it sufficiently forced specialists to continue using the *Ta-yen li* tables in calendrical computations.[70] The *Hsüan-ming li*, adopted four decades after its compilation, was the last Chinese calendrical system adopted by the Japanese for over eight centuries.

Table 6. Calendars Adopted for Use in Japan in Chinese Wave I

Calendrical system	Year compiled in China	Year brought to Japan	Year adopted for use in Japan
Yüan-chia li	443	ca. seventh century	same
Ta-yen li	727	735	764
Wu-chi li	763	780	858
Hsüan-ming li	822	?	862

Source: Nakayama (6), Ch. 6.

[69] It is possible that a fifth calendar system, the *I-feng li*, came to Japan along with the first and was used with it, or afterward in place of it, for less than a century. See Nakayama (6), pp. 69–70, for a discussion and probable solution of this problem.
[70] Ibid., p. 70.

In the second and fourth of these calendars the Japanese had access to the best of Chinese calendrical science. The Japanese officials charged with calendrical duties appear, however, to have concentrated on the practical necessities of introducing and implementing the Chinese date-keeping apparatus, hardly showing their mentors' enthusiasm for constant revision and reform. From the tenth century the office of court astrologer in Japan became hereditary, and astrology and calendrics in general experienced a decline from the tenth through the sixteenth centuries. The Japanese attitude toward calendars was utilitarian, based more on political than scientific concern. And the tradition of portent astrology for state purposes interested them less than it did the Chinese, as we shall see below; the Japanese were drawn, rather, to the arts of individual fate prognostication.

INTELLECTUAL TENDENCIES

The Chinese belief that the emperor conducted the affairs of state under the mandate of heaven, a doctrine rooted in the Chou period and firmly established by Han times, was not always anthropomorphized to make the emperor's mandate simply subject to the "will of heaven." The political world was meant to be concordant with the natural order, with the organismic pattern of nature. That pattern was to be discerned through meticulous observation of natural, and especially celestial, phenomena, in two aspects. On the one hand, regularity in solar, lunar, and planetary movements was regarded as one manifestation of the natural order, or what might be called the "law of heaven," and the task of studying this aspect was institutionalized in China in the Institute of Astrology, the *T'ai-shih chu* (the middle term, *shih*, is the basic one for "history," though it originally also meant "one who makes the calendar"). Producing a calendar based on observed regularities and revising it when significant differences between the calendar and new data were noted, constituted the technical functions of this agency. Production and promulgation of the calendar were among the principal tasks of the emperor, whose title (T'ien-tzu) meant literally "Heaven's son."

Emperors were also quite sensitive to irregularities in the natural order; observing and reporting such apparent anomalies as lunar and solar eclipses, comets, meteors, and the conjunction or occultation of certain planets, or meteorological peculiarities like odd-shaped cloud formations, or even terrestrial abnormalities like earthquakes and floods, were also the responsibility of the Institute. It also had the job of divining the import of these occurrences according to the *yin-yang* art, as the "will" or "warning" of heaven, although this function was supplemented by a separate Bureau of Divination, the T'ai-pu shu. In Japan both the technical and divinatory functions were combined in a single institution, the Institute of Divination (Onmyōryō; lit., "Institute of *Yin-Yang* Arts"). While the combination may have reflected a simple social fact—difficulty in finding enough trained staff for even a small-scale version of the two T'ang agencies—the name chosen for the Japanese institution suggests a preference for divination over astrology, calendrical astronomy, and timekeeping.

Cosmological speculations appear in the oldest Chinese classics, but the first treatise to lift cosmology above sheer fantasy and folklore was the *Chou-pi suan-ching*, with which the *Kai-t'ien* ("sky as a cover") theory was closely connected.[71] Actually the *Chou-pi* classic contained two versions of this theory: a primary or earlier one, thought to have originated sometime between the sixth and fourth centuries B.C., and a secondary or later model originating in the first or second century A.D. Although both models regarded the shape of the sky as circular and the earth as square, the earlier one viewed both heaven and earth as flat and parallel, while the later version viewed them as parallel but curved. Both models were oriented to the central place that the North Pole had in Chinese astronomy, in contrast to the zodiacal framework of the Western cosmos. Though neither model contained rigorous geometrical concepts, they are thought to have expressed well the available information of their times.

[71] Based on Ibid., pp. 24–39.

Another theory, the *Hun-t'ien* ("spherical heaven"), was based on the view that the sky's shape is spherical, and it was closely related to development of the armillary sphere (*hun-i*). Both the theory and the sphere are thought to have been in use by the first century A.D. This theory is known as the "hen egg" theory, because heaven was viewed, by analogy, as corresponding to an egg's shell, and earth to its center, the yolk. But the sphericity attributed to heaven did not lead to a concept of earth's sphericity; the armillary sphere was not followed by terrestrial globes. The egg-yolk analogy had to do with the position, not the shape, of the earth.

Cosmological speculation declined by the T'ang era, when intercultural intercourse peaked, and no lasting influence of Indian or Western cosmologies is discerned. In the Greek tradition, geometric models, especially those involving circles or combinations of circles, were used in calculating the orbits of the sun, moon and planets. But in the Chinese tradition, no cosmological models of any kind—the *Kai-t'ien*, *Hun-t'ien*, or any other—were used in actual calculations. In lieu of such models, though, numerical calculations of astronomical data were highly developed.

Japanese students in the Institute of Divination were required to read the *Chou-pi suan-ching* and the *T'ien-wen chih* (of the Chin dynastic history, *Chin shu*), "the most substantial discussion of early Chinese cosmology." But Japanese specialists were, if anything, less interested than the Chinese in cosmology. The first reference to Chinese cosmology by a Japanese author did not come until A.D. 1414.[72]

Two kinds of astrology entered Japan from China during Chinese Wave I: portent astrology for political purposes and individual fate calculation. The former sought to correlate celestial, meteorological, and seismological portents with the social phenomena related to imperial rule or major political events.[73] The astrological records (*T'ien-wen chih*) of the dynas-

---

[72] The reference was made by a court astrologer named Kamo no Arikata in his *Rekirin mondōshū* (Collected dialogues on the calendar). Ibid., p. 43.

[73] On the Chinese background and Japanese adaptations of portent astrology, see Ibid., pp. 44–54.

tic histories contain voluminous details of portents in chronological order (though after Han times, increasingly less attention was paid to correlation with terrestrial occurrences), ways in which they were interpreted according to *yin-yang* and Five Phases principles, and measures taken to avoid the direst consequences. From these records, certain characteristics are clear. Portent astrology was derived from empirical observations; interpretations were for public, not private affairs; and all data were kept strictly within government control. Lunar eclipses were considered far less serious than solar eclipses, partly because the periodicity of lunar cycles was known before court rituals attained a fixed pattern. Planetary motions acquired astrological significance at an early stage (fourth century B.C.), and even after their cyclic behavior was recognized. Knowledge of mean synodic periods improved from the Han era, but this was not enough to predict planetary phenomena. Their political significance as portents remained.

The Japanese were certainly exposed to Chinese views on portent astrology through the Institute of Divination's use of such texts as the "T'ien-kuan shu" (Record of heavenly offices, i.e., constellations) of Ssu-ma Ch'ien's (135? B.C.–93? B.C.) *Shih chi* (Records of the Grand Historian), and of the astrological treatises of the Former Han and Chin dynasties (*Han shu* and *Chin shu* respectively). During Chinese Wave I the Emperor Tenchi (reigned 661–671) and the Empress Genshō (reigned 715–724) might be singled out as examples of Japanese rulers who manifested some hypersensitivity to heavenly portents. On the whole, though, neither emperors nor the nobility in Japan took such portents very seriously as guides to state affairs. Nonetheless, considerable observational data is found in the official Japanese documents compiled at the time.

The principal source of individual fate calculation in China was the *yin-yang* and Five Phases tradition, which afforded the fundamental concept of correlation to the natural order.[74] Specific indications of one's fate were derived from the Five

---

[74] On Chinese and Japanese uses of fate calculation, see Ibid., pp. 54–57.

Phases, not by the entirely possible but exceedingly difficult correlation with planetary periodicities, but by means of purely numerical cyclical designations, based upon calendrical indexes for the year, month, day, and hour of an individual's birth. The horoscopic tradition of India and the West determined the planetary positions of the year, month, day, and hour of one's birth by charts based upon actual observation. On a lower level, Chinese fortune-telling had already in Han times developed intricate schemes for determining lucky and unlucky days, directions, and other propitious criteria for human behavior. It became even more complicated during the Six Dynasties era (third to sixth centuries).

Japanese students in the divination course of the Institute of Divination studied such Chinese texts as the *Chou i*, or as it more commonly known in Japan and the West, *I-ching* (Book of changes), and other classical texts of Chinese divination. These highly metaphysical treatises do not appear to have been widely read or well understood by the Japanese in Chinese Wave I; it was much later, from the fourteenth century, that divination based on the *I-ching* was once again taken seriously, this time by Zen priests. It was divination based on the calendar, and other cruder everyday practices, that from Chinese Wave I penetrated deeply into Japanese society and remained widely accepted up to and including modern times.

Calendrical divination took the form of an almanac (in Japanese, *guchūreki*; lit., "annotated calendar"), produced by the Institute of Divination, following T'ang practice, by the first day of the eleventh month for the coming year.[75] It was written in Chinese characters, handcopied, and distributed to the court nobility and government officials. The preparation process was twofold: once the purely computational details were completed, hemerological notations were appended to form the almanac. The more technical material contained: the number of days proper to the months, long and short; the

---

[75] The following several pages on prognostication associated with the almanac are based mainly on Suzuki Keishin.

position of any intercalary month; the dates of the vernal and autumnal equinoxes and the summer and winter solstices; tabulation of sunrise and sunset times; and predicted dates of any solar or lunar eclipses indicated. Extracalendrical matter included: predicted fortunes for each day of the year; lucky and unlucky directions relative to each day; the direction assumed by the God of the Year's Virtue (Saitokujin); and various taboos for each day. This extracalendrical content derived from Taoism, Buddhism (especially *Sukuyōdō*), and particularly the *yin-yang* and Five Phases tradition. Because of the pervasive influence of this tradition in Japanese society from ancient to modern times, its basic structure and some of the Japanese adaptations are outlined below (as a general system, without reference to historical development).

The system was based on a sexagenary cycle derived from a method of correlating Ten Celestial Stems and Twelve Terrestrial Branches.[76] The ten stems originally designated the sequence of the days in a ten-day unit, three of which made a month. The names of the ten stems are listed in Table 7. The Japanese *on* readings are adaptations of Chinese readings; the *kun* readings are native Japanese words to which the characters were assigned when borrowed from China. The middle *no* component of the *kun* readings is a connecting term. The prefixes represent the Five Phases: *ki* = Wood, *hi* = Fire, *tsuchi* = Earth, *ka* (*ne*) = Metal, and *mizu* = Water. The suffix *e* means "elder brother," i.e., the *yang* aspect, and *to* the "younger brother," or *yin* aspect of the respective phases.

The twelve branches (see Table 8) designated the twelve months, or roughly a year. The twelve branches also were assigned to the Five Phases, in an order explained later in this section.

The ten stems and twelve branches were coupled to form a sexagenary cycle. Given the correlation of both stems and branches with one of the Five Phases, the sexagenary cycle permitted a twofold phase determination for each year, as

---

[76] For background on the sexagenary cycle, see Needham, 3: 396–398.

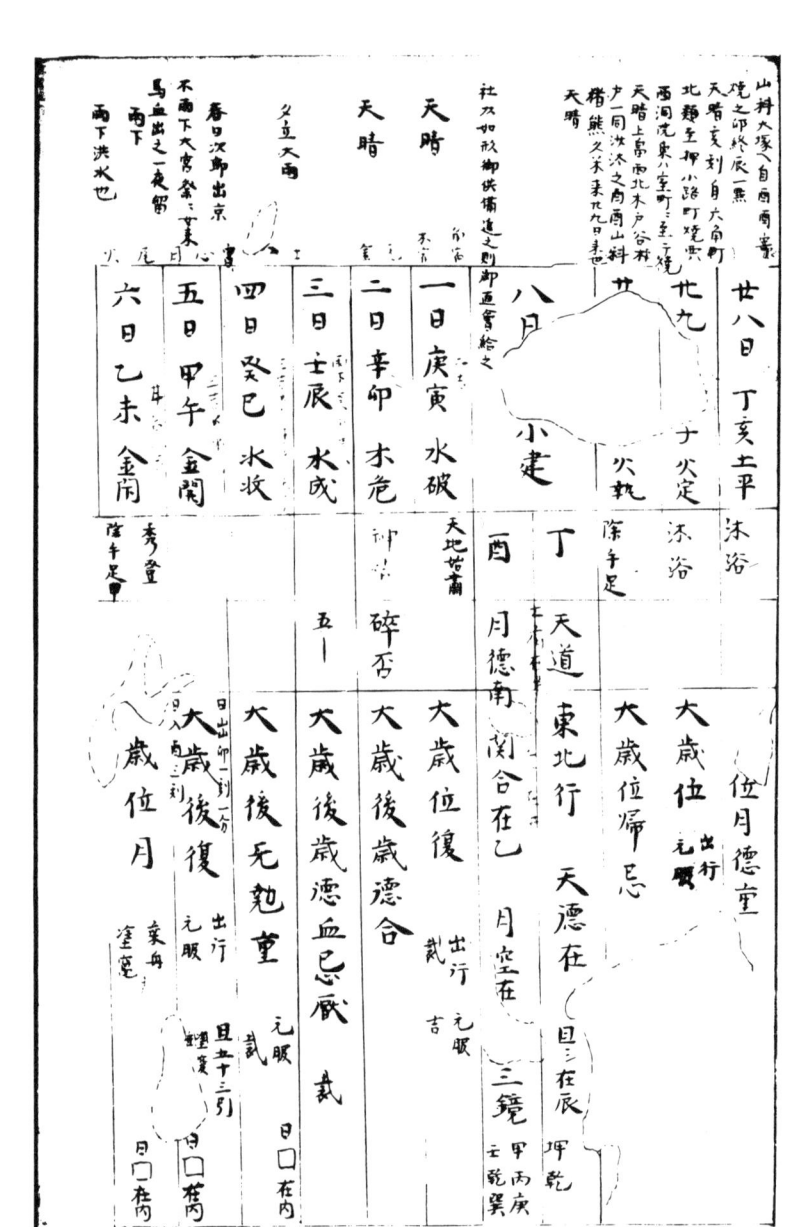

Fig. 5. Handcopied 1795 version of a handwritten annotated calendar (*guchūreki*) for the year 1411; days of the seventh month are shown on the right, those of the eighth month on the left. Small characters just above the top horizontal line indicate days of the week; top characters in the first horizontal column given the days of the month, and just below are the phases of the sexagenary cycle. Lower columns include various

十七日丙午水收	十六日乙巳火成	十五日甲辰火危	十四日癸卯金破	十三日壬寅金執	十二日辛丑土定	十一日庚子土平	十日己亥木平	九日戊戌木満	八日丁酉火除	七日丙申火建
			望	陳年甲	陳年甲	鴻雁来候癸外	沐浴	沐浴	沐浴 淋浴	淋浴
天陰小雨下	天晴	天晴 雨下 駿河上洛	天晴	天晴	天晴	天晴	天晴	天晴但時雨主	洪水也 洪水無之後日間作	近年走程序營炊第

indications of good and bad fortune, good and bad days for various activities, etc. The top margin shows daily diary entries, such as a flood (second from right), "clear with occasional rain" (fourth from right), and records of the comings and goings of persons. (The National Archives, Japan)

Table 7. The Ten Celestial Stems

	(A)	(B)	(C)	(D)	(E)	(F)	(G)	(H)	(I)	(J)
Character	甲	乙	丙	丁	戊	己	庚	辛	壬	癸
Chinese	chia	i	ping	ting	wu	chi	keng	hsin	jen	kuei
Japanese (on)	kō	otsu	hei	tei	bo	ki	kō	shin	jin	ki
Japanese (kun)	kinoe	kinoto	hinoe	hinoto	tsuchinoe	tsuchinoto	kanoe	kanoto	mizunoe	mizunoto

Table 8. The Twelve Terrestrial Branches

	(1)	(2)	(3)	(4)	(5)	(6)	(7)	(8)	(9)	(10)	(11)	(12)
Character	子	丑	寅	卯	辰	巳	午	未	申	酉	戌	亥
Chinese	tzu	ch'ou	yin	mao	ch'en	ssu	wu	wei	shen	yu	hsü	hai
Japanese (on)	shi	chū	in	bō	shin	shi	go	bi	shin	yū	jutsu	gai
Japanese (kun)	ne	ushi	tora	u	tatsu	mi	uma	hitsuji	saru	tori	inu	i
Symbolic animal	rat	ox	tiger	hare	dragon	snake	horse	sheep	monkey	cock	dog	boar

will be clear shortly. First, the sexagenary cycle—applicable to cycles of sixty days or sixty years—was constructed as shown in Table 9, where the first year in the cycle is A_1, or *chia-tzu*, the second year is B_2, or *i-ch'ou*, and so on to the sixtieth year, J_{12}, or *kuei-hai*. To show how these stem-branch couplings gained their significance for divination, an additional word on the Five Phases is needed.

Table 9. Sexagenary Cycle Coordinating the Ten Stems and Twelve Branches.

A_1	B_2	C_3	D_4	E_5	F_6	G_7	H_8	I_9	J_{10}
A_{11}	B_{12}	C_1	D_2	E_3	F_4	G_5	H_6	I_7	J_8
A_9	B_{10}	C_{11}	D_{12}	E_1	F_2	G_3	H_4	I_5	J_6
A_7	B_8	C_9	D_{10}	E_{11}	F_{12}	G_1	H_2	I_3	J_4
A_5	B_6	C_7	D_8	E_9	F_{10}	G_{11}	H_{12}	I_1	J_2
A_3	B_4	C_5	D_6	E_7	F_8	G_9	H_{10}	I_{11}	J_{12}

The Five Phases—Wood, Fire, Earth, Metal, and Water—were thought to characterize everything in heaven and earth. The five planets, for example, were originally regarded as embodying the essence of the energy (*ch'i*) of the Five Phases. By observing the movements of the planets—unpredictable for many centuries—one could know the changes in the energetic cycles that govern all natural phenomena. Many other sets of five (times, things, qualities, etc.) were also correlated with the Five Phases. Table 10 shows some examples.

Table 10. The Five Phases and Correlated Phenomena in Sets of Five

Five Phases	Five Planets	Five Directions	Five Seasons	Five Colors	Five Organs
Wood	Jupiter	east	spring	green	liver
Fire	Mars	south	summer	red	heart
Earth	Saturn	center	end of summer	yellow	spleen
Metal	Venus	west	autumn	white	lungs
Water	Mercury	north	winter	black	kidneys

Assignment of the relations, and many, many others, was quite arbitrary in some instances, though not always. The "end of summer" season is more than a mere convenience; it was the late phase of summer, called the "Greater Heat" (*Ta-shu*), one of the fortnightly periods that made up a full year (Table 11). Likewise, "center" had specific meaning in the context of a circumpolar astronomy.

The functions of the Five Phases were represented by two cycles of mutual production and mutual conquest, whose respective sequences may be diagrammed as follows:

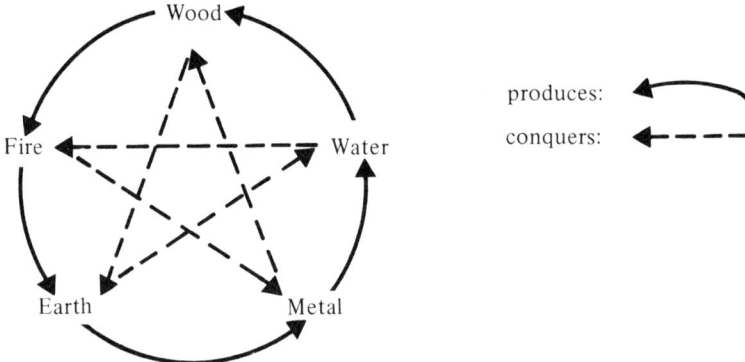

Fig. 6. Representation of the two cyclical functions of mutual production and mutual conquest of the Five Phases.

In correlating the Five Phases with the ten stems and twelve branches, it was necessary to draw on the twofold aspects of *yin* and *yang* so that each Phase could accommodate two stems:

yang	yin	yang	yin	yang	yin	yang	yin	yang	yin
chia	i	ping	ting	wu	chi	keng	hsin	jen	kuei
Wood		Fire		Earth		Metal		Water	

The *yang* quality of *chia* makes it Wood's "elder brother" (superior), and the *yin* quality of *i* gives it a "younger brother" (inferior) position, and the same relations hold true for the stems related to the other Phases.

In correlating the Five Phases with the twelve branches, Earth was assigned four branches, the other Phases two each:

```
         ┌──────────────── Water ────────────────┐
 tzu  ch'ou  yin   mao   ch'en  ssu   wu   wei  shen   yu   hsü   hai
  |    \ /   |    \ /     |    \ /    |   \ /    |    \ /    |
Earth  Wood  Earth  Fire  Earth  Metal  Earth
```

Thus, each unit in the sexagenary cycle (for days or years) corresponds to some combination of two Phases, or a double dose of one. For example, A_1 is *chia-tzu*, hence it corresponds to a Wood-Water combination; B_2 is *i-ch'ou*, a Wood-Earth combination; and so on, to J_{12}, or *kuei-hai*, a Water-Water combination.

These correspondences between stem-branch units and two-fold or double Phase units are used to determine the lucky and unlucky prospects for days and years, according to the cycles of mutual production and conquest of the Five Phases. When the two Phases assigned to a stem-branch unit are in a production relation, such as Wood-Water (as in A_1), the prospects for the year or day are good. When the two Phases are in a conquest relation (as in B_2, where Wood conquers Earth), the prospects are bad. When the Phase unit is the double of one Phase, the energy of the cosmos is considered out of balance, and things on earth do not go well. The fifty-seventh unit in the sixty-phase cycle is *keng-shen*, a stem-branch combination with a Metal-Metal correspondence; consequently, on the fifty-seventh day or in the fifty-seventh year of a cycle, the energy of Metal fills heaven and earth, making everything cold, including human hearts. But the fifty-eighth unit, *hsin-yu* (Japanese, *shin-yū*), also has a Metal-Metal correspondence and is worse than the fifty-seventh unit because it has a *yin*-Metal quality, making things colder than the fifty-seventh's *yang*-Metal quality. The years of *hsin-yu* were regarded as very bad years, both in China and Japan; era names (in Japanese, *nengō*) were usually changed in such years in Japan.

During the period 781–1861 there were nineteen *hsin-yu* years, and in Japan era names were changed in sixteen of them (exceptions: 841, 1561, and 1621). Excluding 1621, the Edo period had four other *hsin-yu* years: 1681, 1741, 1801, and 1861—all of which saw era names changed. The year 1861 was to have been only the second year of the Man'en era; it was made rather the first year of the Bunkyū era. In China, by contrast, only four changes of reign name occurred in *hsin-yu* years between 781 and 1861.

From very early times in China there was a fascination with magic squares or other figures with numbers arranged so that arithmetical operations done in specified directions yielded certain results.

4	9	2
3	5	7
8	1	6

Fig. 7. Representative magic square.

The figure above is a well-known magic square in which addition along any line, column, or diagonal yields the sum 15. It was believed that when the legendary Emperor Yu of the mythical Hsia dynasty (actually prehistoric) quelled the flooding waters of the River Lo, there emerged a turtle with dots distributed on its back in the same positions and numbers as in the diagram. This was received as a heaven-sent omen.

Later on, seven colors were combined with the Five Phases and their related planets in the spaces of the magic square:

4 green Jupiter (Wood)	9 violet Mars (Fire)	2 black Saturn (Earth)
3 blue Jupiter (Wood)	5 yellow Saturn (Earth)	7 red Venus (Metal)
8 white Saturn (Earth)	1 white Mercury (Water)	6 white Venus (Metal)

1-white-Mercury (Water)
2-black-Saturn (Earth)
3-blue-Jupiter (Wood)
4-green-Jupiter (Wood)
5-yellow-Saturn (Earth)
6-white-Venus (Metal)
7-red-Venus (Metal)
8-white-Saturn (Earth)
9-violet-Mars (Fire)

Fig. 8. Magic square with Five Phases Combined with planets and colors

The above diagram was considered the standard arrangement in the nine spaces of the color-planet-Phase combinations. But the places were thought to change with the passage of time sequences (whether year, month, day, or hour), by advancing in a numerically decreasing order from left to right, top to bottom, as shown in the sets below:

```
  (1)        (2)        (3)
4 9 2      3 8 1      2 7 9
3 5 7      2 4 6      1 3 5
8 1 6      7 9 5      6 8 4

  (4)        (5)        (6)
1 6 8      9 5 7      8 4 6
9 2 4      8 1 3      7 9 2
5 7 3      4 6 2      3 5 1

  (7)        (8)        (9)
7 3 5      6 2 4      5 1 3
6 8 1      5 7 9      4 6 8
2 4 9      1 3 8      9 2 7
```

Fig. 9. Magic square representation of sequential changes of phase-planet-color combinations

The way of assigning the nine-place sets to the sexagenary cycle, say for years, was to assign set (5) to the first year, and sets (6), (7), (8), ... and so on to succeeding years. After completion of a full cycle of 60 years, the first year in the second cycle corresponds to set (2); and after another 60-year cycle, to set (8), and so on. Only after 180 years would the original year-set combination recur; hence, it was thought that "everything happens in 180-year cycles." Application of the set sequences to months and days as well as years began in the T'ang period; inclusion of hours in the scheme was started in the Sung period. In Japan, the sets were usually used only for years and days.

Dangerous directions were also derived from the sets. Some applied generally; for example, the related direction of the 4-green-Jupiter set (east) and its opposite (west) would be dangerous directions toward which no action should be taken in years (or on days) of set (4). Such discouraged actions might include travel, constructing buildings, etc. For an individual, the central position on the set for the year (or day) of one's birth was considered his "life-determinant" and the direction indicated by that particular number-color-planet unit at any given time (year, month, day, or hour) was regarded, along with its opposite direction, as most unlucky for him.

Another mode of day-counting was based on a six-part sequence which, after introduction to Japan, was reordered, renamed, and reinterpreted. The Japanese names and sequence are:

Senshō: Will win if you go first.
Tomobiki: Do not hold a funeral, or you will soon have another.
Senpu: Will lose if you go first.
Butsumetsu: "Buddha's death," a very bad day for doing anything.
Taian: Very good day for anything (many weddings are still scheduled on this day).
Shakku: Good for doing things in midday, but not in the morning or evening.

Senshō is made the first day of the first and seventh months, the others following daily in sequence, the sequence repeating itself until the end of each month. The assignments to first days of the other months are: *Tomobiki* in the second and eighth months; *Senpu* in the third and ninth months; *Butsumetsu* in the fourth and tenth months; *Taian* in the fifth and eleventh months; and *Shakku* in the sixth and twelfth months.

Another mode of determining daily fortunes was based on the twenty-four fortnightly periods (see Table 11); twelve of these were called "medial *ch'i*" (*chung-ch'i*), and between them were interspersed the remaining twelve, called "nodal *ch'i*" (*chieh-ch'i*), analogous to the nodes of the bamboo. Correlated with the nodes were the twelve branches, which were given special names in Japan: *tatsu, nozoku, mitsu, taira, sadamu, toru, yaburu, ayau, naru, osamu, hiraku,* and *tozu*.

Tatsu is assigned to the first day of *tora* (branch 3) after the *ch'i*-node fortnight of the first month, followed by the rest in sequence. The first day of *u* (branch 4) after the *ch'i*-node fortnight of the second month begins another sequence starting with *tatsu*, and so on. *Tatsu* is assigned to the successive order of the twelve branches following the *ch'i*-node fortnight of each month. The *tatsu, nozoku, mitsu,* ... series had its own implications for good or bad fortune. A few examples:

Tatsu: Good day, but avoid weddings, medical care, etc.
Nozoku: Good day for revisions (such as haircuts, cutting one's nails, starting medical care, etc.), but bad for other things.
Mitsu: Good for ceremonies and prayers to the gods or Buddha, etc., but not for other things.
Taira: Good day, but not for initiating activity (such as medical care, seeding, etc.).

The daily advice based on these correspondences was written in various columns of the almanac. Reactions to the advice are recorded, as it was not uncommon to use the margins of the almanac as a kind of personal diary. Absorbed widely in all segments of Japanese society from the late Heian era on, some of these various modes of fate prognostication are still accepted today by some as conventional guidelines for decisions of particular importance, such as selection of propitious days for weddings, funerals, business deals, trips, the direction a new house should face, and the like.

Table 11. The Twenty-four Fortnightly Periods

Number[a]	Character	Chinese	Japanese	English equivalent	Beginning[b] (Gregorian calendar)	
1	冬至	Tung-chih	Tōji	Winter solstice	Dec.	22
2	小寒	Hsiao-han	Shōkan	Lesser cold	Jan.	6
3	大寒	Ta-han	Daikan	Greater cold		20
4	立春	Li-ch'un	Risshun	Beginning of spring	Feb.	4
5	雨水	Yü-shui	Usui	Rain waters		19
6	驚蟄	Ching-chih	Keichitsu	Awakening of creatures (from hibernation)	Mar.	6
7	春分	Ch'un-fen	Shunbun	Spring equinox		21
8	清明	Ch'ing-ming	Seimei	Clear and bright	Apr.	5
9	穀雨	Ku-yü	Kokuu	Great rain		21
10	立夏	Li-hsia	Rikka	Beginning of summer	May	6
11	小滿	Hsiao-man	Shōman	Lesser fullness (of grain)		22
12	芒種	Mang-chung	Bōshu	Grain in ear	June	6
13	夏至	Hsia-chih	Geshi	Summer solstice		22
14	小暑	Hsiao-shu	Shōsho	Lesser heat	July	8
15	大暑	Ta-shu	Taisho	Greater heat		23
16	立秋	Li-ch'iu	Risshū	Beginning of autumn	Aug.	8
17	處暑	Ch'u-shu	Shosho	End of heat		24
18	白露	Pe-(pai-)lu	Hakuro	White dews	Sept.	8
19	秋分	Ch'iu-fen	Shūbun	Autumn equinox		23
20	寒露	Han-lu	Kanro	Cold dews	Oct.	9
21	霜降	Shuang-chiang	Sōkō	Descent of hoar frost		24
22	立冬	Li-tung	Rittō	Beginning of winter	Nov.	8
23	小雪	Hsiao-hsüeh	Shōsetsu	Lesser snow		23
24	大雪	Ta-hsüeh	Taisetsu	Greater snow	Dec.	8

Sources: Needham, 3: 405; Yabuuchi (2), pp. 446–451. English equivalents are adapted from Needham; the order of periods and approximate beginning dates according to the Gregorian calendar follow Yabuuchi.

[a] All odd-numbered periods are medial *ch'i* (*chung-ch'i* 中氣); even-numbered periods are nodal *ch'i* (*chieh-ch'i* 節氣).

[b] All dates are approximate and may vary one day before or after.

The purely Indian astrology that came to China first in the translation *Mo-teng-ch'ieh ching* (A.D. 230) used the twenty-eight lunar mansions (or "lunar lodges," as some now say, because the planets do not reside, but merely spend the night there). It was not until the fifth or sixth century that the Indians were able to calculate planetary positions by Greek methods. In 759 Amoghavajra (Pu-k'ung) translated the *Hsiu-yao ching* (Canon of lunar mansions and planets), in which horoscopic practice was explicitly though sketchily explained. This canon was supplemented by the more detailed *Ch'i-yao jang-tsai chueh* (Formulas for avoiding calamities according to the seven luminaries), an early-ninth-century work by the Buddhist Chin Chü-ch'a.[77] It contained planetary tables that could actually be used for casting horoscopes. The Japanese practice of Indian astrology, based on the *Hsiu-yao ching* (in Japanese, *Sukuyōkyō*) and called *Sukuyōdō*, began late in Chinese Wave I under Buddhist promotion and aristocratic patronage but flourished only after the end of the period, as explained in the next chapter.

LOGICAL ASPECTS

From the time of its earliest known use in Shang times to the time of its replacement by the Gregorian calendar when the last (Ch'ing) Chinese dynasty collapsed early in the twentieth century, the Chinese calendar was a lunisolar calendar. Such a calendar is far more complicated than one based solely on either solar or lunar cycles, as it must attempt correlation of two basically irreconcilable periods: the tropical year and the synodic month. Possession of a fairly accurate lunisolar calendar by the Chinese in early times presupposes an accumulation of considerable observational data, as well as frequent and persistent efforts at evolving and improving calendrical methodology.

Historically, calendars have been constructed by making various adjustments to such natural units of time as the rotation of the earth (day), the waxing and waning of the moon

[77] Ibid., p. 62.

(month), and the round of the seasons (year). In the ancient cultures of China, Babylon, India, and Greece, calendars were based on combined lunar and solar calculations, that is, one full phase of the moon's undulation was made one month, while one year was made to coincide with one full cycle of the seasons, or in our terms, the time of one full revolution of the earth around the sun. As an average lunar cycle is 29.53 days, a purely lunar calendar can be easily arranged with 6 long months of 30 days and 6 short months of 29 days, allowing the days of the months to coincide almost perfectly with the successive shapes of the waxing-waning moon and a particular day of the month to coincide with a particular shape of the moon, as when the moon is invisible on the first and full on the fifteenth. A lunar calendar's 12-month year, however, has only 354 days [(30 × 6) + (29 × 6) = 354], or 11.2422 days less than the tropical year of 365.2422 days. In Islamic countries today, where a calendar of this kind is used, New Year's Day moves constantly backward through the seasons in a cycle of about 32.5 years. An occasional 384-day year (with 7 long months) can be added often enough to yield a long-term average closer to 365 days.

The Chinese method for approximating the tropical year while retaining approximation to the synodic month was largely achieved by the time of the *Ssu-fen li* calendar of the fourth century B.C. Adding intercalary months seven times in the basic 19-year cycle (*chang*) yielded 235 months [(12 × 19) + 7 = 235]. Correlated to a 365.25-day year, this also assumed an average month of 29.53 + days, since

$$\frac{365.25 \times 19}{235} = 29\frac{499}{940} = 29.53 + .$$

The number of days in a 19-year cycle, though, did not come out in whole days (365.25 × 19 = 6,939¾), making a new cycle start one-fourth of a day, or six hours, earlier. After the Chinese gained greater precision in their constants to detect the difference, four 19-year cycles were combined in the 76-

year cycle (pu) to give a cycle that began and ended at the same hour ($6,939\frac{3}{4} \times 4 = 27,759$).[78]

By such measures the lunisolar calendar was kept generally in coincidence with both lunar and solar periods. While the methods of calculation were simple, a large volume of observed data extending over a long time, as well as developed theories and techniques, were needed to reach relative precision. Although high precision for solar and lunar cycles was reached quite early, endless burnishing of the imperial charisma prompted repeated reviews of procedures and many minor revisions. Another consequence of the close tie of the lunisolar calendar to monarchical authority was that the general public could not be sure whether the calendar would have an intercalary month until it was officially published.

Approximation of lunar and solar cycles was not the only motive behind the tireless efforts of Chinese astronomers. Astrological concerns for eclipses and planetary motions were at least equally strong. While the details cannot be elaborated here, the long and arduous search for reliable knowledge of

[78] The calendrical reform of 46 B.C. under Julius Caesar gave the Roman calendar an average year-length of 365.25 days by adding an extra day every fourth, or "leap" year, to its normal 365-day year:
$$\frac{(365 \times 3) + (366 \times 1)}{4} = 365.25.$$
Slightly longer than the tropical year of 365.2422 days, this calendar was one full day early in 128 years. By 1580 the vernal equinox fell on March 11, or ten days earlier than it should have. The Gregorian reform of 1582 largely corrected this by eliminating the leap year's extra day three times in each 400-year cycle. This meant not adding intercalary days to February in century years not divisible by 400 (e.g., the year 1600 was a leap year, but 1700, 1800, and 1900 were common years). The Gregorian calendar has an average year of 365.2425 days:
$$\frac{[365 \times (300 + 3)] + [366 \times (100 - 3)]}{4} = 365.2425,$$
or a difference in relation to the solar year of 0.0003 days (or about 26 seconds), and therefore requires 3,300 years to vary one full day. While this difference increases about 0.53 seconds per century (because the solar year shortens slightly), the calendar would not fall one full day ahead of the sun until the year 4316, but periodic adjustments using an atomic clock are now made so that this will never happen.

mean and true conjunction, of precession, and of anomalistic lunar and solar motion derived partly from these concerns. It was in relation to the use of interpolation formulas for dealing with anomalistic motions of the sun and moon that mathematical treatment of astronomical data reached its highest peak in Sui and T'ang times. In his computation of lunar and solar anomalies (see p. 48), Liu Ch'o employed the interpolation formula for equal intervals of the following type:

Given values for $f(\alpha), f(\alpha + w), f(\alpha + 2w), \ldots, f(\alpha + nw)$, the value of $f(\alpha + x)$ can be determined.

A century later the Buddhist monk I-hsing advanced to use of the interpolation formula for unequal intervals of the following type:

Given values for $f(\alpha), f(\alpha + w_1), f(\alpha + w_1 + w_2), \ldots, f(\alpha + w_1 + w_2 \ldots + w_n)$, the value of $f(\alpha + x)$ can be determined.

This improved the ability to determine the true motion of the sun so necessary for accurate predictions of solar eclipses, which are far more difficult than calculations of lunar eclipses. Thus, on this basis, the computation of solar eclipses was much improved by the time of the *Hsüan-ming li*, the last calendrical system adopted by Japan (though these technicalities were not in evidence in the calendrical treatises themselves; simple step-by-step instructions and tables made it possible to calculate complete ephemerides using a counting board, with knowledge of mathematical operations and no understanding of the astronomical significance of the procedures).

In China the calendar was part of the paraphernalia of dynastic legitimacy, and its revision a confirmation of the mandate's revivification. All this was beside the point in Japan, where the single, unbroken imperial line rendered the mandate of heaven theory superfluous. Hence, the Japanese court was content if its Institute of Divination carried out the routine work of annual issuance of the calendar, without delving deeply into calendrical theory and techniques. The *Hsüan-ming li*, called *Senmyōreki* in Japanese, remained in continuous use, without revision, from the year 862 until renewed stimulus from continental science in Chinese Wave II (1401–1854) enabled Japanese specialists to carry out suitable revisions in 1684—a time lapse of 823 years! While the Chinese amassed data and developed theories and techniques (for whatever reason), the Japanese failed to engage in research. Even their ability to predict eclipses (in which the nobility had some interest) declined from late in Chinese Wave I.[79]

SOCIAL FACTORS

Advanced knowledge of astrology and calendrical astronomy was a state prerogative in China, and its monopoly of these sciences was at times quite thorough, preventing access to the latest knowledge by private citizens. Even when private citizens were permitted to work on astronomy, it was usual for someone who submitted an important calendrical reform or who made a spectacularly confirmed astrological prediction to be immediately appointed to the Institute of Astrology. In appropriating astrology and calendrics as state functions, Japan kept the same formal bureaucratic controls. Though there was, at this stage, no occasion for private initiative in calendrical reform, there did emerge, as we shall see, competition in eclipse prediction. But the wide spread into private circles of astronomical and calendrical knowledge that finally occurred in China in the Sung period (tenth to thirteenth centuries) was not paralleled in Japan until much later, in the seventeenth century.

As the official instrument for measuring and marking time, it was imperative to have a single calendar for an entire state, and desirable for a common cultural realm. The Japanese adoption of the Chinese system was not only technically a derivative operation, but in Chinese eyes an act of cultural subordination common among states on China's periphery. After Japan withdrew from formal relations with China, the

---

[79] To their credit, the Japanese specialists did carry out faithfully over a long period the regular tasks of careful observation; e.g., in both China and Japan, as well as in Korea and the Middle East, reports of "new stars" (great stellar explosions) in 1006 and 1054 were preserved. On the contemporary importance of these records, see Needham, 3: 426–429, esp. note *b* on p. 429.

capabilities of her specialists declined. Eclipse predictions were fairly accurate when Japan adopted China's best and latest calendars; but as time passed, the adopted calendars increasingly needed revisions which the Japanese could not make. Hence, from late in Chinese Wave I near and complete misses in predictions occurred with greater frequency. Decreased proficiency in the official Institute of Divination created a kind of vacuum which drew others into the business of eclipse prediction. The Institute's monopoly was partially broken as mathematicians trained in the University and Buddhist practitioners of *Sukuyōdō* vied with each other with irregular success. The competition in eclipse prediction was patronized by court circles as only one of the many forms of fate calculation favored.

By far the most pervasive and enduring impact on society from the realm of astrology and calendars was the wide acceptance of the practice of calendar-related divination described earlier. Its main competitor (or, at times, collaborator) was the practice of *Sukuyōdō* by Buddhist priests with aristocratic patronage. These practices were to escalate, while interest in the technical side of calendrical astronomy declined, in the succeeding Semiseclusion Era.

MATHEMATICS

HISTORICAL TRENDS

From an early start around the fourteenth century B.C. in numerical notations, addition and subtraction, Chinese mathematics had by Han times progressed considerably to include multiplication, division, place-value concepts, knowledge of square and cube roots, fractions, negative numbers, and rudimentary geometric and algebraic processes. Though progress continued to T'ang times, when the Chinese had developed some skill at indeterminate analysis and even cubic equations, there appeared a tendency for bureaucratic preoccupations to restrict the scope and achievements of mathematical inquiry, until the more urbane and commercial atmosphere of the Sung period stimulated new creativity.

The focus of Chinese Wave I falls primarily upon a classical corpus of "ten mathematical manuals" (*Suan-ching shih-shu*) edited by Chinese government officials in A.D. 656 for use as textbooks in the mathematical course of the T'ang university (Kuo-tzu-chien).[80] The ten texts represent two historical phases: the collection of pre-Han achievements into two Han-period compilations, with the remaining eight texts incorporating the specific gains made by early T'ang times. The older of the two Han texts was the *Chou-pi suan-ching*, the cosmological classic; it is thought to be a Former Han compilation of materials dating as far back as the Warring States era. A computation of the dimensions of the universe, from assumptions about its shape, this text included such mathematical operations as the four arithmetical processes, a means for finding the square root of a number, a right angle theorem (for the special case $3^2 + 4^2 = 5^2$), and a rough value of 3 for π. The other Han text, the *Chiu-chang suan-shu* (Nine chapters on the mathematical art), was the foremost classic of its time in Chinese mathematics. Compiled with commentaries by Liu Hui in A.D. 263, it represents a much more advanced stage of mathematical knowledge, and a more practical range of applications, than the *Chou-pi* text. It deals with the rules of addition, subtraction, multiplication, and division, with measurement of areas and volume, with percentages, proportions, and ratios, with square and cube roots, and with the properties of the right-angled triangle. It also includes one quadratic equation. The Nine Chapters came to be regarded as the standard work in mathematics, and later mathematicians sought more to extend its possibilities than to look for new ones.

To his commentaries on the Nine Chapters, Liu Hui added an appendix entitled "Sea island mathematical manual" (*Hai-tao suan-ching*) after its first problem. No longer than one of the chapters of the original book, it was concerned entirely with measurements of heights and distances determined by

[80] Needham, 3: 18.

similar right-angled triangles simulated with tall surveyor's poles and horizontal bars. The method of working out the values of its geometric problems was essentially algebraic in character, but written as usual in discursive, not symbolic form. "The mathmatical manual of Master Sun" (*Sun-tzu suan-ching*), written probably between the late third century and the fifth century A.D., gave explicit details of the four arithmetical processes implicit in the Nine Chapters and opened a new line of inquiry into indeterminate analysis. The first text to deal with true indeterminate analysis was the "Mathematical manual of Chang Ch'iu-chien" (*Chang Ch'iu-chien suan-ching*), composed perhaps in the sixth century A.D. It also gave two formulas for arithmetical progression and made some clever applications of the type of quadratic equation seen first in the Nine Chapters.

The quality of the ten classics was uneven; indeed, the worst of them involved nothing more than multiplication and division, and many of its results were only rough approximations or definitely wrong. The only work thought to have significantly surpassed the Nine Chapters was the *Chui shu* (lit., Coupling Method) by Tsu Ch'ung-chih (A.D. 430–501), compiler of the *Ta-ming li* calendar, which was famous for its fairly accurate calculation of the upper and lower limits of π (see pp. 81–82). This work, however, was lost, and what is known comes mainly from the dynastic history of the Sui (*Sui shu*). The only T'ang period product in the ten classics was the "Continuation of ancient mathematics" (*Ch'i-ku suan-ching*), written about A.D. 625 by Wang Hsiao-t'ung, in which one problem is solved by a cubic equation—the first known advance over quadratic equations since the Nine Chapters.

Indian mathematics filtered into China before the T'ang period, but the high point was the A.D. 718 translation of the Indian astrological work (*Chiu-chih li*) in which Indian numerals, the dot symbol for zero, and a simple and limited table of sine functions were included. As noted earlier, these elements of Indian mathematics were not adopted by the

Chinese, nor were any other essential Indian (or Western) influences assimilated.[81]

Mathematics was incorporated into the Chinese administrative system much later than were medicine, astrology, and calendrical astronomy, all of which had been included in government institutions in the Han period. Coming first under government sponsorship sometime in the Six Dynasties era (after the Later Han, from A.D. 222 to the beginning of the Sui, 589), its place in the T'ang administration was lower than that of the other sciences. Nonetheless, it was definitely a part of the T'ang system copied by the Japanese.

Many other mathematical treatises were written from Han to T'ang times (many of which were lost) besides the ten classics. There were also other interests outside the scope of officially sponsored mathematics—notably puzzles, and particularly magic squares. Numerical schemes accompanied many divination practices, as in the Book of Changes, and in other developments of *yin-yang* and Five Phases thought. But these interests were taken up in Japan very late in and after Chinese Wave I, when rational activity declined; hence, treatment of these is postponed to the following chapter.

Rudimentary arithmetical skills had entered Japan only very late in the preparatory period. The sudden introduction in Chinese Wave I of the developed mathematics of the ten classics tradition based on the Nine Chapters presumably preempted the possibility of a mathematics truly Japanese in style. This would have mattered little, had Japan enjoyed even moderate success in adopting the Chinese tradition. Though the texts used in the mathematics course of the *ritsuryō* University were the ten classics, the course itself was isolated from the main curriculum of the University, and very likely from its best students. Given the relatively low level of Japanese interest in calendars and astronomy, there was no stimulation from those quarters. It is hardly surprising, then, that advanced

---

[81] Yabuuchi (2), pp. 128–129.

Chinese mathematics did not take root in the intellectual soil of Japan at this time. (Only after the Japanese turned their attention anew in the fifteenth and sixteenth centuries to the reintroduction of practical mathematics for commercial and technical purposes was a foundation laid for higher mathematics, which developed in the seventeenth century in the distinct form known as *wasan*; see Chapter 4.)

INTELLECTUAL TENDENCIES

In the Chinese, as in other cultures, several kinds of mathematical interests can be seen. In varying degrees one finds a straightforward *utilitarian* interest, responding to fiscal, commercial, and technical demands; a *logical* impulse that strives toward some system of abstract generality and logical precision, often with close affinities to philosophy; a purely *recreational* playing with numerical and geometrical relations; and finally, a *numerological* bent of mind which exerts itself in fields from fate calculation to alchemy. In the Western mathematical experience there was, to be sure, a considerable body of utilitarian material developed in answer to commercial and technical demands, as well as games and mystic numerology; but there was also a body of more strictly "pure" mathematics, notably Euclid's geometry, which enjoyed special regard as a norm for other branches of mathematics, as a close companion to philosophy, and as a complement of formal logic. The more strictly logical emphasis was not found in the mathematical enterprise of China. Chinese mathematics was, above all things, essentially utilitarian, though games and numerology were not lacking. But on the whole, mathematics was valued not so much for its relations to other academic fields as for its social uses. It was not viewed as an indispensable element in the cultivation of higher learning as such. In the *Chou li* idealization of government and its uses of learning simple arithmetic is given a place, it is true, among the Six Arts, as one of the skills, that is, needed by those expected to work in government offices. Here it is its utility that is valued, not its claim to a place in the household of learning.

The low value accorded mathematics in Japan during

Chinese Wave I has already been suggested. When numerological devices won a place in the surge of magical and superstitious practices in the post-Chinese Wave I era, its intellectual respectability sank even further. Even so, it was regard for utility that reopened the door to Chinese mathematics among fifteenth- and sixteenth-century merchants—though purely pastime interests, much more abstract in character, were to become the dominant concern of some leisured samurai in the seventeenth century.

During Chinese Wave I proper, the intellectual position of mathematics in Japan inevitably conformed, at best, to that held by mathematics in the Chinese intellectual spectrum. Its inclusion in the academic agencies of the *ritsuryō* system was strictly for maintaining that system itself. The teaching of mathematics in the University was directed entirely toward the bureaucratic needs of an adopted administrative system, e.g., for fiscal affairs, land-surveying, and large-scale construction projects. Human computers were produced, not scholars of mathematics.

LOGICAL ASPECTS

The scope of achievements in Chinese mathematics has already been alluded to in the historical notes, especially with reference to the Nine Chapters and its commentaries. Here certain distinctive features need to be reviewed. For instance, all calculating—whether addition, subtraction, multiplication, or division—was generally done with counting rods placed in appropriate rows on a counting board. Place-values were indicated by positioning in proper columns, and a blank space denoted zero. Red rods were used for positive numbers, black ones for negative numbers. While counting rods were known in pre-Han times, references to them become more frequent from Han times. Methods for multiplication and division on the counting board were more complicated, of course, than for addition and subtraction. Though references from a lost mathematical text (*Shu-shu chi-i*, reported in the Chin dynastic history) suggest some early forms of the abacus, known use of this more advanced calculating instrument comes much later,

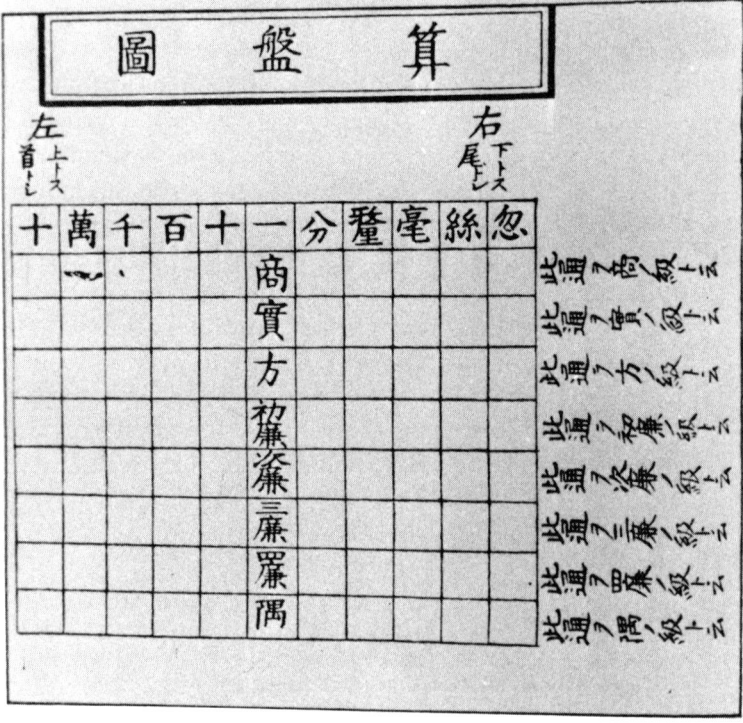

Fig. 10. General form of the counting board (*sanban*); from a 1698 work.

possibly in the Sung era, and the first detailed description of the abacus as known today is dated 1593.[82]

Fractions were known quite early (at least as early as the calendar year of $365\frac{1}{4}$ days), though they were consistently written out, a fact that possibly delayed popularization of decimals. A decimal system was known much earlier than the first reference in the *Sun-tzu suan-ching*, judging from the place-value system in use; and a weight system of decimals, based on descending tens, was put to use in A.D. 992.[83] Ancient

---

[82] Ibid., pp. 127–128; but a later work, Yabuuchi (7), cautions that historical material to support this theory is scarce. See also Needham, 3: 75.
[83] Needham, 3: 12, 85.

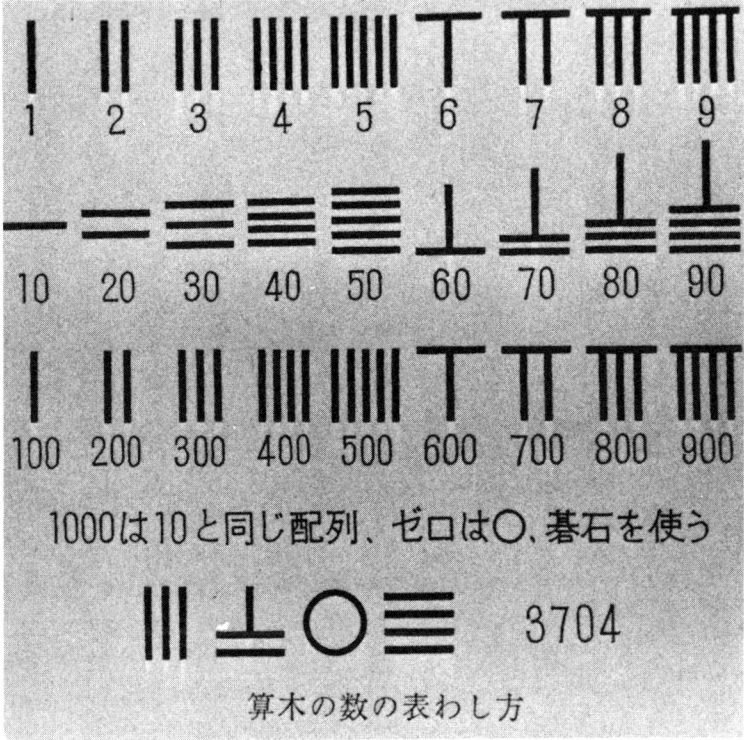

Fig. 11. Arrangements of counting rods to represent numbers. These arrangements were placed in matrices on the counting board to work problems.

handling of negative numbers and of processes for extracting square and cube roots was essentially the same as today. Algebraic equations were always formulated with numerical coefficients, as in equations of the type $5x^2 + 3x + 2 = 0$, and not of the type $ax^2 + bx + c = 0$. These selected instances serve to indicate the high academic level reached in Chinese mathematics by T'ang times, as do certainly the interpolation formulas for equal intervals by Liu Ch'o and for unequal intervals by the Buddhist I-hsing in their respective calendrical calculations (see p. 72). Finally, the calculations of the upper and lower limits of π by Tsu Ch'ung-chih, who used inscribed and circumscribed polygons of 3,072 sides to arrive at $3.1415926 < \pi < 3.1415927$, came remarkably close to the correct value of 3.1415926535....[84]

As the Nine Chapters formed the trunk of which later advances were the branches, it merits further remarks. Basically it was a collection of problems—246 in all—divided into nine chapters. These were problems encountered by government officials in their daily work. While it gave step-by-step instructions for solving each problem, which was stated in terms of specific numerical values, there was no general solution which would also hold for problems in which the parameters had other values. All of the problems were quite concrete, never abstract; thus it was perfectly natural that algebraic equations always had numerical coefficients. Treatment of geometrical figures consisted entirely of numerical calculations of length, area, and volume. The calculations were not treated in general propositions, such as "the sum of the angles of a triangle is equal to two right angles." Ironically, whereas a large proportion of the problems arise from geometrical situations—mensuration of variously shaped fields and the like—it is precisely the geometrical character of the Nine Chapters that is the weakest, compared to its arithmetical and algebraic qualities.[85] The search for a system with generality and rigor that has characterized Western mathematics after the model of Euclid's geometry is lacking in the Nine Chapters and in the tradition based upon it.

As with the *ritsuryō* system as a whole, so with mathematics: the Japanese were preoccupied with adoption of an established tradition, not with improvement of its norms or procedures. How much the full tradition was needed in the actual work of *ritsuryō* officials, or to what extent they really understood its value, much less its deficiencies, cannot be answered specifically. Ordinary arithmetical calculations were certainly used widely during and after Chinese Wave I, but in the subsequent

---

[84] Yabuuchi (2), p. 117.
[85] Ibid., p. 113; also Needham, 3: 91–168, esp. 153, 156.

era of semiseclusion from China, Japanese interests and abilities in the subject matter of higher mathematics drastically declined. Afterward, it was in the seventeenth century that the first Japanese-authored mathematical treatise appeared (see Chapter 3). From this alone, it is difficult to avoid the conclusion that Japanese understanding of Chinese mathematics during Chinese Wave I was minimal.

SOCIAL FACTORS

For the *ritsuryō* University's course on mathematics there were two teachers, who generally received only very low court ranks.[86] Initially there were thirty students, though the number dropped to twenty in the A.D. 808 reform of the University curriculum. At first there were actually two courses, one of them using only the now lost *Chui shu*[87] and the T'ang period *Ch'i-ku suan-ching* (see p. 76), which were probably considered the most difficult texts. Whether there was little interest in this course, or it was too difficult, is not clear; but it was soon dropped, leaving only the more routine course using the eight remaining texts of the ten-classics corpus for training prospective bureaucrats.

As with the rest of the University faculty, the mathematics professors concentrated on teaching, without performing regular official business. Top-level mathematics was a government monopoly, and there was as yet no merchant class with either the financial or intellectual resources to develop such sophisticated pursuits. Apart from rudimentary uses by artisans, the land-surveying, fiscal, and construction needs of government and of large temples and shrines were obviously met without creating any strong demand for sustaining, much less upgrading, the advanced mathematics taught in the University.

---

[86] In China professors of mathematics received usually the lowest court ranks. Cf. Yabuuchi (2), p. 130. But in Japan they received court ranks a little above the lowest, i.e., Junior Seventh Rank, Upper Grade. For the court rank system in Japan, see Reischauer and Fairbank, p. 476.
[87] The *Chui shu* was lost in China by the end of the eleventh century. It existed in Korea and Japan at least up to the thirteenth century. Cf. Yabuuchi (2), p. 121.

84   Science in Japan's First Cultural Transformation

As interest in the University itself waned, and rational interests diminished, the decline of mathematics was sharpest.[88] In the latter half of the Semiseclusion Era (i.e., 1192–1401), a distinct merchant class began to emerge; and in the early phase of Chinese Wave II (i.e., 1401–1639) some merchants took up the pursuit of, first, practical and later higher mathematics, thus laying the groundwork for the seventeenth-century flowering of *wasan*. The Early Chinese Wave II reintroduction of Chinese mathematics represented a totally new development, however, not a continuation of the Chinese Wave I academic program.

MEDICINE

Chinese medicine came of age in the more rationalistic Confucian world of the Han era, from which time all extant treatises date. Though the history of the origins and transmission of the principal texts, with numerous commentaries and insertions, is only approximate for the ten centuries from Han to Sui times,[89] the documents themselves reflect considerable effort to systematize Chinese medical experience according to the *yin-yang* and Five Phases theories. After medicine was incorporated into the Han administrative system, some of the principal treatises came to be regarded as authoritative and not only set the standards for subsequent medical scholarship in China but also constituted the model which Japan sought to adopt through its Institute of Medicine (Ten'yakuryō).

HISTORICAL TRENDS

The most important treatise was the "Yellow emperor's inner classic" (*Huang-ti nei-ching*), compiled between the third and first centuries B.C. The *Nei-ching*'s explanation of the human body, its dynamics, and its relations to the psyche and the cosmic environment as the interplay of a number of functional

---

[88] Documents in the National Repository (Shōsōin) in Nara give quantitative descriptions of the composition of alloyed metals, but after the end of Chinese Wave I such records as these also vanished. Yoshida (1), pp. 71–72.
[89] Porkert, p. 4.

systems was "the first systematic description of the system of correspondences"[90] that formed the theoretical basis of classical Chinese medicine. This classic consists of two parts: the *Su-wen*, which contains the fundamental elaboration of physiological and pathological theory based on *yin-yang* and Five Phases concepts; and the *Ling-shu*, which discusses mainly therapy—mostly centered on medicinal prescriptions, but also including physical therapies such as bonesetting and breathing exercises, and stimulation treatments such as acupuncture, moxibustion, and massage. The *Nei-ching*'s great value—beyond bringing much order to the medical knowledge of its time—lay in its functional approach to the body as a whole, as an integrated organism, in contrast to the more analytic, organ-centered medicine of today, though its theories rested on largely speculative grounds.

No less speculative yet more clinical in character was the treatise that came to be regarded as the standard work on therapy, the "On cold damage disorders" (*Shang-han tsa-ping lun*), compiled early in the third century (end of the Later Han era) by Chang Chung-ching. The oldest extant text in its field, it systematized and classified various symptomatic syndromes in terms of the specific prescriptions used to cure them. Its list of ninety-two prescriptions includes seventy-one herbal, twelve animal, and nine mineral ingredients. This determinative work on methods of diagnosis and prescriptions was in time followed by various treatises on special therapeutic topics, such as the "Classic on acupuncture and moxibustion" (*Chia-i ching*) by Huang-fu Mi (223–282) and the "Classic on pulsation" (*Mo ching*) by Wang Shu-ho (fl. third century).

Heavy reliance on medicinal preparations called forth a series of works on materia medica (*pen-ts'ao*) that gave theoretical classifications, descriptions, places of production, and therapeutic effects of an increasingly wide range of herbal, animal, and mineral substances. The first in the series was a Later Han compilation called "The Divine Cultivator's classic

---

[90] Ibid. Chapters 1–34 of the *Nei-ching* are translated in Veith.

on materia medica" (*Shen-nung pen-ts'ao ching*). A sixth-century improvement of this work by Tao Hung-ching (456–536), entitled "Collected annotations on materia medica" (*Pen-ts'ao chi-chu*), became the standard source book on materia medica. It listed 730 substances.[91] In 659 a "Newly revised materia medica" (*Hsin-hsiu pen-ts'ao*) was produced under imperial order by Su Ching. It contained an expanded list of 850 entries, including some from India and from what is described only as the "western region"[92]—presumably Central Asia and the Hellenistic world—additions that are not surprising in the cosmopolitan T'ang period.

More than any other work, the *Nei-ching* was subjected to continuous study, quotation, annotation, and editing. The most important of these efforts during the Chinese Wave I period was a 762 edition by Wang Ping in which he introduced a new formulation of the system of correspondences. In his interpretation, termed "phase energetics" by Manfred Porkert, the correspondences were redefined in terms of "Five Circuit Phases" (*wu-yün*) and "Six Energetic Configurations" (*liu-ch'i*) to elaborate and specify the manifold macrocosmic (i.e., meteorological, climatic, immunological) influences on microcosmic (i.e., physiological) processes.[93] Highly rationalistic yet largely speculative, Wang Ping's interpretation did not attain an authoritative place in Chinese medicine until the Sung era[94] and was not part of the Chinese medical tradition introduced into Japan during Chinese Wave I.[95]

None of the major classics was completely free from popular or Taoist influences. Taoists were particularly active in pur-

---

[91] Ishihara (1), p. 60.
[92] Yabuuchi (2), pp. 281–282.
[93] Porkert, Ch. 2 on macrocosmic dimensions, and Ch. 3 on microcosmic dimensions.
[94] Ibid., pp. 56–59.
[95] "Phase energetics" was reflected in the limited introduction of Chinese medicine by Japanese Buddhist priest-practitioners in the Semiseclusion Era (Ch. 2) but was first introduced as a system in Early Chinese Wave II (Ch. 3). It became a major target of attack by some Japanese medical scholars in Late Chinese Wave II (Ch. 4).

suing the search for drugs to assure long life and immortality, and developments from Han to T'ang times saw a significant increase in Taoist influences on Chinese medicine. While their activities led to a kind of alchemy, they also contributed, as did eventually Buddhist activities, to an accelerated emphasis on various rites and exorcistic incantations of popular origin as auxiliary modes of tending the sick. As a result, by T'ang times there was an expanded store of practical knowledge of symptoms and medicines organized theoretically according to naturalistic assumptions in physiology, pathology, and therapy, and on the other hand a promotion of more magical and religious methods of treatment. Both of these developmental lines were reflected in the T'ang medical institutes, in curricula that included an assortment of courses on internal medicine, surgery (for minor lacerations and swellings), acupuncture, moxibustion, massage (including bonesetting), and exorcistic incantations. For all the complex variety of Chinese medicine at this stage, it was not in the least inferior to its contemporary counterpart in the West, which had its own speculative theories based on four elements (air, fire, earth, and water) and Galen's pathology of humors. The European art, it has been noted, was much more isolated from currents of popular medicine than the Chinese and only meagerly supplemented by them between the time of Dioscorides (first century A.D.) and the Renaissance.[96]

How much and what kinds of Chinese medical practice first filtered into Japan through medical practitioners sent from Korea and through mainlanders settling in Japan before Chinese Wave I is not clear. In any case, early influx was piecemeal and irregular until two young Buddhist priests, Enichi and Fukuin, accompanied the third *kenzuishi* mission to China in 608 to study medicine and returned in 623 to promote the systematic adoption of Chinese medicine by Japan. And only after the founding of the government's Institute of Medicine could real work toward this goal begin.

[96] Nathan Sivin, private communication.

Medicine had a much broader base in Chinese society than in Japan, where its study and practice were largely separated from society in general. The *ritsuryō* Institute of Medicine was intended to serve only the imperial household and officials in the capital cities, and the medical courses in the provincial colleges likewise were for the benefit of provincial officials. Even so, the prospects for medicine proved brighter than for astrology, calendrical astronomy, and mathematics, presumably due to the strong demands for it. For example, the first compilation of Chinese medical sources by a Japanese appeared early in the ninth century. Japanese-authored texts in other sciences occurred much later: the first book related to astrology, for example, appeared in 1414 (see p. 198), while the first known monograph on mathematics is dated 1622.

The earliest known medical work produced in Japan was a short manual on materia medica, the *Yakkei taiso* (Outline of materia medica), of which only a later edition exists. It was produced during the reign of the Emperor Kanmu (reigned 781–806) by Wake no Hiroyo. Working under order of the Emperor Heizei (reigned 806–809), who feared all indigenous practices might be forgotten because of enthusiasm over Chinese medicine, Abe no Manao and Izumo no Hirosada in 808 compiled an extensive record of treatments and medicinal preparations employed in Japan from antiquity; the work was entitled *Daidō ruijūhō* (Classified prescriptions [collected in the] Daidō era), and only a forgery of a later edition exists.[97] After this work, nine medical books were compiled by Japanese (none are extant) before the *Ishinpō* (sometimes *Ishinhō*; Essential medical prescriptions), the oldest extant Japanese medical text still preserved in full. Begun in 982 by Tanba no Yasuyori, a court physician of Chinese descent, and completed in 984, it is a compilation of excerpts from many authoritative Sui and T'ang sources, many of which no longer exist. Thus, its historical importance is great, not least as an indication of the level of medicine introduced into Japan during Chinese

[97] Fujikawa Yū, pp. 50–53.

Fig. 12. Sketch and text in a section on pregnancy in the *Ishinpō* (Essential medical prescriptions; 982–984); sketch shows the "Leg-colon-bladder" channel with vital points (in the left leg) and a fetus in upright position in the womb. (The National Archives, Japan)

Wave I, though the book itself appeared after the wave ended and does not cite a single document of Japanese origin.[98]

INTELLECTUAL TENDENCIES

Confucian thought as such cannot be said to have contributed directly to the formation of medical theory and practice in China. It would be shortsighted, however, to overlook its role in shaping the rational style of all scholarship and in leading the way for incorporation of learning into government in the Han era. Medicine, like other fields, benefited greatly from imperial orders to collect and compile extensive materials into standard texts.

The Taoist involvement was more direct.[99] Out of the shamanistic milieu there emerged in China's Warring States era a kind of practitioner (*fang-shih*) whose practice was a combination of exorcism, fortune-telling, and the art of preparing immortality drugs. The last activity in particular contributed greatly, from the Han era on, to a wider knowledge of drugs in general, of the chemical properties of materials, and of the instruments of research. Central to the search for immortality drugs was the development of techniques for refining alchemical elixirs of immortality, called *tan* after the common ingredient of many of them, cinnabar (red mercuric sulfide). Taoists were fascinated by the observation that the vermilion-colored cinnabar turned a bright silver color when heated (yielding mercury, or "quicksilver"), and that, when heated further, it oxidized and the original vermilion hue (this time embodied in mercuric oxide) returned. This remarkable sequence of changes must have perfectly met their fancy that the robust health of youth could be reclaimed from the ravages of old age and disease. Many of their drugs following this discovery were mercuric compounds, though for that very reason just as often poisonous. Taoists also engaged in aurifaction, though the "gold" produced artificially from base metals was used, not for profit-making ventures as in the West, but

[98] Ibid., p. 6.
[99] Yabuuchi (2), pp. 26–33.

for making vessels presumed to have special qualities suitable for the preparation of immortality drugs.[100] Not all Taoists confined themselves to immortality drugs and alchemy; many turned eventually to studies of more orthodox medicine. For example, Tao Hung-ching, compiler of the "Collected annotations on materia medica," was the organizer of the Mt. Mao sect of Taoism, which was the state cult for a while during his lifetime and the most powerful sect of Taoism for centuries afterward.

The extent of Indian influence, which came mostly through Buddhist channels, is not yet clear,[101] but many new medicinal materials seem to have reached China by this route. The Chinese Buddhists certainly rendered extensive health services to the masses, though their efforts had the concomitant effect of strengthening esoteric and exorcistic tendencies. By T'ang times there was a plethora of magical and religious methods and taboos of both Taoist and Buddhist connections, which in actual practice were seldom distinguished.

The determinative influence of *yin-yang* and Five Phases concepts on Chinese medicine can hardly be overemphasized. These concepts provided the theoretical framework for all physiology and pathology, as well as the guidelines for prescribing particular drugs to treat specific syndromes. This enabled medical scholars to organize their experiences—and thus to perceive—according to naturalistic assumptions, with no need for recourse to supernatural causes and evil spirits. The disadvantage, as with all speculative systems, was that, when the data became too voluminous and complex, explanations became strained and less useful in practice.

The basic assumption of Chinese medicine was that the universe is the macrocosm of which the human body is the microcosm, and that disorders come about when the cyclical functioning of the latter moves out of phase with that of the former. As the *yin-yang* and Five Phases are the fundamental

---

[100] Ibid., p. 30. For a discussion of Chinese alchemy in T'ang times and a translation of one text, see Sivin (1).
[101] Yabuuchi (2), p. 37.

principles for understanding cyclical behavior in the natural order, they thus are the primary tools for understanding the human body and the causes and remedies of its malfunctions. The assignment of *yin-yang* and Five Phases principles to all natural phenomena was a means of analyzing complex wholes in space and time (processes or configurations) into two or five interdependent aspects whose relations to each other were constant—and to show how individual phenomena corresponded to some aspect of a whole process or configuration.

Each *yang* quality was subdivided into *yin* and *yang* aspects, and each of these were so differentiated again and again. The same differentiation also applied to each *yin* quality, making

Table 12. *Yin-Yang* Correspondences

Yin	*Yang*
cosmic	
Earth	Heaven
moon	sun
night	day
darkness	brightness
water, rain	fire
moisture	dryness
autumn, winter	spring, summer
things small and weak	things big and powerful
quiescence	movement
bodily	
female	male
interior	exterior
lower part	upper part
right side	left side
ventral side	dorsal side
visceral orbs	bowel orbs

Source: Adapted from Porkert, pp. 24ff.

N.B. Terms in this table refer not simply to physical conditions but to qualitative standards by which natural, including medical, data were analyzed and organized for describing the body's functions, diagnosing disorders, and prescribing remedies.

Table 13. Five Phases Correspondences

Fivefold Group	Phase				
	Wood	Fire	Earth	Metal	Water
cosmic					
planets	Jupiter	Mars	Saturn	Venus	Mercury
directions	east	south	center[a]	west	north
seasons	spring	summer	summer's end	autumn	winter
climates	windy	hot	humid	dry	cold
colors	green	red	yellow	white	black
flavors	sour	bitter	sweet	acrid	salty
bodily					
viscera	liver	heart	spleen	lungs	kidneys
orifices	eyes	ears	mouth	nose	anus and urethra
forms	muscles	veins	flesh	skin and hair	bones
emotions	anger	pleasure	cogitation	sorrow	fear
fluids	tears	sweat	saliva	nasal secretions	saliva

N.B. Generally speaking, the items in any given column above are some of the key qualitative standards for defining a specific functional system of the body in terms of the correspondences inside and outside the body.
Sources: Needham 2: 262–263; Porkert, pp. 43ff.
[a] Reflects circumpolar design of Chinese cosmology.

possible extensive gradations and overlaps of *yang*-in-*yang*, *yang*-in-*yin*, *yin*-in-*yin*, and *yin*-in-*yang*.[102] Coordination of these abstractions with all sorts of natural phenomena was also associated with the calendrical system, so that, for instance, the hours and seasons and azimuth directions in which the sun expands its energy—starting in the morning, in the east, and in spring, then peaking at noon, in the south, and in summer—were qualified as active *yang* aspects; while the waning of the sun in the evening, in the west, and in autumn, or its apparent disappearance at midnight, in the north, and in winter, were qualified as quiescent, stable *yin* aspects. Various

[102] Needham, 2: 276/7, Plate XVI; Porkert, pp. 31–34.

arrangements of the Five Phases were also available to further diversify the qualifications; one of them used in the *Nei-ching*[103] associates Wood with the east and spring, Fire with the south and summer, Metal with the west and autumn, and Water with the north and winter in a circular geometrical design that has Earth at the center in a neutral position. In the early stage of Chinese medical theory, *yin-yang* polarities were dominant; from T'ang times on, the cyclical alternations of both *yin-yang* and Five Phases aspects were increasingly emphasized in medicine after Wang Ping first gave them definite formulation, and particularly after his interpretation became authoritative and further elaborated from the Sung era on.[104]

In the Japanese Institute of Medicine this tradition was conveyed in books at the heart of the curriculum, but it does not appear to have been taken with great seriousness until reintroduced during Chinese Wave II in the greatly elaborated form of Wang Ping and his Sung, Chin, and Yüan successors. Medical practices, more than theories, were adopted in Chinese Wave I, and it was Buddhist practitioners more than court-related personnel who kept them alive and made them available to the public for many generations. But, as we shall see in the next chapter, the conceptual framework of medicine in Japan was colored very little by Buddhist ideas. Part of the Buddhists' role in medical services was performed by wandering priests, who stressed exorcism and magic as much as the more rational modes of Chinese medicine proper. In their practices, Taoist and Buddhist elements received from China became indistinguishably mixed with popular Japanese religious practices of pre-Chinese Wave I origin. Practicing far from major cities and without access to medical literature, the wandering priests were particularly prone to eclecticism.

LOGICAL ASPECTS

To insist that the Chinese medical tradition from Han to T'ang times was not inferior to that elsewhere is not to overlook its

---

[103] Porkert, p. 47.
[104] Ibid., pp. 43, 51, and esp. Ch. 2.

unique and distinguishing features. Its most distinctive characteristics stemmed from the use of the *yin-yang* and Five Phases concepts to achieve order and clarity.

Chinese physiology held that in the human body there are two sets of functional systems, which only very loosely correspond to our anatomical organs. One is a set of five, associated with and named for the "Five Viscera" (*wu-tsang*), which included the liver, heart, spleen, lungs, and kidneys. Theoretical discourse about these extremely loose associations was about activity within the economy of the body rather than about organs and tissues. The function of the "splenetic" system, for instance, would correspond in Western medicine to the physiology and pathology of digestion, absorption, material metabolism, balance of bodily fluids, and certain aspects of blood circulation.[105] The second set is associated with and named for the "Six Bowels" (*liu-fu*), that is, the stomach, large intestine, small intestine, gall bladder, urinary bladder, and the *san-chiao* (meaning unclear in the earliest literature, where it had no clear spatial location or anatomical substratum; for several centuries from the Later Han on efforts were made to locate the three *chiao* in regions roughly above, around, and below the stomach, but from T'ang times this attempt was abandoned in favor of simply defining the *san-chiao* function, by some as the central coordinator of all energy flow[106]). The first five systems (or "functional orbs," as Manfred Porkert calls them[107]) store refined *yang* energy (*ch'i*), which is distributed throughout the body by a sort of pneuma also called *ch'i*. The systems of the Six Bowels have the functions of fermenting food, separating the clear product from the turbid, and transporting the dregs to be excreted. All the orbs are connected with the extremities and the surface of the body, and with each other, by twelve main channels (*ching mo*) that run vertically along the torso and limbs near the surface of

---

[105] See, for example, the summary in a 1972 Peking textbook of traditional Chinese medicine, *Hsin pien Chung-i-hsüeh kai yao*, p. 12, translated in Sivin (2).
[106] Porkert, pp. 158–162.
[107] Ibid., Ch. 3.

the body. In these channels flow *yin* energy (*hsueh*), carried by blood (also called *hsueh*), as well as *yang* energy. The twelve main channels are connected by many branch channels (*lo mo*), with further interconnection and ramification by several lesser types of channels. At various points along the main channels are distributed 365 vital "holes," that is, invisible and histologically indistinguishable sensitive points.[108] By stimulating them, Chinese medicine sought to regulate the flow of energy between the two sets of functional systems by selective stimulation and retardation. Hence, the points are the loci for application of acupuncture, moxibustion, and massage, and at the same time, "openings" through which disease agents can enter the body.

Chinese knowledge of the somatic functional systems must have been based upon anatomical observations made in antiquity. Though there is a modicum of accurate information in describing the internal organs which constituted the physical substrata of the functional orbs,[109] given the emphases of Chinese medicine, there was no occasion to attempt more than a rough sketch of the inside of the body. The scheme of organ substrata was not tested by direct observation until some dissections were performed in the Sung period. The orbs themselves were regarded as interrelated and as connected by the energetic circulation system (which does not correspond to the modern vascular system), their functions being both physical and psychical. For example, the orb associated with the heart produces *yin* energy and controls the state of the mind generally; the liver orb stores *yin* energy and controls reflection; or

[108] Porkert observes: "The sensitive points provide the positive empirical and historically primary data on which the theory [of channels and regulatory points] is based; the conduits [channels], on the other hand, are only the result of systematic speculations"—a theoretical system he likens to "the lines of force of a magnetic field or to the orbits of planets as defined by gravitation and mass." Ibid., p. 198.

[109] The Chinese understanding of the anatomical substrata is summarized in part IV.8 of the description of each orb in Ch. 3 of Porkert—though his descriptions are largely based on interpretations that became authoritative only from the Sung era on.

the *yang* orb associated with the gall bladder, the "bowel whose contents are clear," is particularly connected with courage.

Basic to pathology was the idea that disease results from any lack of equability in the cyclical functions of the body or in their harmony with the regular cycles of nature (as analyzed using the twofold *yin-yang* and fivefold phase concepts), and recovery of the lost balance between the various functions is necessary to cure disease. The proper balance might be upset by an outer cause, or by an internal factor such as emotional excess, or by both, resulting in an excess or deficiency of *yin* or *yang* energy in the general circulation, or a pathological change in the relation of a single orb to the others. These relations are normally analogous to those of the Five Phases themselves. Etiology in Chinese medicine was not always consistent with the highly abstract character of this pathology. The train of physiological changes that led to medical disorders was often traced back to demonic possession, or to sin as defined by popular religion, or to Buddhist *karma*; for instance, certain disorders symptomatically close to the modern view of leprosy were sometimes regarded as the result of bad deeds in a previous life. It was probably because of this contradiction that etiology normally played no part whatever in diagnosis. Regardless of cause, classical medicine was consistent in its application of theory to diagnose disorders by systematic examination and questioning of the patient and to explain the development of disorders (even possession disorders) by sequences of naturalistic events within the body once the pathological agents had entered.

Therapy aimed at restoring balance by stimulating any indicated increase or decrease of *yin* or *yang* energy, so that each function could proceed according to its place in the theoretical scheme. The main emphasis was upon the use of drugs, and prescriptions were often directed to some part of the body peripheral to the affected functional system. The classification of drugs in the earliest schemes was made according to intended effects, not the qualities of substances themselves. In "The Divine Cultivator's classic on materia

medica," for example, there were three basic kinds of drugs:[110]
"Upper": 120 kinds, for immortality
"Middle": 120 kinds, for sustaining health
"Lower": 125 kinds, for curing diseases.

While the total of 365 drugs suggests the extent to which therapy was accommodated in theory to cosmology, later books on materia medica further extended the list, while working toward taxonomies based on the sources of the medicines. The primary place given to immortality drugs in the earliest scheme also indicates the Taoist influence on the formation of Chinese medicine.

The medicinal preparations were taken in the form of liquids, powder, or pills. Very often acupuncture or moxibustion were applied along with drugs. In particular cases, massage, physical exercises, breath control techniques, and other methods were also utilized. Surgery was generally limited to treatment of boils and wounds. There is a legend (probably Indian in origin) that in the Three Kingdoms period a famous doctor named Hua T'o performed a variety of major operations, including amputations under partial anesthesia. In subsequent Chinese medical history no such major operations were reported, and traditional surgery had only a meager existence, poor in variety and skills compared to the extensive use of drugs and auxiliary therapies. Finally, exorcistic therapy was sometimes practiced in conjunction with drugs and other methods, though only in folk medicine was it used alone to any extent.

The main weaknesses of Chinese medicine were not unusual for the historical stage but need to be mentioned for their relevance to later chapters. Medicine shared a certain insularity with all Chinese scholarship; while adoption of non-Chinese materials surely contributed to the widening of the range of materia medica from the fifth to the seventh centuries, the fundamental theories remained unchanged for centuries.[111]

[110] Yabuuchi (2), pp. 32–33.
[111] Ibid., pp. 38–43.

The functional and energetic approach of physiology became too schematic, and too little effort was expended on empirical verification of its theories; thus anatomy, incidental to such theories, was negligible. Failure to be more than superficially critical up to modern times (and well into the Edo period in Japan) definitely limited further development. That the metaphysically oriented pathology made little reference to deities or demons (while exorcism was practiced, even among the gentry clientele, its role was minor and not integral to pathology) must be counted an advance for the historical stage concerned. That pathological theory congealed at the level of early assumptions, failing to push on to empirical testing and revision, or even rejection of some positions, must be judged to have set limits to progress. Finally, the research methodology was rather limited and remained so. Acknowledgment of the definite accomplishments of Chinese medicine cannot subdue the observation that up to modern times it was methodologically no more able than its medieval European counterpart to transcend the basic approach of prescribing various drugs for known syndromes. Throughout premodern times it remained an accumulation of practical experience ordered by an increasingly elaborate body of theory, but theory in which correlations were only assumed and demonstrations rarely sought.

It is not enough to say that similar weaknesses were shared all over the world in ancient times. Western medicine was indeed slow to overcome its own inadequacies in physiology, anatomy, pathology, and research methodology, but it did finally begin making radical conceptual breakthroughs in the sixteenth century in anatomy and blood circulation, and in methods of observation, which gradually were organized into verified systems. But when the improved knowledge came first to East Asia, medicine was still fixed in its traditional forms in both China and Japan.

The Japanese appropriation of Chinese medicine was designed to include all levels of medical theory and practice: internal and external medicine, and the auxiliary therapies of acupuncture, moxibustion, and massage; the wide range of drugs; and exorcistic therapy. As noted already, the propensity was more to practice than theory. Since the Japanese were disinclined to delve deeply into metaphysical concepts, much of the basic physiology and pathology was not readily assimilated. It cannot be concluded, therefore, that the *ritsuryō* program accomplished much more in medicine than the formal introduction of the basic material, and the Japanese could hardly be expected to have overcome points of weakness in the Chinese tradition. On the other hand, the response and results were in fact greater in medicine than in the other sciences, as evidenced partly by the much earlier appearance of Japanese texts in the field of medicine—though, with the exception of the *Daidō ruijūhō*, they were all compilations of excerpts from Chinese classics. Moreover, following Chinese Wave I, Chinese-style medical practice was gradually diffused into many sectors of Japan, thanks to Buddhist efforts, on a scale not seen in the other sciences. Critical scholarship that fully understood and sought to adapt theories and therapies did not emerge until the seventeenth-century intellectual renaissance (see Chapter 4), and only after the late eighteenth-century introduction of modern Western medicine could certain defects of Chinese medicine be definitely confirmed (see Chapter 5).

SOCIAL FACTORS

In both China and Japan medical services and education were on several levels formally quite similar. The scope of the respective programs, however, differed vastly. At the highest administrative level were special bureaus to serve the courts and officials of each country. Whereas the T'ang court maintained two separate bureaus for the emperor and crown prince respectively, the Japanese court had only one for the whole imperial household. The combined staff of the two T'ang bureaus, including administrators, physicians, pharmacists, and specialists in acupuncture, moxibustion, massage, and exorcism, ran well over eighty, excluding clerical personnel.

The Naiyakushi bureau serving the imperial household in Japan had a staff of only sixteen.[112]

On the next level was the T'ang medical institute (T'ai-i shu), after which the *ritsuryō* Institute of Medicine was patterned. Both were to provide medical care for government officials in addition to educating future practitioners. The teaching staff, students, and attendants of the T'ang institute numbered over two hundred, while the Japanese counterpart included about half that number. Education and health services for provincial officials reveal a similar gap. T'ang China was divided into fifteen (initially ten) circuits (*tao*), which were subdivided into prefectures (*chou*), and the T'ang plan called for a medical institute in each prefecture. The prefectures were graded upper, middle, and lower, and the professor, assistant professor, and student numbers in each of these was 1-1-15, 1-1-12, and 1-0-10 respectively. There were also three urban prefectures, each with one professor, one assistant professor, and twenty students, as well as some medical personnel attached to local garrisons.[113] In Japan medical care in the provinces was to be provided through medical courses and staff in the provincial colleges (Kokugaku; Fugaku in Kyushu), to which one-fifth of the students were to be assigned; though, as noted earlier, the actual results are difficult to assess.

Both governments encouraged temple clinics for treating the general public, and it may be surmised that here too T'ang services were more extensive than those of the Japanese. The extension of medical care beyond official circles to commoners in Chinese Wave I did, however, reflect the close alliance of Japanese Buddhism with the imperial court. When the Shitennōji temple was built in the late sixth century in Osaka, four medical facilities were attached to it. Apart from the Keiden'in, which was to care for sick and aged priests, there were three facilities for sick and poor commoners: the Hiden'in,

---

[112] Ibid., pp. 216–268.
[113] Ibid., p. 283.

Ryōbyōin, and Seyakuin. In 723 facilities named Seyakuin and Hiden'in were established at Kōfukuji temple (clan temple of the Fujiwara household) in Nara. Again, the Empress Kōmyō (701–760), who came from the Fujiwara line, founded facilities called Seyakuin and Hiden'in for treatment of commoners. From the beginning, these facilities received support from both the imperial court and the Fujiwara household. In 825 a government organ named Seyakuin, independent of the Institute of Medicine, was legally instituted. Its function was to supervise the Hiden'in facilities for commoners and to perform funeral services for members of the court (including preparation of the corpse and burial). The directorship of this agency went to members of the Fujiwara household who were not themselves medically trained. Much else is not clear, but the indication clearly is that official support did much to strengthen the social base from which the Buddhists, motivated by their own virtue of compassion, provided the bulk of medical services to the Japanese people for the next half-millennium.

2

Five Centuries of Indigenous Development

The Semiseclusion Era: 894–1401
The isolation policy of the Tokugawa era (seventeenth through mid-nineteenth century) is so well known it is often forgotten that the Japanese government at the end of the ninth century also drastically limited foreign contact—a policy maintained for the next half-millennium. During this long interim of "semiseclusion" the impetus for social change shifted from imitation of a foreign model to pursuit of domestic priorities. This shift became quite apparent in the three centuries from the last *kentōshi* mission of 894 to establishment of the Kamakura shogunate in 1192—the early phase of the era. This phase was dominated politically and culturally by the court nobility, who retained many formalities of the *ritsuryō* system but in practice undermined its spirit and power. Through their literary and aesthetic activities they raised the general cultural level but in the process downgraded rational activities, and thus the sciences. Changes in the late phase of the era—the entire Kamakura period (1192–1336) and half of the Muromachi period (1336–1467)—went even further. Dominated by Japan's new military class, this phase saw the formation of feudalism as well as new stirrings occasioned partly by agricultural, commercial, and technical advances, and partly by the introduction of new forms of Buddhism from China. The import of these later changes for the revival of science in Japan was realized only after continental contacts were renewed at the beginning of the fifteenth century and the Semiseclusion Era came to an end.

SINO-JAPANESE RELATIONS

The early phase of the Semiseclusion Era coincides roughly with most of the Sung period (960–1279), a long period of internal growth and stability in China during which commerce flourished and overflowed into large-scale maritime trade, marking China's first emergence as a major seafaring nation. Though Arabs and Persians had earlier handled most trade to the south and west of China, and Koreans that to the east with Japan, the Chinese now increasingly took over their own overseas trading. In the process, China's orientation to the outside world became less confined to the northern steppes and western land routes across Central Asia as it was extended to embrace also the new world over sea routes to Southeast Asia and to the West by way of the Indian Ocean.

From late in Chinese Wave I merchant ships from China had visited Japanese ports in Kyushu rather frequently. The Sung dynasty, not active in expanding its territory, also welcomed foreign ships to Chinese ports. After cancelation of the *kentōshi* missions in 894, however, Japan not only ceased sending missions abroad and exchanging official letters, but also prohibited Japanese ships from voyaging abroad.[1] Chinese merchant vessels kept calling at ports in northern Kyushu but were subject to severe restrictions imposed by Japan on the number of ships and the amount of cargo per year. Moreover, only after agents of the Japanese imperial household and court nobility made their purchases were ordinary merchants free to buy up the remainder. One notable exception during the era's early phase was Taira no Kiyomori (1118–1181), a samurai who governed most of Japan briefly before the Kamakura shogunate arose. He actively promoted trade with China by welcoming Chinese ships and sending abroad Japanese vessels.[2]

In 1125 the Jürched clans from the steppes north of China gained control over northern China, founding the Chin dynasty

---

[1] Mori (2), pp. 212–213.
[2] Kiyomori once received a letter from a provincial governor in China, which was addressed to "the king of Japan," and in return he sent a letter and gifts in an effort to increase trade. Export items to China at this time included swords and armor, gold and sulphur, folding screens and folding fans, and Japanese paintings (*Yamato-e*); imports from China included silk cloth, books, copper coins, paper, black ink-sticks and ink-stones, and spices procured in the South Seas.

(1126–1234), and Sung rule was limited to central and southern China (Southern Sung: 1127–1279). In the middle of the thirteenth century both dynasties fell before the onslaught of Mongol power which established the Yüan dynasty (1279–1368) over all China, extended its dominion westward to envelop much of the Islamic world and seriously threaten the Christian realm in Europe, then turned eastward to devastate and conquer the Korean peninsula and demand Japanese recognition of its suzerainty. By this time the inept Japanese nobility had yielded power to the Kamakura military regime, which bluntly refused recognition. Mongol invasion forces of twenty-five thousand and one hundred forty thousand troops, dispatched in 1274 and 1281 respectively, were repelled by stiff Japanese resistance aided by foul weather. Defenses against a possible third assault (actually contemplated by the Yüan) were maintained for more than two decades, but it never came.

Yüan aggressiveness in territorial expansion did not prevent the encouragment of foreign trade. Despite failure to bring Japan into its dominion, and even though formal state relations were not established, the Yüan government permitted Chinese merchant ships to frequent Japanese ports before, during, and after the invasion attempts. Moreover, on several occasions Japanese ships were authorized by the Kamakura and Ashikaga shogunates to go to China on trading expeditions; the profits were to be used for large temple or shrine construction in Japan. For example, ships were sent on behalf of the Kenchōji temple in Kamakura (1325), the Sumiyoshi Shrine in Osaka (1332), the Tenryūji temple in Kyoto (1341), and the Ryōbyōin medical facility of Shitennōji temple in Osaka (1367).

Apart from the few authorized ships, there were, after the Mongol invasions failed in the late thirteenth century, the rather large-scale activities of the pirate-traders, *wakō*, who operated first along the Korean and later along the Chinese coasts as well. *Wakō* were not simple pirates but avid entrepeneurs (merchants and local samurai) hampered by unfavorable national and international conditions. The Japanese continued generally to prohibit overseas ventures, and the Ming dynasty, successor to the Yüan from 1368, completely reversed the Chinese posture toward the outside world. From about 1371 it inaugurated and continued to issue maritime prohibitions (*hai-chin*) forbidding overseas trading by Chinese ships and allowing only foreign ships with an appointed envoy bearing official documents and tribute to enter Chinese ports.[3] Tally coupons issued to identify authorized foreign vessels were used in the Tally Trade Agreements formalized with more than fifty states. Japan was included in this arrangement mainly to control the illegal *wakō* activities—but, in effect, it signaled the end of the era of partial seclusion.

By far the chief cultural value of the irregular shipping variously conducted throughout the Semiseclusion Era was the transport provided to Buddhist priests who wanted to visit China. Apart from the exceptions noted above, they alone, during the entire era, were granted court permission to visit abroad. In the Yüan period particularly, many priests went over to the mainland, and a few Chinese priests came to Japan. In addition to new forms of Buddhism, they also introduced, as we shall see, certain contemporary forms of Chinese learning and science. It was they who made seclusion "semi-," though cultural influx into Japan was definitely more limited in scope and character than in Chinese Wave I.

CHINESE CULTURAL DEVELOPMENTS

Sung culture derived partly from a renaissance of classical ideals and partly from the evolution of new urban, commercial, and civilian values. The ideal of unchallenged supremacy of the emperor had been only partially fulfilled in the T'ang dynasty, because aristocrats had retained strong political powers. In Sung society aristocratic powers were successfully

---

[3] Tanaka (2), pp. 180–181.

pruned, largely by abolishing independent local militia and putting all troops under imperial command, and thus imperial supremacy was established. Military officers who held high government posts were transferred or retired to make way for civilian entry into political positions. To assure selection of civilians with both ability and loyalty, the civil examination system was fully reinstated. This not only made the bureaucracy the main force in government but also helped to restore moral and intellectual criteria for leadership in state affairs.

In social and economic developments there was much not wholly anticipated in classical ideals. Along with greatly increased productivity in agriculture—rice crops seem to have doubled in the eleventh and twelfth centuries[4]—there was rapid growth in commerce, related, as noted above, to large-scale maritime trade. Commercial activity also produced a highly developed money economy which, together with great population growth, gave rise to large commercial cities—a functional evolution beyond earlier large cities in which political functions were primary. Technical innovations included navigational use of the magnetic compass from the twelfth century on (though magnetic polarity had been discovered at least as early as the fourth century by the Chinese, and declination was described about 1088 by Shen Kua[5]) and the making of gunpowder. Enjoyed mostly as fireworks, military use of gunpowder in the Sung era was limited to incendiary flares, until the Jürched invaders used it also in explosives and the Mongols developed primitive cannon and grenades to frighten cavalry mounts.

All learning benefited from the dual Sung thrusts, traditional and innovative. Study of the classics was revived, and many classical texts were revised and printed. Though late in the T'ang era and during the following Five Dynasties period (907–960) woodblock printing gradually had become rather

---

[4] Reischauer and Fairbank, p. 212.
[5] Needham, 4/1: 249–250, 278; biography by Sivin in *Dictionary of Scientific Biography*, s.v.

common, printing of books in the Sung era was much more extensive.[6] The major impetus to printing was the government-sponsored reprinting of classical texts in history, literature, Confucianism, and science in general, especially medicine. A vast array of books also came from nongovernment ventures responding to a widened public market, including a newly literate merchant class.

Along with urban trends toward secularization and sophistication in artistic, literary, and scholarly life, the best minds of the age forged a new synthesis of Confucianism and natural philosophy in which terms and concepts from the Buddhist and Taoist traditions were freely incorporated. Called Neo-Confucianism by Western scholars, the new synthesis provided an updated metaphysical framework for traditional ethics, partly by trying to take into account the whole range of natural knowledge, which by Sung times had expanded considerably; in the process, Confucian thought was accommodated to the developed cosmologies of China. The work of many men, but principally of Chu Hsi (1130–1200), the Neo-Confucian synthesis exercised a pervasive influence on medical theories in the Chin (Jürched) and Yüan (Mongol) periods.

To some extent Chinese cultural progress was retarded in the Yüan period, when the urbane, secular, and sophisticated

---

[6] Movable type made of baked ceramics was invented by this time but was unsuitable for large-scale printing due to the fragility of the type and technical difficulties in cutting and baking; not made with molds, it was not easily adapted to mass production. Metal cast-type was invented in Korea sometime in the early half of the fourteenth century (possibly earlier), and this technique was developed in China in the Yüan era. Movable bronze type was commissioned by the Korean Yi dynasty in 1390 and the first issue was in 1403; about a hundred years lapsed before this development reached China. Metal movable type was never used widely in China and Japan, though, mostly because of difficulty in embedding the type so that the printing surface was flat. Wooden movable type, appearing in the Yüan era, was found more satisfactory. Korea, however, lacking enough good wood, solved the problem of embedding metal type properly by standardizing types (modulae) and improving the setting in beeswax in bronze trays, a solution that proved much cheaper than woodblocks for limited editions. Cf. Jeon, pp. 167–184. On Chinese printing, see Carter.

Chinese were ruled by the essentially nomadic Mongols whose cultural heritage was rooted in military prowess and shamanistic religion. The Mongol lords, however, successfully recruited both Chinese and non-Chinese collaborators for administrative positions and also encouraged intercourse with the Islamic and Christian realms.[7] Confucianism was fostered, partly to retain the loyalty of Chinese officials, but greater attention was shown Buddhism, and even more so Islam. Generally, Confucian scholars serving as administrators enjoyed less access to top positions in government than before, and proportionately less influence. Their scholarship did not flourish, but neither did it die. Throughout the Yüan era the main traditions of Chinese culture were little affected by the Mongol presence or by the Islamic influence it encouraged.

When purely Chinese rule was restored in 1368 by the Ming dynasty, high priority was placed on elimination of the "barbarian" aftermath of Mongol domination and on the revival of classical Chinese culture in its purity. By that time, however, the bureaucracy had become quite conservative, and from Ming times on learning and science in China lost their creativity. Jealous of its position in East Asia, the Ming dynasty sought aggressively to reestablish tributary relations with its neighbors; thus, the main impact of Ming rule on the Semiseclusion Era was to provide the conditions for bringing it to an end.

JAPAN'S SHIFT TO FEUDALISM

Had Japan, like Korea, been subjugated in the Yüan era and Mongol suzerainty been acknowledged, it is possible that the infant Kamakura feudal order would have crumbled, and further that the Chinese-style political, social, and academic

---

[7] This was the dynasty that welcomed Marco Polo and many others from the West; during the Yüan genuine Chinese discoveries such as paper and printing, gunpowder, and the magnetic compass were transmitted to the West. Their use there was later associated with much more rapid social change than in China, where these and other technical improvements tended to be absorbed into existing social and political structures.

institutions would have been revived. Korea, in the Koryŏ dynasty (935–1392), had also, in 1170, experienced a military takeover similar to that of Japan by the Kamakura regime; but from 1231 Korea was forced to accept Mongol domination until 1356. When the Yüan dynasty was replaced in 1368 by the indigenous Ming rule, the Koryŏ too was soon (1392) supplanted by the Yi dynasty (1392–1910) which, like the Ming in China, reinstituted Confucianism (with emphasis much altered in practice) as the guiding philosophy of government. Korean history thereafter consistently followed the Chinese dynastic pattern, though without a change in dynasty, while Japan deviated from that model to evolve its own feudal system. Korea honored Confucianism over Buddhism[8] much earlier than did Japan, where Buddhism remained a vital force for at least two more centuries and Confucian ideology did not flourish until early in the seventeenth century.

Capitulation to the Mongol lords late in the thirteenth century would have surely brought Japan once again under heavy and direct continental influence. Whether this would have restored Japan to the dynastic line, however, cannot be easily assumed, because the deviation did not derive solely from diminished contact with China. Early in the Semiseclusion Era an attempt was made, in fact, to implement once more the *ritsuryō* system, an attempt codified in the 927 document *Engishiki* (Procedures of the Engi era; an elaboration of the Yōrō Code of 718). But the temper of the times ran against this effort. The court nobility continued to relish luxury imports from China, it is true, and they were not totally disinterested in certain formal features of Chinese culture, such as honorary titles of the *ritsuryō* system, including those in academic fields, which were retained for many centuries. But they also undermined *ritsuryō* institutions by monopolizing government positions and particularly by amassing huge, tax-free estates—an activity equally engaged in by the imperial house-

---

[8] Buddhism remained active on a popular level, though not with official support, even in the Yi dynasty. Cf. Ledyard, and Hatada (1), pp. 63–64.

hold and larger temples and shrines. As more and more lands were, often by devious means, removed from the tax registers, incomes of the central and provincial governments drastically declined. Thus, the weakening of the *ritsuryō* system had a three-centuries head start before the Mongol invasions even posed the possibility of restored continental tutelage.

Weakened central controls left the provinces in a general state of lawlessness. Local managers who ran the estates for absentee noble landlords were forced to form armed units of peasant-warriors who, by the end of the era's early phase, emerged as a kind of "rural military aristocracy."[9] Political competition among the nobility produced a growing series of court intrigues into which the local land-based militia was increasingly drawn. Eventually the local militias had little recourse other than to take a stronger hand in national politics; in time competition among military leaders led to a final showdown between the two most powerful groups, the Taira (or Heike) and Minamoto (or Genji) clans. In 1185 the Genji won a decisive victory and moved the political and military functions of government away from the intrigue-ridden capital of Kyoto to their own seat of power in Kamakura in eastern Japan, which was formally designated a shogunate in 1192. This shift of power from the nobility to the new samurai class inaugurated the second, or late phase of the Semiseclusion Era. The pattern of effective rule by a military regime, without abolishing the traditional forms of the imperial court or the status system of the nobility, continued for nearly seven hundred years up to the Meiji Restoration. Initially, landholdings of the imperial family, the court nobility, and of the major temples and shrines were not confiscated (though severe land losses were suffered later in the Warring States era, 1467– 1603). Nor were the traditional *ritsuryō* administrative posts fully abolished by the Kamakura government. But the earlier substance and spirit of these positions were already gone.

The estates of the early phase were largely self-sufficient

---

[9] Reischauer and Fairbank, pp. 522f.

economic units, and there was very little exchange of commodities among them.[10] Under the relative security provided by the Kamakura regime, a growing commodity-exchange economy emerged which, in turn, gave rise to new commercial centers at ports and highway station towns. There appeared a small but sometimes wealthy class of merchants. Economic growth was, however, supported largely by increased agricultural production. From the middle of the Heian era (794–1192) to the Muromachi era (1336–1467) total land area under cultivation in Japan is said to have nearly doubled, from 500,000 *chō* to 950,000 *chō* (approximately 1,220,000 to 2,330,000 acres),[11] and there was a concomitant increase in knowledge of intensive farming and more effective methods of planting and harvesting.

The techniques introduced in Chinese Wave I were, excepting a few special cases, assimilated into the major population centers by the early phase of the Semiseclusion Era. The formation of large estates diffused these techniques into the provinces. Artisans particularly developed great finesse in responding to the demands of the nobility and later the samurai; sword craftsmanship, for example, reached a peak in the Kamakura period. Moreover, regionalization of the economy stimulated a new degree of specialization in skills related to commodity production and marketing, money and exchange procedures, coastal shipping, and in mining, irrigation, and land reclamation, to mention a few. But it was the rise of enterprising merchant households during the late phase that brought forth the most significant technical innovations: the introduction of printing, and generally increased skill in practical mathematics as a sheer necessity. Merchants, though, like farmers and artisans, remained subordinate classes in a society oriented in the early phase to the court hierarchy, and in the late phase to military regimentation. By far the most equalizing forces were the development of a Japanese syllabic

[10] Yazaki, p. 76.
[11] Toyoda, p. 302.

script called *kana* and the introduction of new forms of Buddhism from China.

The script, in use by the tenth century, was a rather easily memorized set of forty-seven truncated Chinese characters, each of which represented a syllable of spoken Japanese. Writing could be done, then, all in *kana*, all in Chinese characters, or (from the eleventh century) by mixing the two. By bringing literacy within closer range of the public, the script did much to diffuse a common culture among the Japanese. An indigenous literature emerged that included not only the famous novel *Genji monogatari* (The Tale of Genji; early eleventh century) but also many histories, dictionaries, and encyclopedias. The effect of the script on learning and science was more limited, as it served to facilitate improvements in expression, not in thought as such. Nevertheless, the new script was of great importance in the spread of learning when priests in the late phase began to make certain scientific treatises available in this medium (p. 146).

"New Buddhism," so called to distinguish it from the "Old Buddhism" of Chinese Wave I, included the "Pure Land" (*Jōdo*), the "Lotus Sutra" (*Hokke*),[12] and the Ch'an (Zen) sects. No longer so closely identified with the nobility as was "Old Buddhism," the new sects were well received by all social classes—noblemen, samurai, and commoners—and thus gained a much broader popular base, partly because these sects were in fact as much created as adopted by the Japanese. Temples came to be built by local communities, not only by powerful leaders to promote an imported social system. Most importantly, the overseas travel involved in introducing these sects undoubtedly helped some of the leading priests to sense the changes taking place in East Asia and thus become the main agents of the intellectual and cultural ferment that

---

[12] The Lotus Sutra was actually introduced into Japan in Chinese Wave I; it was Nichiren's contribution to popularize it in the thirteenth century as the basis of a Japanese mass religious movement more typical of the late phase of the Semiseclusion Era. See, e.g., Nakamura, pp. 395–396, 429, 443, 451, 464, 511, 568, 582.

eventually led to the second great wave of cultural influx from China—Chinese Wave II.

Learning

THE EARLY PHASE: AFTERMATH OF CHINESE WAVE I

In Chinese Wave I the Japanese court nobility had engaged vigorously in the actual tasks of governing. They had sought not only to improve domestic conditions but also to comprehend Japan's position in East Asia. They had had a keen sense of responsibility as officials serving the nation as a whole and for this purpose had introduced Chinese learning and science. In the early phase of the Semiseclusion Era the nobility lost interest in the East Asian scene and in service to Japan as a whole, becoming absorbed in court ceremonials, political intrigues, expansion of their own personal privileges and property, and in the aesthetic and amorous life of the capital city. They evolved a cultured, courtly life-style (ōchōfū) that stressed personal skills in Chinese letters, art, music, and the Japanese verse form, familiarity with Chinese classics and Buddhist canons, and mastery of court ceremonials and precedents. But they were increasingly disinterested in administrative competence and the basic education necessary for it. Moreover, beneath their cultured refinement of literary and artistic tastes ran a strongly irrational and superstitious impulse. They feared evil spirits of the living and the dead, as well as various taboos, curses, and demonic forces—the dire consequences of which they sought to avoid by divination, exorcism, prayers, and supplications. In practice they, like the commoners, made little distinction between the intellectually dissimilar notions of ethnic Shinto, Chinese-style divination (onmyōdō), and Buddhist fate calculation (sukuyōdō).

Though the nobility carried literary and aesthetic refinement to extremes, few paid comparable attention to the more ethical and rational standards of Confucian studies and law, much less science, and virtually none tried to keep up with developments in learning and science in Sung China and afterward. Inevitably, these radical changes in attitude led to serious alterations in the *ritsuryō* institutions of learning—changes particularly visible in the alternatives to the University that emerged (see Table 14).

ALTERNATIVES TO THE UNIVERSITY

The decline of the University from the middle of the tenth century was due partly to a loosening of internal discipline, but equally to gradual loss of the nobility's patronage. Sons of the lower nobility had increasingly less hope of high placement in government, even if they went to the University, as higher administrative posts were monopolized by the upper nobility. The latter were assured appointments to high office regardless of academic record and, because they had other educational alternatives, had less need of the University. As the University gradually lost its raison d'être, its enrollment and caliber declined, including the once popular literature course. When the University building was destroyed by fire in 1177, it was never rebuilt, though nominal appointments to the University were continued.

The academic alternatives that evolved were of three kinds. One was the Monjōin, developed as the literature "department" of the University from about 810, with its East and West sections, and its own lecture hall and dormitory. Begun as part of the University, these sections of the Monjōin later came under the hereditary management of the Ōe (East section) and Sugawara (West section) households, and in time were operated as the private institutes of these households. The students came from leading noble families.[13]

Another alternative grew out of the University's affiliate schools (Daigakuryō bessō). As noted in the previous chapter, these schools were founded during the ninth century by ranking families among the nobility to prepare their sons for the University's entrance examination and to supplement its curriculum after enrollment. After the oldest, the Kangakuin, founded in 821 by the Fujiwara household, was given status as an affiliate school of the University in 872, others followed suit.

[13] Hisaki, pp. 84–87, 110–114.

The Shōgakuin, founded by the Arihara household in 881, was made an affiliate in 900, and the Tachibana household's Gakkan'in, founded around 845, was accorded affiliate status in 964.[14] These affiliate schools rose to prominence during the tenth century as the University went into decline; they remained active throughout most of the early phase of the Semiseclusion Era.[15] The Gakkan'in and Shōgakuin went out of existence sometime early in the twelfth century. The Fujiwara household's Kangakuin began to decline also in the twelfth century but continued to operate until late in the Kamakura period. The Monjōin lost prestige as these affiliate schools prospered, but it also survived until late in the Kamakura era.

The third development that afforded some real alternative to the University was a modification of one of its original functions, namely, professorial appointments. Late in the ninth century the Ozuki household had acquired hereditary rights to the professorship in mathematics, and by the end of that century the Ōe and Sugawara households assumed hereditary control of the literature professorships, for which they selected and trained their successors.[16] From the middle of the tenth century, concurrently with the decline of the University and the rise of the affiliate schools, the tendency for professorial appointments to be made hereditary became quite strong. This tendency was not, however, limited to the University alone; by the middle of the eleventh century virtually all of the major disciplines of the University and of the institutes of Divination and Medicine had been incorporated into a system in which one or more noble households had hereditary rights

---

[14] Ibid., pp. 181–184.
[15] A related activity was the founding of a sort of library called Kōbun'in by the Wake household around the end of the eighth century. It placed no particular emphasis on educating students and did not survive beyond the end of the ninth century; hence it did not constitute a real alternative to the University. The name was revived, however, by the Tokugawa shogunate in 1663 when it was given to the Hayashi academy (est. 1630) of Confucian studies. Cf. Hisaki, p. 151.
[16] Ibid., p. 100.

and responsibilities for specific disciplines. This arrangement was called "House Learning" (*Kagaku*).

HOUSE LEARNING

The assignment of hereditary professorships in particular fields to specific households took place gradually. Beginning with mathematics and literature in the late ninth century, other assignments were made during the tenth and eleventh centuries, and the last, in law, appears to have been made in the twelfth century.[17] By that time not only the University but also its affiliates were losing their importance. Hereditary professorships in the House Learning system were continued in most academic fields long after the Semiseclusion Era ended, though without creativity and not infrequently without real mastery of the inherited disciplines.

Once the pattern of House Learning was established, the officially designated scholars educated their sons (natural or adopted) in their assigned specialties, lest the scholarly tradition be lost, and with it the court appointment. This meant that, despite fluctuations in social or intellectual demand, or in stimulation from outside the country, the barest minimum of each discipline had some chance of being preserved. On the other hand, this system made mobility from one specialty to another impossible and set formidable barriers between scholars and the public. Indeed, to enhance the authority of their respective fields, scholars tended to stress the more esoteric rather than the more general aspects of their studies. Consequently, scholarship became a formality, genuine understanding was impeded, and intellectual creativity seriously impaired. Closed to new talent, those disciplines that could not draw upon strong social or intellectual demand declined most rapidly in quality and productivity. This was especially true of the more conceptually rigorous disciplines; thus, in the sciences, mathematics suffered the most, astrology and calendrical astronomy next, and medicine the least. All in all, the academic caretakers manifested little motivation to do research

---

[17] Ibid., pp. 194–196.

Table 14. Learning in the Early Phase of the Semiseclusion Era

	600	700	800	900	1000	1100
					————Semiseclusion Era————	
			————Ninth century————	894	Mid-tenth century	1177
University (Daigakuryō)		ca. 670 University started	Peak of University		Decline of University	Burned; never rebuilt
University affiliates (Daigakuryō bessō)			By 810: Monjōin ——————————→ Virtually private institution of Ōe and Sugawara households			
				821— 872— Emergence of private academies	Recognition as University affiliates	Decline of University affiliates
Hereditary professorships and House Learning (Kagaku)				By end of ninth century First hereditary professorships: Mathematics, Ozuki	Mid-tenth century—Mid-eleventh century —————→ Extension of hereditary professorships to most fields	
				Literature, Ōe and Sugawara		Systematized as House Learning

and even less apparent interest in contact with China through Buddhist priests who traveled there. It is not surprising, then, that learning in the hands of the nobility became largely nominal by the second phase of the Semiseclusion Era and had very little to do with the revival of Chinese learning and science in Chinese Wave II.

THE LATE PHASE: NEW STIRRINGS IN LEARNING

The samurai founders of the Kamakura regime had had no share in the previous course of learning, and their simple, austere, and disciplined life-style stood in conscious contrast to the leisured and luxurious life of the Kyoto aristocrats. But they did not cut themselves off totally from the learned traditions of the nobility. Some of the lower nobility with House Learning assignments were welcomed to Kamakura to assist the shogunate. From early in the thirteenth century low-ranking court physicians also moved to Kamakura to serve the shogun, his family, and top shogunate officials, and one head of the Institute of Medicine visited there to offer his services.

A more important alliance was that developed by the new military class with Buddhist priests, who gave the samurai their first grasp of the larger East Asian scene. Especially after the Ashikaga shogunate assumed power, with its headquarters in the Muromachi township of Kyoto, learned priests, and particularly those of the Rinzai sect of Zen, were quite active as political and diplomatic advisors. Some were appointed envoys to China, and some served as supervisors for officially authorized trading ships to Yüan China. As the only sustained link with China, they alone maintained a somewhat wider perspective of regional affairs.

Samurai generally received their primary education at local Buddhist temples (source of the term *terakoya*, "temple schools"), and with the help of learned priests some of the leading samurai formed their own academic facilities such as the Hōjō household's Kanazawa Library (Kanazawa Bunko), and the Uesugi family's Ashikaga School (Ashikaga Gakkō) in the northern part of the Kantō plain. Leading samurai were attracted to the cultured tastes of courtly life in Kyoto, and developed their own aristocratic samurai culture. They were not inclined to revive the *ritsuryō* academic system and, like the nobility, had little interest in science.

BUDDHIST LEARNING

Learned priests undoubtedly made the most significant contributions to learning in the Semiseclusion Era, though not necessarily in its early phase. The Old Buddhism of that phase was too tied to the secular concerns of the nobility—their health, wealth, and power. It was the learning of the privileged class, concerned with points of doctrine and remote from the personal religious experiences of ordinary people, except for the appeal of esoteric rites, exorcism, and fortune-telling.

Buddhist learning underwent great changes, though, in the late phase of the Semiseclusion Era after the introduction of New Buddhism. The new sects gained greater contact with all sectors of the populace, and Zen priests, particularly those of the Rinzai sect, won a new place in the affairs of state. Most important, priests of New Buddhism took advantage of their contacts with China to explore also non-Buddhist intellectual developments, especially Neo-Confucianism, which became part of their study programs in Japan. Old Buddhism had its centers of study and training, such as Mt. Kōya and Mt. Hiei, outside the metropolitan centers. The most prominent institutions of learning founded by New Buddhism were the temple complexes called *gozan* (lit., "five mountains"), a group of five (and later more) temples built first in Kamakura and afterward in Kyoto when the shogunate moved there. Founded by the Rinzai sect, partly at the instigation of the shogunate and in imitation of the Ch'an models of China, the *gozan* temples became centers for the introduction of contemporary Zen styles in temple-building, and for the zealous promotion of Zen disciplines both in everyday life and in intellectual pursuit. Along with Buddhism proper, classical Chinese learning and Neo-Confucianism were at the heart of their academic activity.

The sciences were not included in the *gozan* studies (or those of other temples), nor were scientific concepts viewed as integral to, or relevant extensions of, their learning. Even so, medicine was the chosen avenue of service for many priests. Inclined more to practical treatment than to theoretical studies or attempts at systematization of medical knowledge, a number of priests did in the late phase perform notable service in the compilation of medical texts. Very late in this phase some priests went to China to do specialized medical research. Apart from medicine, though, the Buddhist contribution to science was meager. Yet the *gozan* temples in both Kamakura and Kyoto did serve to keep alive interest in Chinese culture during the era's late phase and thus to prepare the way for the renewed activity in learning and science in the next wave period. Not surprisingly, medical scholars with priestly backgrounds were to be in the vanguard of that revival.

The Specific Sciences

Given such unfavorable conditions as the general isolation from scientific activity on the continent, except for the limited Buddhist contacts, the merely formalistic preservation of Chinese Wave I sciences in the House Learning system, and the isolation of House Learning from the general public, as well as the lack of strong patronage from any major power bloc in society, the course of science was very much left to chance, and in fact was highly uneven. The overall course is marked by decline in the early phase of the era and by only faltering steps in the late phase toward a revival which finally occurred with the reintroduction of Chinese sciences in the fifteenth and sixteenth centuries.

ASTROLOGY AND CALENDRICAL ASTRONOMY

Detailed observations of celestial phenomena were continued in China throughout the Sung, Yüan, and Ming periods for the two traditional purposes: to determine the astronomical constants and epochal positions of the sun, moon, and planets needed to make an accurate calendar; and to search for anomalies that could be interpreted as having implications for imperial rule. Frequent calendar changes continued to take place as increasingly conventional accompaniments to changes of emperor; few were improvements except in details. Calendar changes during the early half of the Sung period (Northern Sung) number nine, with nine more in the Southern Sung period.[18] Japanese Buddhist priests visiting China in these times acquired some competence in both traditional computational methods and in matters related to *sukuyōdō*, enabling them to compete with official specialists in the early phase of the era.

The most remarkable features of the Yüan period were the production in 1280 of the *Shou-shih li* by Kuo Shou-ching and the welcome given Islamic scholars by the Mongol government. The *Shou-shih li*, considered the best of all astronomical systems calculated by traditional methods, was backed by a vast amount of precise observational data and by use of improved observational instruments. As was conventional, it included step-by-step instructions for computing ephemerides (i.e., predicting all predictable astronomical phenomena). It was thus not a calendar but a generator of calendars. It is known that in 1267 the Muslim Jamal al-Din (Cha-ma-lu-ting in the Historical Records of the Yüan Dynasty) used some astronomical instruments that he brought from Persia—an armillary sphere, celestial and terrestrial globes, plane sundials for equinoxes and solstices, and another vaguely described as for celestial observation (probably an astrolabe).[19] These are thought to have been used for observations for the *Shou-shih li*. But the mathematical and conceptual approaches used in compiling it were traditional Chinese methods.[20] An Islamic Observatory was built in 1271 in Peking, and Islamic astronomers performed their own independent observations, using their own instruments, references, and means of calculation; their data were then submitted to the Chinese specialists. In the Ming period those Muslims remaining in China were

[18] Yabuuchi (4), p. 90.
[19] Needham, 3: 372–378.
[20] Yabuuchi (4), pp. 101–105; Needham, 3: 381.

incorporated into the Chinese astronomical bureau as a separate section. But during the three centuries of the Ming era, only one calendar, the *Ta-t'ung li* (a slight modification of the *Shou-shih li*), was used.[21]

In Japan, Kamo no Yasunori, who held appointments as professor of astrology and as head of the Institute of Divination, died in 987, after entrusting the work of astrology to his ablest disciple, Abe no Seimei (921–1005). Calendrical astronomy, considered a less honorable and exacting task, was entrusted to Yasunori's son Kamo no Mitsuyoshi. Subsequently, astrology became the House Learning of the Abe family (and later, its descendants, the Tsuchimikado household), and calendrical astronomy the House Learning of the Kamo household (later, through a nonlineal family, called Kōtokui). While the functions of both households remained official ones, the Abe family enjoyed a more privileged status at court; it had access to the hall of the imperial court, which the Kamo family did not. Also, members of the Abe line were subsequently appointed to the post of director of the Institute of Divination, the Kamo line supplying the vice-director.[22] The Kamo family later became extinct, and from the middle of the sixteenth century both astrology and calendrical astronomy were taken over by the Tsuchimikado household (descendants of the Abe household), which controlled both fields up to the Meiji period when the Western (Gregorian) calendar was adopted.

Even after the University virtually vanished, the Institute of Divination, along with the Institute of Medicine, continued to exist up to the Meiji Restoration (1868). Like the University and the Institute of Medicine, the Institute of Divination possessed its own lands (*shokuden*) which provided revenue for its activities. The Institute's lands were regarded as the private property of the Abe and Kamo households after the House Learning system went into effect. Buildings of the Institute of

[21] For Islamic astronomy in the Yüan and Ming periods, see Yabuuchi (5), pp. 202–234, and Needham, 3: 372.
[22] Nakayama (6), p. 22.

Divination in Kyoto were destroyed by fire in 1127, at which time all but a few of the records, references, and instruments were lost. These materials were treated as highly secret government property, and therefore private ownership of them was prohibited. In this respect the loss was much greater than when fire destroyed the University in 1177, as private ownership of the Confucian and other Chinese classics used in the University was common among its professors and other learned men.

The major texts of the Chinese *yin-yang* art were required reading in the Japanese Institute of Divination, and it may be assumed that the Abe family was knowledgeable in *yin-yang* studies, and particularly the modes of calendar-related divination. To Abe no Seimei is traditionally attributed authorship of the *Hoki naiden* (Ritual implement tradition), a *ying-yang* (*onmyōdō*) text.[23] It dealt mainly with hemerological concerns but drew also on Buddhist, Shinto, and possibly Taoist sources. This text was less theoretically developed than the Chinese classic in the field, the *Wu-hsing ta-i* (Fundamental principles of the Five Phases); it was typical of the eclectic tendencies of the nobility and of their turning from public affairs to matters of private fate and fortune.

Practitioners of *onmyōdō* trained in the Institute of Divination were called *onmyōji* (or *on'yōshi*). They served as government officials at the Institute itself or were assigned to Dazaifu, to some other provincial center, or to one of the military garrisons (*chinjufu*) in eastern Japan. In Chinese Wave I their function was to engage in divination and related rituals in matters of national interest. In the early phase of the Semiseclusion Era the *onmyōji* increasingly served the personal interests of the court nobility, a reflection of the latter's contracted world perspective and gradually lowered rational propensities. On a more popular level, many of the *onmyōji* assigned to the provinces eventually settled down and spread *onmyōdō* ideas and practices among the general populace in a much de-

[23] Ibid., p. 63.

generated form. That is, their services performed for individual clients became increasingly mixed with Buddhist *sukuyōdō* elements and various indigenous taboos and exorcistic rites common among noblemen and commoners alike throughout the Semiseclusion Era. Afterward, around the end of the fifteenth century, there appeared a number of practitioners who, without official appointment or approval, called themselves *onmyōji*; their practices grew so indiscriminate by the Tokugawa era that the shogunate in 1684 put them all under the formal control of the Tsuchimikado household. Even in China, *yin-yang* divination, though replete with cosmological theory, was never brought into a functional relation with calendrical astronomy; the *onmyōdō* practices in Japan veered even further away from exact scientific work.

The upper hand in fortune-telling in the early phase of the Semiseclusion Era was held, however, by the Buddhist practitioners of *sukuyōdō*, who were called *sukuyōji*. The practice had initiated late in Chinese Wave I with the introduction of the *Sukuyōkyō* canon (*Hsiu-yao ching*) and the *Ch'i-yao jang-tsai chueh*, a more detailed text that included planetary tables for casting horoscopes (see p. 69).[24] The earliest known horoscope in Japan is one cast for a young boy in 1112; the document that records it, the *Sukuyō unmei kanroku* (A record of fate prognostication according to the mansions and planets), quotes the above texts as well as a Chinese translation (*Tu-li-yü-ssu ching*; ca. 800) based on Hellenistic zodiacal astrology, without reference to Indian sources. This horoscope and the second oldest known, for the year 1269, were both cast by consulting the *Fu-t'ien li*, an unofficial Chinese calendar compiled by Ts'ao Shih-wei between 780 and 783[25] and brought to Japan in 957 by a Buddhist monk. The horoscopic art of Indo-Hellenistic background could, in principle, have been a stimulus to astronomical observation and study in Japan; the opposite seems to have been closer to the truth. As Nakayama

[24] Ibid., p. 62.
[25] Yabuuchi (2), p. 459; also Nakayama (6), p. 62.

Shigeru asserts, "Although practitioners quoted these works, there is virtually no record that they also observed the courses of heavenly bodies for the purpose of casting horoscopes. This art, therefore, did not help in stimulating astronomical activities but was merely absorbed into the underworld of Chinese divination practices."[26]

Buddhist activity in general was more given to dealing with the many superstitions and openly secular interests of the dominant court nobility in the early phase of the era, particularly the performance of esoteric rites. For example, on days when solar eclipses were predicted (indicated each year in the civil calendar), court buildings were covered with coarse straw sheeting and court business for the day was canceled. Several leading temples in Kyoto then performed rituals by court order to ward off suspected evils. If the eclipse failed to materialize, the emperor sent messengers to the temples to express gratitude for the successful rituals. The possibility of faulty predictions seems to have been beside the point.

The *sukuyōji* began early in the eleventh century to predict solar and lunar eclipses on their own, competing with the Institute of Divination's specialists in calendrical astronomy. That the ruling nobility viewed eclipses with a sense of grave foreboding appears evident from the excessive number recorded —more than could possibly have been observed in Japan at that time.[27] While the Institute's specialists no longer had direct contact with China, many Buddhist priests did. Their accuracy in predicting eclipses that actually occurred, and in not predicting ones that failed to materialize, often surpassed the record of the government specialists. Men trained in the University's course in mathematics later entered the competition and occasionally excelled over one or the other of the competitors.

The *sukuyōji* lost their patronage after the first phase of the Semiseclusion Era, largely because the samurai class had tastes

[26] Nakayama (6), p. 62.
[27] Ibid., pp. 51–52.

different from those of the declining nobility, but also because they patronized the priests of New Buddhism instead. Many *sukuyōdō* practices survived by being absorbed into *onmyōdō*, along with indigenous elements, and thus were diffused throughout society. A new sort of activity in fate prediction arose among the priests of New Buddhism in the second phase of the era. It came from the *gozan* monasteries where the Book of Changes was studied as one of the "outer canons" (i.e., Chinese classics, as distinguished from the "inner canons" of Buddhism proper).

The Chinese system of divination developed around the Book of Changes (*I-ching*; Japanese, *Ekikyō*) is thought to have been based originally on a loosely organized corpus of divination texts, folklore, and peasant omen interpretations from the early Chou dynasty or even earlier. By the end of the Han period a number of Confucian appendices were added that transformed the corpus into a repository of concepts so abstracted from concrete experience as to be applicable to almost every conceivable natural or human situation. By Sung times the forces and phases represented by stereotyped trigrams and hexagrams used in this practice came to be viewed "not only as abstract formulations of all kinds of natural processes, but as invisible operators and causative factors."[28] These ideas had been correlated to calendrical phases as well as to the apparatus, substances, procedures, and propitious times for carrying out various alchemical processes. Eight trigrams were combined into sixty-four hexagrams that were made to represent a vast array of dialectical processes of advance and retrogression, increase and diminution, etc. For fate calculation, a handful of sticks were tossed to reveal the appropriate hexagrammatic patterns for a given situation.

Eki divination, as it was called in Japan, reached its peak in the post–Semiseclusion Era time of civil wars when Zen priests used it in their role as advisors to military leaders. The Ashikaga School gained a special status as a kind of professional

---

[28] Needham, 2: 329.

training center for *eki* practitioners. Zen priest-advisers did not, however, confine themselves solely to *eki* methods but also used portent astrology to interpret unusual celestial and terrestrial phenomena. The last word is, though, that neither *sukuyōdō* and *eki* as practiced by Buddhists, nor for that matter, *onmyōdō* practice, led to objective study of the heavens; nor did any of these methods of fate calculation enjoy any direct interrelationships with astronomical science.[29]

Finally, with regard to calendars, it has already been noted in Chapter 1 that the *Hsüan-ming li*, used from 822 to 892 in China and in Japan from 862, was a good one. But by the early phase of the Semiseclusion Era its secular errors had accumulated to the point where calculations based on it were increasingly inaccurate—frequent failures in eclipse predicting made the Institute of Divination specialists an easy target of criticism by the *sukuyōji* and others. Cut off from technical improvements in China, specialists at the Institute were unable to devise means for correcting their calendar, much less construct a new astronomical system; they were too proud or too disinterested to adopt contemporary Chinese calendrical systems that had been brought up to date. The almanac (*guchūreki*; see Chapter 1), on the other hand, occupied a prominent place in the lives of the nobility; many of them used it as a diary by writing in the margins. Originally written in Chinese characters and handcopied, a *kana* version (*kanagoyomi*) emerged in the Semiseclusion Era. It too was at first copied by hand but later was printed by woodblocks. The date of initial appearance is uncertain, though the *kanagoyomi* surely existed by the thirteenth century, presumably in manuscript form; the oldest extant printed copy is dated 1331. Once distributed only among government officials, its use spread far and wide during the Semiseclusion Era. In Japan's Warring States era (1467–1603) the annotated calendars were commercially printed in many places in Japan.

[29] Nakayama (6), pp. 59, 63.

MATHEMATICS

In the Sung reprinting of many classical texts, mathematics was no exception. Of the "Ten mathematical manuals," used earlier as text-books in the T'ang university's course on mathematics, all were reprinted, once in the Northern Sung and again in the Southern Sung, except for the most advanced, the *Chui shu*, which had already been lost.[30] The reprints did much to provide a basis for progress in the late Sung and Yüan periods, while encouraging advances to proceed along traditional lines. These patterns of achievement can best be seen in the work of four leading mathematicians: Ch'in Chiu-shao (1202–1261), Li Yeh, Yang Hui, and Chu Shih-chieh.

Ch'in Chiu-shao, who lived in the Southern Sung, in 1247 wrote his *Shu-shu chiu-chang* (Mathematical treatise in nine sections), which dealt with eighty-one problems.[31] Neither a revision of nor a commentary on the famous Nine Chapters (pp. 75, 82), it covered most of the problems and methods of that earlier classic but concentrated on more difficult problems such as the indeterminate analysis developed in the work of the Buddhist I-hsing. For this Ch'in Chiu-shao made use of the "heaven-origin method" (*t'ien-yüan shu*; i.e., a method of coefficients for solving algebraic unknowns). He used variously 3, $\frac{22}{7}$, and $\sqrt{10}$ for π. He treated problems of the equipartition of a trapezoid, the first instance of various problems of the same nature which interested later Chinese and Japanese mathematicians; and one such problem was solved according to an equation treated as one of the tenth degree. Before Ch'in Chiu-shao there is no known instance of equations of degrees higher than the third. The practical orientation of the mathematical tradition to which Ch'in was heir hardly required equations

---

[30] Yabuuchi (4), p. 54.
[31] Libbrecht's book is an exhaustive study of this treatise, with comparisons of its major achievements to similar processes in India, Islam, and Europe; it provides biographical data on Ch'in Chiu-shao as well as significant information on artisanal, economic, administrative, and military affairs related to Ch'in Chiu-shao's problems.

higher than the third degree. He went to great pains to cast the practical problems of his day (to be sure, more complex than those dealt with in the Nine Chapters which shaped that tradition) in ways that did require equations of such high degree in order to satisfy his pioneering desire to solve them. This resulted in very complicated procedures—"an example not of inelegant solution but of experimentation in a direction not easily accommodated by the traditional character of Chinese mathematics."[32]

Whereas Ch'in Chiu-shao gave detailed descriptions of the practical processes of root extraction so important in his work, he failed to describe the setting up of determinate equations. Among extant documents, such algebraic considerations are explained for the first time in Li Yeh's *Ts'e-yüan hai-ching* (Sea-mirror of circle measurements; 1248) and his *I-ku yen-tuan* (New steps in computation; 1259).[33] While Ch'in used the "heaven-origin method" in indeterminate problems, he did not use it in numerical equations, which Li Yeh did, writing *yüan* beside the unknown to distinguish it. *T'ai* marked the extreme limit, i.e., the absolute term. Thus, the notation[34] in Figure 13 represents the equation $2x^3 + 15x^2 + 166x - 4460 = 0$. Positive and negative numbers written in red and black ink respectively proved rather inconvenient, so Li Yeh used a diagonal stroke through the right-hand figure (an "O" for zero) in the bottom row to denote a negative, a habit adopted and used by Chinese mathematicians from his time on.

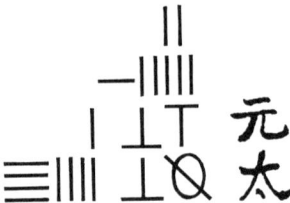

Fig. 13. Algebraic equation by the "heaven-origin" method (元 = *yüan*, 太 = *t'ai*).

[32] Libbrecht, p. xiii, in the foreword by Nathan Sivin.
[33] Li Yeh lived in the Northern Chin under Jürched and Mongol rule, and it is improbable that he had contact with Ch'in Chiu-shao. Needham, 3: 41.
[34] Ibid., p. 45; see also Smith and Mikami, p. 50.

Rules that lead to formulae for the summation of arithmetical progressions appear in the 1261 work *Hsiang-chieh chiu-chang suan-fa tsuan-lei* (Detailed analysis of the mathematical rules in the Nine Chapters and their reclassification) by Yang Hui of the Southern Sung. His greatest achievement, however, was the solution of linear equations with four or five unknowns, and he gives one case of a biquadratic equation. But Yang Hui did not employ the "heaven-origin method" in solving equations with several unknowns; this came half a century later in the work of Chu Shih-chieh of the Yüan era. His two chief works were the *Suan-hsüeh ch'i-meng* (Introduction to mathematical studies) and the *Ssu-yüan yü-chien* (Precious mirror of the four origins), written in 1299 and 1303 respectively.[35] Since it was an elementary textbook, the former contained nothing new beyond the results of his predecessors. It is of interest chiefly because of the strong influence it was later to exercise on the development of Japanese mathematics (see Chapter 4).

Prior to Chu Shih-chieh "heaven-origin" algebra was carried out only for cases of one unknown; Chu extended it to cases of four unknowns.[36] The four unknowns were distinguished as heaven, earth, man, and thing, the latter three being arranged to the left, right, and above heaven on the counting board. The process—essentially the same as what was called Horner's Method in Europe—was to take trial solutions and from them to generate arbitrary intermediate unknowns other than the ones sought, and then from known relations given by the data of the problem, to improve the trial solution one order of magnitude at a time, using at each step new intermediate unknowns. The algebra of the "heaven-origin" coefficients is said by those who recorded it in the thirteenth century to have

---

[35] Both of these texts were lost in China. The "Introduction" was restored in 1839 by Lo Shih-lin's (?–1853) discovery of a Korean reprint made in 1660; the "Precious mirror" was recovered only slightly earlier. Cf. Needham, 3:46.
[36] Chu Shih-chieh was a contemporary of Kuo Shou-ching, compiler of the *Shou-chih li* calendrical system, for which Islamic observational instruments were presumably used. While some relevant texts remain to be carefully analyzed, thus far no concrete Islamic influences on Chinese mathematics have been discerned. On this problem, see Needham, 3:48–50, and Mikami (1), p. 108.

descended from earlier ages, though it was never practiced widely among Chinese mathematicians. But it definitely sprang from the older practices of extracting square and cube roots by manipulations on the counting board—a peculiarity of Chinese mathematics.[37]

Traditional Chinese mathematics exemplified by the "Ten mathematical manuals" inherited from the T'ang period and carried to its peak by the Yüan period, subsequently declined in the Ming era. From beginning to end it was cultivated primarily for the use of administrative officials; and while they treated mostly practical problems, their interests often lay far away from the everyday problems of ordinary merchants and artisans. Nonetheless, the elementary mathematics necessary to the work of commerce and crafts developed steadily during the Southern Sung and Yüan periods and was the context for the emergence of the abacus. No established theory of the date of the first use of the abacus stands, but it was certainly in widespread use in China by the Ming era and possibly was in use as early as the late Southern Sung period.[38]

Although some of the Sung or Yüan works may have come to Japan in Buddhist hands during the Semiseclusion Era, there is unhappily no discernible sign of their influence throughout the era. The reprinted "Ten mathematical manuals" were never brought over as a corpus to Japan, even in later periods, and the higher mathematics as well as that of merchants and craftsmen—including the abacus—did not reach Japan until Chinese Wave II was under way. The Sung and Yüan achievements in mathematics contrast strikingly with the sharp decline of mathematics in Japan during the Semiseclusion Era.

Hereditary professorships occurred earliest in the field of mathematics, the first appointment going to the Ozuki household late in the ninth century.[39] The Miyoshi household acquired similar privileges in this field by the end of the tenth century. Only the names of those receiving formal appointments as court mathematicians are recorded, however, with

[37] Wang Ling and Needham.
[38] Needham, 3: 74–79.
[39] Hisaki, p. 100.

no indication as to their achievements in mathematics. It is known, as was noted in the previous section, that some of them challenged astronomers in the Institute of Divination, as well as *sukuyōji*, in predicting eclipses in the early phase of the Semiseclusion Era.

Practical mathematics of use to *ritsuryō* administrators had been central to the University's mathematical course in Chinese Wave I. Following the transfer of the capital from Nara to Kyoto, no further moves were made, and construction and civil engineering projects of the government tended to diminish in scale. Political power passed first to the nobility (early phase) and then to the land-based military cliques (late phase), and the need for practical mathematics of a high level to be used in finance and engineering seems to have declined in the process. While claiming no special progress in the Semiseclusion Era, calendrical astronomy, and to a lesser degree astrology, had at least some regular jobs to perform; medicine had the stimulus of social and personal needs to deal with human illness, no matter what the social and political situation. Mathematics as kept in the custodial households could count on no such demands or necessity for its existence. The court-appointed mathematicians had nothing to do with the commercial and technical requirements of commoners. The potential stimulus of new developments on the continent during the Sung and Yüan periods remained outside their range of vision and interest.

According to a book catalogue entitled *Nihonkoku kenzaisho mokuroku* (Catalogue of books seen in Japan), compiled around 890, some of the Sui and T'ang books on mathematics were brought to Japan and preserved up to at least the end of the ninth century. These books went out of use and most appear to have been lost by or during the turmoil of Japan's Warring States period; and, as already indicated, there is no clear sign that the Sung reprints of the "Ten mathematical manuals" and other Sung and Yüan texts reached Japan.[40] The picture one gets from existing records is one of serious decline of mathe-

[40] Sawada, pp. 42–43.

matical interests and capabilities, more serious than in any other scientific field.

A glimpse of the real situation can be gleaned from texts written for the elementary education of sons of the court nobility in the early phase of the era, and for sons of the samurai in its late phase. Such a text is the *Kuchizusami* (lit., "singing to oneself," i.e., impromptus), compiled in 970 by Minamoto no Tamenori. It consists of nineteen sections, the last of which is a "miscellaneous" section that includes several items related to arithmetic. Some of the items involved rudimentary instruction; e.g., a multiplication table that begins with 9 × 9, and works down to 1 × 1, and a way of counting large numbers in decimally arranged units: 10, 100, 1,000, 10,000, 100,000, 1,000,000 and 10,000,000 (*to, momo, chi, yorozu, soyorozu, momoyorozu,* and *chiyorozu*).[41] Another example is a method for finding the total number of bamboo rods in a bundle, given the number of rods in the circumference.

Only step-by-step computation is recorded, with no explanation of the method it exemplifies. The text also includes some numerological schemes. One is for predicting the sex of an unborn child: add 12 to the age of the pregnant mother, then

Fig. 14. Simple arithmetical problem posed by a bundle of rods.

[41] Smith and Mikami, p. 4.

subtract in succession 1, 2, 3, 4, ... until further subtraction becomes impossible; if the remainder is an odd number, the child will be a boy; if even, a girl. The influence of *yin-yang* thought is obvious in the association of odd numbers (*yang*) with male fortunes, and of even numbers (*yin*) with female fates. (A similar exercise appears in the *Sun-tzu suan-ching*.) Another example is a method for making prognoses in cases of serious illness: add 93 to the patient's age and then divide by 3; no remainder means death for a female, life for a male; some remainder has the opposite implications. In all these examples, whether straightforward instruction, games, or magic, the learning was mainly by rote and the content entirely of an elementary nature. They certainly suggest that the image of mathematics prevalent among the nobility was that rudimentary arithmetic is enough for life, and the rest is games or magic.

While it was extremely rare for any of the Buddhists in touch with Chinese scholarship to take an active interest in high-level Chinese mathematics, there is some fragmentary record of priests who studied the mathematical textbooks introduced in Chinese Wave I. According to the autobiography of Engetsu (1300–1375), twelfth chief priest of the Kenninji temple (Rinzai sect) in Kyoto, he studied the Nine Chapters in 1311 at the age of twelve, along with study of the Classic of Filial Piety (*Hsiao ching*; in Japanese, *Kōkyō*) and of the Analects (*Lun yü*; in Japanese, *Rongo*) of Confucianism, besides studies of the Buddhist canons. In the late phase of the Semiseclusion Era many Zen priests accompanying Japanese trading ships to China made the actual transactions. Presumably they had some knowledge of practical mathematics, acquired in or out of the temples. Further information on mathematics in the era's late phase derives from the fact that a small number of priests incorporated bits of very practical arithmetic into *ōraimono* (didactic exchanges of letters) for use initially in the elementary education of children of the court nobility and high-ranking samurai. Texts in the *ōraimono* style appeared in the eleventh century (late Heian era) and continued to be produced until the end of the Edo period, their use being extended, of course, to commoners in the latter period. The compilers were usually members of the court nobility in the early part of the Semiseclusion Era, Buddhist priests from the Kamakura to the Warring States periods, and literary writers and professional teachers in the *terakoya* schools in the Edo era.

A Zen priest named Kokan Shiren (1278–1346) of Tōfukuji temple (one of the Kyoto *gozan* group) published an *ōraimono* in 1346 entitled *Isei teikin ōrai* (Extraordinary exchange of letters) that contained a number of mathematical games. One of these, called *Mamakodate* (*mamako* = stepchild, *date* = standing), supposes a mother who has fifteen true children and fifteen stepchildren standing in a circle. Starting from some one child and counting clockwise, the problem is to eliminate every

Fig. 15. The *mamakodate* problem as illustrated in Miyake Kenryū's *Shojutsu sangaku zue* (1795 ed.).

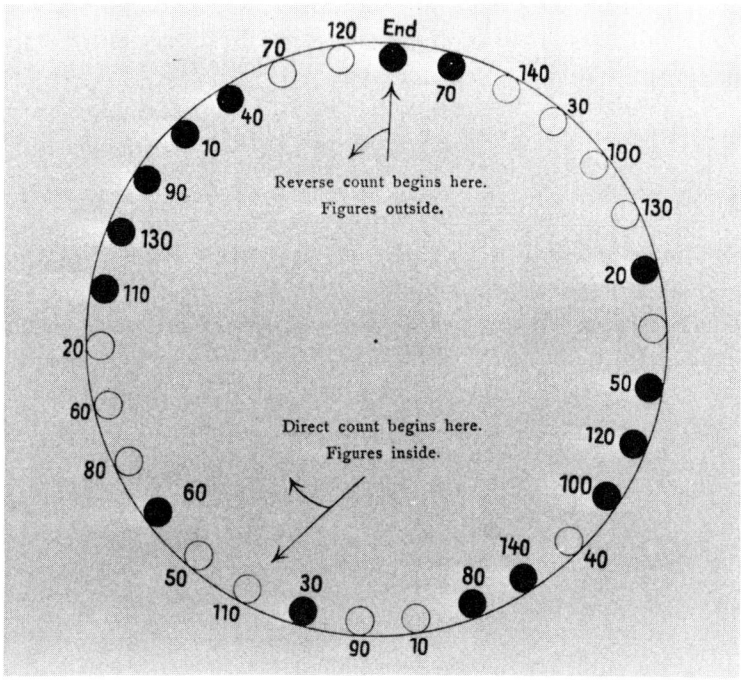

Fig. 16. Diagram of the solution for the *mamakodate* problem. (Smith and Mikami, p. 83)

tenth child so as to leave only the true children standing in the circle. In what order should they be lined up?[42]

Neither the court-appointed mathematicians nor priests interested in mathematics provided the preparation for, or the transition to, the revival of higher-level mathematics in a later age. That role was played by commoners. Practical mathematics useful in the work of artisans and merchants gradually permeated the daily lives of ordinary citizens in the Semiseclusion Era. It was presumably quite unsophisticated at this stage, and its content is difficult to assess. Yet it appears to

[42] Nihon Gakushiin (4), 1: 160; Smith and Mikami, pp. 80–84; according to Needham, 3: 62, there is as yet no known instance of this puzzle in a Chinese mathematical work.

have eventually served as the major springboard for the development of mathematics in the seventeenth-century intellectual renaissance and to have created a commercial market for printed books on mathematics. The custodial mathematics of House Learning was to have no role in that development.

MEDICINE

The two thrusts of Sung culture—to revive classical learning and to create a new, contemporary synthesis of knowledge—had their greatest effect on medical scholarship of all scientific fields. The printing of medical texts under imperial auspices was particularly extensive, both in sheer quantity and in the range of materials. Interest focused partly on the reprinting of medical classics, especially the Han text, the Yellow Emperor's Inner Classic (*Huang-ti nei-ching*) and the third-century text, On Cold Damage Disorders (*Shang-han tsa-ping lun*), reprinted in 1027 and 1065 respectively. New compilations were revised and expanded from the tenth through the thirteenth centuries as well. Massive collections of prescriptions for clinical use were compiled by imperial order. An imperial handbook which set out 297 prescriptions for given syndromes, without reference to underlying physiological and pathological theories, was the *Ho-chi-chü fang* (Official prescriptions), printed in 1106. Revised and printed again and again, this manual was not only widely used in China but also became the principal medical handbook in Japan from the Kamakura period to the middle of the Edo era. It and other simple, schematic books on prescriptions and materia medica did much to standardize medicinal preparations, making it possible to prepare and stock large quantities of drugs in advance.[43] The simplified manual, along with standardized medicines, helped make medical care more widely accessible to the masses in both China and Japan. The principal danger was that practitioners might not carefully check symptoms before choosing a prescription. The large-scale printing of a

[43] Yabuuchi (4), pp. 141–142.

varied medical literature made it rather easily accessible not only to medical scholars and practitioners, but to other literati as well. Thus, familiarity with medical thought spread in the Sung period further than before beyond circles of professional medical men. This tendency made it rather easy for many of the reprinted classics and new books to reach Japan in the hands of Buddhist priest-practitioners.

While Neo-Confucian thought became differentiated into various schools in Sung times, and even though there was considerable mutual influence between it and medical thought, Chinese medicine did not undergo a similar differentiation into schools organized around distinct systematic principles until the Chin and Yüan eras. The Liu and Chang schools that emerged in the Chin and the Li and Chu schools of the Yüan were each founded upon different emphases within the general framework of "phase energetics"—differences sometimes due to circumstances of time, place, and class. Besides these four schools, there evolved from the many users of the Official Prescriptions manual a school called *Chü-fang-p'ai* (Official Prescriptions school) in which medical practice was simple and straightforward but lacked guiding principles.[44] Later, in the Ming period, the Liu and Chang schools of the Chin era and the Li and Chu schools of the Yüan era were to be criticized as much too metaphysical to be really useful in practice, and a cry would be raised to return to the classics of ancient times, especially the canon On Cold Damage Disorders.

Though partial transmission of Chinese medicine to Japan occurred during the Semiseclusion Era, Li and Chu medicine were not introduced as complete systems until two centuries later through the efforts of Tashiro Sanki (1465–1537) and his student Manase Dōsan (1507–1594). The earlier Chin schools of Liu and Chang were introduced even later, in the middle of the seventeenth century, by Aeba Tōan (1615–1673). Thus it was well into the period of Chinese Wave II—and after Chinese medicine had lost its creativity—before the main

[44] Ishihara (1), p. 82.

schools of Chinese medicine were dealt with as whole systems of theory and practice by Japanese medical scholars. And it was only after sophisticated levels of understanding were reached in the seventeenth century that there arose, in the eighteenth century, a critical attitude toward received theories, an attitude based on detailed research into specific problems.

Gradual dissolution of the *ritsuryō* institutional basis for the Japanese scientific enterprise had its effect on medicine too. Yet we find relatively more productivity among the nobility during the early phase of the Semiseclusion Era in medicine than in the other sciences, notably compilation of medical books from Chinese sources available up to the T'ang era. The formal arrangements were the same as for other sciences; by the end of the tenth century medicine became the House Learning of the Tanba and Wake households, who dominated the Institute of Medicine since they had custody of medical scholarship and responsibility for the care of the imperial family and ranking nobility.

Tanba no Yasuyori, a specialist in acupuncture, was compiler of the most important extant medical document of the early phase, the *Ishinpō* (Essential medical prescriptions; compiled between 982 and 984). As noted in Chapter 1, it preserves portions of many now lost Sui and T'ang medical treatises (quoting more than two hundred Chinese, but no Japanese, sources). Though mainly concerned with clinical treatment, its physiology and pathology, like those of its Chinese sources, are a mixture of ideas from traditional Chinese medicine as influenced to a small extent by Indian medical theories appearing in various Buddhist scriptures, as well as references to Taoist drugs and sexual hygiene. It hardly reflects the real state of medicine in Japan early in the Semiseclusion Era. The *ritsuryō* program for medical training in the capital's Institute of Medicine and the provincial schools (Kokugaku) was already in serious decline by this time, and it is doubtful that understanding of the *Ishinpō*'s varied and complex background was very profound. In fact, it proved too extensive for everyday use; a handier manual was needed. Tanba no Masatada (1021–

1088), a great-grandson of Tanba no Yasuyori, supplied the need in 1081 by compiling the *Iryakushō* (Selected therapies), a collection of emergency treatments excerpted from the *Ishinpō*.[45]

One source[46] lists twenty-nine medical texts produced in the Heian period (794–1192), twenty-two of which fall in the early phase of the Semiseclusion Era. Authorship, except in the earliest stage, is attributed in most known cases to members of the Tanba and Wake families. The House Learning of these households was cut off from China and essentially restricted to court circles; it worked only with sources available up to the Sui and T'ang eras. Though some of the court-related doctors went to Kamakura to serve the military officials in the era's late phase, leadership in medicine in that phase was relinquished to Buddhist priests, whose own literary activity surpassed that of the nobility.

Temple clinics fostered by the *ritsuryō* government in Chinese Wave I declined in the Semiseclusion Era's early phase, during which Buddhist medical care centered on exorcism of the sick —a practice quite popular among the nobility—and only one medical book was prepared by a Buddhist priest in the early phase. One striking example of medical services performed by sects of Old Buddhism even in the era's late phase was the Gokurakuji, or "Paradise Temple," built in Kamakura with shogunate support in 1259 by the priest Ninshō of the Shingon sect. With many large facilities, including a hospital and a leprosarium, it is reported to have accommodated about fifty-seven thousand patients in a twenty-year period.[47] Burned in a 1333 battle when the Kamakura shogunate fell, it was rebuilt, but its services declined after the new Ashikaga regime was set up in Kyoto in 1336 and Kamakura's population dropped drastically.

Clearly it was the priests of New Buddhism who exercised greatest leadership in medicine in the late phase of the Semi-

---
[45] Hattori (2), pp. 150–151.
[46] Sugimoto (2), pp. 67–68.
[47] Hattori (3), p. 77.

seclusion Era. They were particularly active in introducing contemporary Chinese medicine, and among them a new class of priest-practitioners was born. Like their predecessors among the nobility, priest-practitioners were limited in their comprehension of Chinese medical theories; but visits to China by priests gradually increased, and most medical books produced in the late phase came from the hands of priest-practitioners. Because of the relevance of their work to this and later historical stages, it may be helpful to digress briefly and comment on some characteristics of "Buddhist medicine."[48]

Buddhist doctrine was rooted in the view of all temporal life as empty of permanent meaning; this could be found only in nirvana. Life on earth passed through four stages: birth, illness, aging, and death. Of these four, only the condition of illness could be mitigated; and the Buddhist virtue of compassion led many to devote much effort to medical care. Throughout many generations definite ideas on physiology, pathology, and therapy evolved, though references to these ideas are scattered over a wide range of sutras. Buddhist medicine was auxiliary to Buddhist learning in general, not an independent discipline. Pathological ideas found in various sutras often conflicted in certain respects; but the axiomatic basis of all is that nature consists of four elements: earth, water, fire, and wind. This theory, Greek in origin, was taken for granted in Indian thought. The human body, then, is composed of the four elements and remains healthy as long as they are in harmony. A dysfunction of any one element, though, can cause 101 varieties of disease; with four elements, the possible range of diseases runs to 404. Dysfunctions of the water or wind elements cause chilling diseases, while earth or fire elemental dysfunctions cause feverish illnesses. Beyond this, a few examples must suffice to suggest the barest minimum of Buddhist medical theory.

The activities of all living creatures are characterized by

[48] Based mainly on Hattori (4), p. 36f.

five psychophysical principles: physical existence (the body); perception (mental activity to accept things as they are); moral judgment (inner reaction to good and evil); deeds (all overt behavior); and discernment (mental activity to recognize and judge things). As these principles might suggest, there can be *outer* causes of disease which are four in number (wind, heat, cold, and "miscellaneous") and also *inner* causes, again four (greed, anger, stupidity, and pride). The theory of causality, based on belief in reincarnation, holds that in addition to causes arising in this life, illness can also occur from causes relating to a previous life.

Traces of Buddhist medical ideas can be seen in Chinese texts, through which they were first transmitted to Japan. Such ideas exercised greatest influence in Chinese Wave I and the Semiseclusion Era. References to the four elements and the causality theory appear in the *Ishinpō*, but merely as uncritical excerpts—these ideas were firmly imbedded in odd corners of Chinese medicine long before Tanba no Yasuyori's time. Buddhist medical practitioners, though they referred to Buddhist pathology while discussing diseases, utilized traditional Chinese drugs and prescriptions when treating the sick.

The priest Eisai (1141–1215), who twice visited Sung China and first introduced the Rinzai sect to Japan, in 1168 brought back tea and initiated its cultivation in Japan. In a short work entitled *Kissa yōjōki* (Guide to good health through tea drinking), dedicated to Shogun Minamoto no Sanetomo (1192–1219) in 1214, he discussed the remarkable medicinal effects of tea and gave a detailed explanation of Chinese and Indian medicinal theories pertaining to tea as he understood them.

The productive labors of the priest Kajiwara Shōzen (1266–1337) perhaps best represent the conscious effort to improve Japanese medicine through introduction of contemporary Chinese medical books. Among the thirty sources quoted throughout the many chapters of his *Ton'ishō* (Selected medical cures), compiled in 1303, some are from the Sui, T'ang, and Five Dynasties periods, though most are of Sung origin.

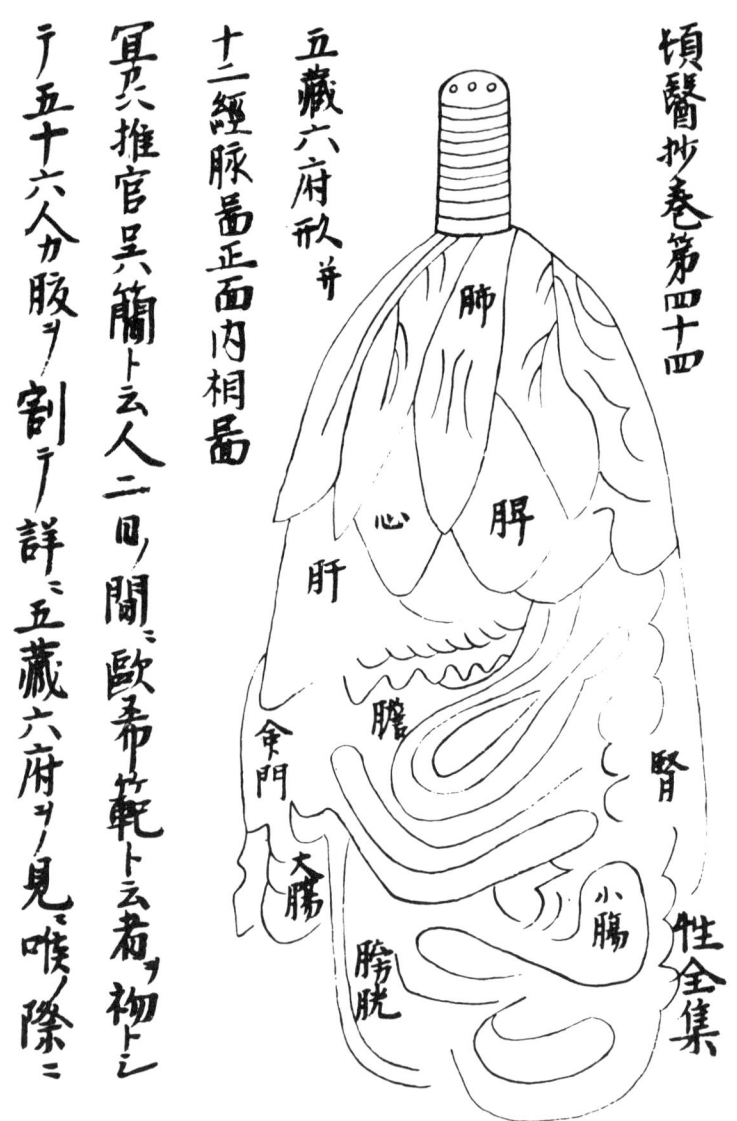

Fig. 17a. One of the sketches of the "five viscera" and the "six bowels" in Kajiwara Shōzen's *Ton'ishō* (Selected medical cures; 1303). Chinese characters to the left indicate that this sketch came from a late Northern Sung text. (The National Archives, Japan)

Fig. 17b. Sketch showing one of the twelve major channels with vital points; from the *Ton'ishō*. (The National Archives, Japan)

The overwhelming majority of prescriptions are drawn from a few Chinese works of the preceding six and a half centuries.[49] Shōzen nourished a desire to make the best medical care widely available, keeping no traditions secret; his *Ton'ishō* was for public use. At the same time, there was a limit to those who could read Chinese characters freely, so he put this text in the mixed *kana* and character style for a broader reading public. But there was more to Chinese medicine than this work covered, so in 1315 he produced another, the *Man'anpō* (Prescriptions for all occasions[?]), an expanded version of the *Ton'ishō* written in Chinese (*kanbun*) with his own commentaries, for his invalid son Fuyukage to use in his medical practice.[50]

The Zen priest Yūrin in about 1368 compiled a medical encyclopedia entitled *Fukudenhō* (Prescriptions gathered for [Buddha's] blessing) that was also based on contemporary Sung works but utilized an original classification of diseases into twelve groups and described them in terms of cause, symptoms, diagnosis, prognosis, and therapy—a scheme that approximates modern treatises.[51] This work was also written in the mixed *kana*-character style but went a step further to give familiar Japanese (*kun*) equivalents for terms in addition to the more literary Chinese-style (*on*) readings for Chinese characters (e.g., *yamai* for "sickness" in addition to *byō*). This was a big step forward in the diffusion and assimilation of Chinese medicine in Japan.

Perhaps the most conspicuous aspect of the diffusion of Chinese medicine in Japan was the spread and use of books on materia medica. Early in Japanese history, before Chinese Wave I, primitive medicines made from simple and easily obtainable substances (grasses, roots, barks, etc.) prevailed. Chinese Wave I brought a much more extensive range of drugs (reflecting the more highly developed commerce of a

[49] Hattori (3), p. 122.
[50] Ibid., pp. 138–141.
[51] Ishihara (2), p. 63.

state many times larger than Japan). Even for the privileged classes, the new drugs were costly. This led to a search for suitable materials in Japan. In the early phase of the Semiseclusion Era a number of books on materia medica (in Japanese, *honzō*) were produced that gave the Japanese names and places of production corresponding to the materia medica (*pen-ts'ao*) required by Chinese medicinal prescriptions. One such book was the *Honzō wamyō* (Japanese names of drugs) compiled around 920 by Fukane Sukehito. Of the 657 authentic Chinese materials listed, 520 are said to have been produced in Japan. The *Engishiki* (see p. 110) refers to about two hundred medicinal substances which were to be sent annually to the central government from the provinces.[52] In the late phase of the Semiseclusion Era many Chinese-style medicines came into wide use in Japan. Most of them were domestic; but as demand exceeded production, medicinal materials continued to be imported from China in large quantities throughout the era.

Finally, two embryonic movements merit mention. Genuine specialization in Japanese medicine occurred after the Semiseclusion Era; but by the end of the fourteenth century Majima Seigan (? –1379), a Buddhist priest who distinguished himself in the treatment of eye diseases, established his own school for that specialty. Tanba no Kaneyasu specialized in mouth diseases during the Nin'an era (1166–1169). Later, in Japan's Warring States period, specialization in "combat medicine" was to arise in response to great demand, and specialization was also to appear in gynecology and pediatrics. Another trend, appearing very late in the Semiseclusion Era, involved visits to China by Buddhist priests who studied medicine independently of religion. This was a harbinger of the general dissociation of medical activity from religious activity that would occur in Chinese Wave II.

[52] Books I–X of the fifty books (*kan*) of the *Engishiki* have been translated into English by Felicia Bock; for a list of some of the medicinal substances, see Bock, Books I–V, pp. 178–179, and Appendix VII.

3

Pressures toward Modern Society

Early Chinese Cultural Wave II: 1401–1639

Chinese Cultural Wave II had its formal beginning in 1401 when the Tally Trade negotiations reopened trade relations with China for the first time in half a millennium. The agents implementing this new intercourse were mainly merchants and Buddhist priests. Slow in getting under way, this wave continued as a whole for four and a half centuries, up to and throughout the period of isolation (1639–1854) under the Tokugawa regime. In this chapter we are concerned only with its preisolation phase (1401–1639), or Early Chinese Wave II.

During this phase internal pressures built up toward a post-medieval social order. Relatively autonomous local military cliques engaged in continuous warfare until unity was restored to the nation by Oda Nobunaga (1534–1582), Toyotomi Hideyoshi (1536–1598), and finally Tokugawa Ieyasu. Despite chronic civil strife, domestic commerce and industry flourished, and fairly large cities—and with them a new townsmen's culture—emerged as part of an overall cultural fluidity that overseas took shape in the adventurous expansion of Japanese merchants into most of East and Southeast Asia. It was in the context of this fluidity that both Buddhist priests and merchants became the active agents of a new influx of continental learning and science that set the stage for a great awakening of intellect in the seventeenth century, soon after isolation began.

The last century of Early Chinese Wave II overlapped with

Table 15. Early and Late Phases of Chinese Wave II and Western Waves I and II

1400	1500	1600	1700	1800	1900

1401 —— Early Chinese Wave II —— 1639 —— Late Chinese Wave II —— 1854
                                            1639 —— Isolation Period —— 1854
              1543 —————— 1639   1720 —————————— 1854
                   Western Wave I   Western Wave II
   (————— Chapter 3 —————)   (Chapter 4)   (——Chapter 5——)

the initial wave of cultural influx from the West between 1543 and 1639, or Western Cultural Wave I. Traders and Catholic missionaries took part in a broad cultural venture including multiple scholarly, scientific, technical and religious aspects that could, in this transitional age, have wrought far-reaching changes in the Japanese social and intellectual makeup. The final Tokugawa decision for isolation in 1639 not only brought an abrupt end to Western Wave I but also altered drastically the conditions under which Chinese Wave II continued. What might have been Japan's second great era of modernization ended, rather, in a return to tradition and a closing of the country to most foreign intercourse.

Early Chinese Wave II and Western Wave I both fell within the scope of the Ming dynasty (1368–1644) when Chinese society was quite stable but in many respects stagnant. They also coincided with the time when many Western European nations were undergoing their own transitions into modern societies.

THE EAST ASIAN BACKGROUND: CHINESE CULTURAL WAVE II

Ming society developed partly out of reaction to Mongol rule and thus was characterized by efforts to prevent future foreign domination. Its policies toward that end achieved sufficient national strength and stability to yield 276 years of relative peace (until China once again came under external control, by the Manchu dynasty, in 1644). From the early half of the fifteenth century a lucrative overseas trade developed. Domestically there was growth in population, cultivated land area, and production of goods. Reaction to Mongol rule was also evident, however, in efforts to restore the purity of Chinese culture—sustained by renewed confidence in China's cultural superiority and by hostility to things foreign. This insularity was further strengthened by reinstatement of the civil examination system, which, by confining itself to the Four Books and Five Classics (the standard curriculum of the late Sung period), tended to exclude unorthodox ideas and, of course, guaranteed the privileged place of the literati in society. At the same time, private academies flourished throughout China, and literary production, both official and commercial, was extensive. The principal philosophical achievement was the formation of the idealistic Confucian school of Wang Yang-ming (1472–1529) in reaction to the more rationalistic emphases of the Sung era school of Chu Hsi and others.

The Ming government was aggressive, not in territorial expansion, but in formalizing tribute-trade relations to stabilize its diplomatic environment. The Japanese had been interested in promoting trade with China from late in the Semiseclusion Era, as evidenced by the eruption of early *wakō* activity. The Ming dynasty, by drawing Japan into the Tally Trade Agreement, intended to get rid of the *wakō*.[1] The first efforts to persuade the Ashikaga shogunate to suppress the *wakō* failed; but Yoshimitsu, the third Ashikaga shogun, did not feel purely symbolic recognition of Chinese primacy in East Asia an ex-

Fig. 18. Stylized scroll drawing of a Tally Trade ship sent to Ming China, equipped with with woven rice-straw sails. The ladder-like structures alongside were for oarsmen to stand while sculling. (Shinsho Gokurakuji temple, Kyoto)

[1] Tanaka (2), pp. 42–44.

cessive price to pay for access to the lucrative Chinese trade. In 1401 he sent an envoy to the Chinese court in response to renewed overtures made by the third Ming emperor and in 1404 received a golden seal as feudatory "king of Japan" for use in stamping the tallies issued to identify officially commissioned ships to the Ming court; the first ships were dispatched in the same year.[2]

Actual trading took place on three levels. First, there was the shogunate's official "tribute" to the Ming emperor, and the imperial "gift" sent back to the shogun. Secondly, there was official trade in which the Ming government purchased goods brought by the shogunate's representatives, the ships' captains, and by Japanese merchants permitted to accompany the ships with their own merchandise. Finally, there were private transactions, in which those on board the Japanese ships dealt directly with Chinese merchants. The first two categories were confined to the capital of Peking, but the third occurred at the port of call, the Ming capital, and en route between port and capital.

The Ashikaga shogunate could not, and did not try to, monopolize the total operations or profits. They were quite willing, for payment, to share the tallies with other powerful groups in Japan. Hence, in a particular operation there might be several ships; e.g., one of the shogunate, another of a large shrine or temple, and perhaps one or more belonging to strong provincial clans. Except for the initial Tally Trade ship, the envoy and vice-envoy were always Zen priests of the Kyoto *gozan* temples. They wrote and bore to the Ming court the official letters, which were dated according to the contemporary Chinese calendar (the *Ta-t'ung li*) given to Yoshimitsu in 1402 by the Ming emperor.[3]

The Tally Trade arrangement encountered serious problems on both sides of the China Sea. The Japanese were ever eager

---

[2] Ibid., pp. 54, 65.
[3] Ibid., p. 64.

to increase trade, but the Chinese were reluctant to permit its expansion. Though the original agreement authorized only one ship, the Japanese gradually increased the number, in 1451 sending as many as nine with about twelve hundred persons aboard and huge cargoes.[4] Afterward the Ming court relaxed the conditions to permit three ships every ten years. On the other hand, the limits of Ashikaga power over the Japanese principals in the Tally Trade were exposed in 1469 when the powerful Ōuchi household in Western Honshu, in the confusion of the civil wars begun in 1467, seized the tally coupons and gained a monopoly over the Tally Trade which the shogunate openly acknowledged in 1516. The last Tally Trade ship from Japan sailed in 1547, and the Ōuchi family was overthrown a decade later by its own vassals, ending a century and a half of official trade relations.[5]

Even before the Tally Trade ended, *wakō* activity was resumed around the 1520s. Known as the "later *wakō*" to distinguish them from those active before the Tally Trade, the Japanese smugglers found many cooperative counterparts along the Chinese coasts. The commodity economies of both countries were too developed to allow enforcement of the prohibitions on private trading that were supposed to protect the Tally Trade Agreement. Hence, early in the sixteenth century, despite the Ming prohibitions on overseas trade in effect from 1371 to 1567, the activity of Chinese merchant ships in East and Southeast Asia was remarkable, and at the peak of "later *wakō*" activity in the 1550s only 30 percent of the *wakō* (lit., "Japanese pirates") are said to have been Japanese. In 1567 the Ming government relaxed its policy against private trading overseas somewhat, though with some conditions: Chinese ships could sail to the Southern Seas, but not to Japan, and exports from China of nitrates, sulphur, copper, and iron were prohibited.[6]

[4] Ibid., pp. 94–95.
[5] Tally Trade ships were sent from Japan to China a total of seventeen times between 1404 and 1547. Tanaka (2), pp. 65–66.
[6] Ibid., pp. 208–209.

Not long afterward, Hideyoshi made attempts at conquest on the Korean peninsula. In 1592 and again in 1597 he sent large military expeditions to Korea that were repulsed, at great cost to the Japanese and even greater damage to Korea, by the allied forces of the Ming and Yi dynasties. As Hideyoshi's schemes ultimately included China, official Sino-Japanese relations were ruptured once again. Though frustrated militarily, Hideyoshi did promote foreign trade. It was he who finally suppressed the *wakō* by establishing a system of authorized shipping that enabled powerful fiefs and the merchant households they protected to mount successful trading ventures not only to China but throughout Southeast Asia as well. The letters of accreditation issued to these ships were stamped with

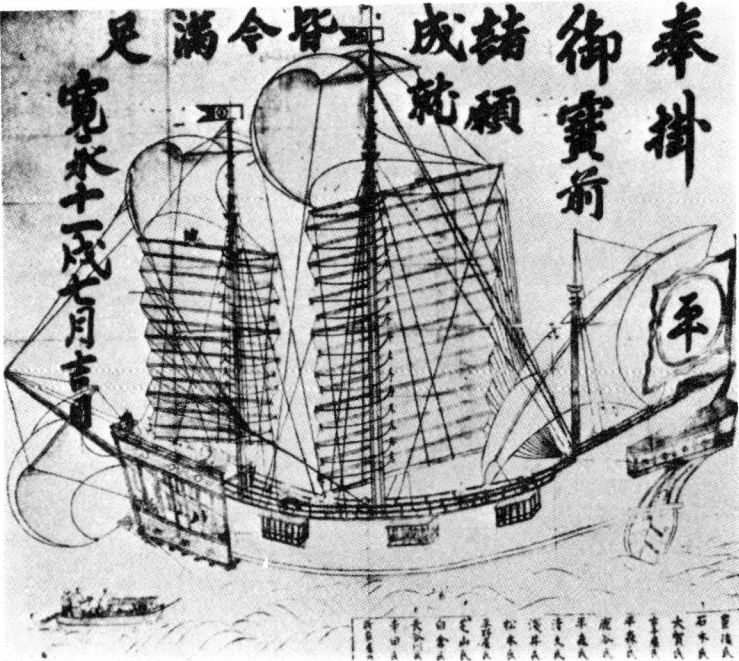

Fig. 19. Sketch of a "Red Seal Ship" drawn on a large votive plaque, dedicated to a temple in Nagasaki. The names of sixteen merchants who chartered this ship are listed at lower right.

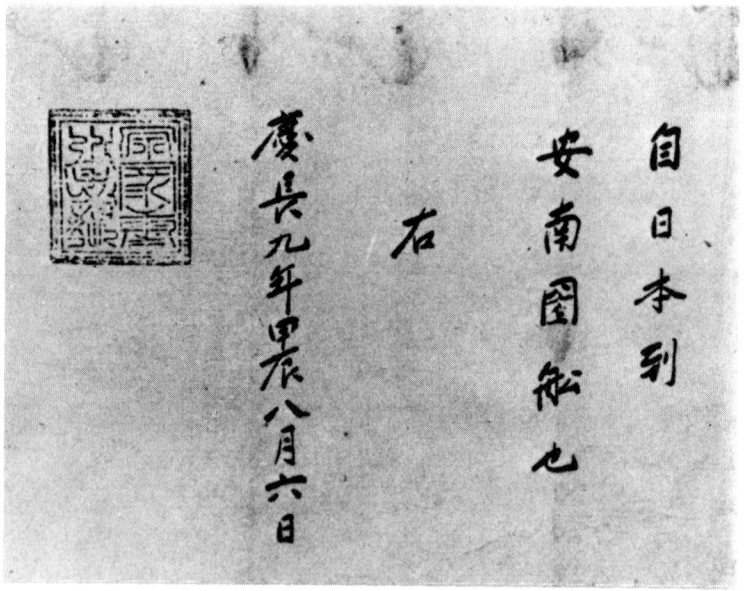

Fig. 20. "Red Seal Ship" certificate, with the vermilion seal of the Tokugawa shogunate (upper left); issued in 1604 to a ship bound for Annam (Vietnam). (Shōkokuji Temple, Kyoto)

the bright vermilion seal of Hideyoshi (and later, the Tokugawa shoguns) giving them the name "Red Seal Ships" (*Shuinsen*).

In response to the Red Seal Ships system, large numbers of Japanese from various social strata ventured into Southeast Asia and established autonomous trading communities, called "Japanese towns" (*Nihonmachi*), that generally enjoyed extraterritorial privileges.[7] Founded in more than twenty places, in the Philippines, Vietnam, Siam, Burma, and so on, the largest towns are said to have had sizable populations in their peak periods.[8] The Japanese community in Manila around the 1620s, for example, is reported to have had some three thousand Japanese residents (probably one-half of them Catholics), and

---

[7] Iwao (2), pp. 186–187. In many of these places there were also Chinese traders' settlements; like the Japanese towns, they were often independent of local governments.
[8] Ibid., p. 106.

Ayutthaya in Siam is thought to have had about fifteen hundred Japanese in the same period. Other trading posts founded by the Japanese late in the sixteenth century included one in Da Nang, Vietnam, and another in Arakan, Burma.[9]

These overseas trading communities emerged rather spontaneously—that is, they were private economic ventures of commercial households and not the result of action taken by local or central governmental agencies. Only at the point of accreditation did central authority touch on the Southeast Asian trade; but there were no formal diplomatic relations between Japan and the other countries. Yet, this trade handled little of direct importance to the development of learning and science—except through transactions at Southeast Asian ports with European traders,[10] who by this time were also engaged in offshore trading along China's coasts and, more importantly, at Japanese ports in Kyushu (see below). The combined Southeast Asian and European trade did, of course, help powerful commercial households amass great wealth and thus strengthened the nongovernmental pressures toward a modern society. The isolation decrees of the 1630s, however, narrowed the European trade down to the single contact with the Dutch, and the Red Seal Ships were completely abolished. The Japanese residents of the "Japanese towns" were left stranded all over Southeast Asia—and faced the death penalty if they attempted to return home. Both the Chinese and the Japanese lacked the integrated strategy of the European trading companies (backed by home governments), and their ships and navigational techniques were inferior. Without the interven-

[9] Ibid., pp. 111, 112, 116, 136, 139f.
[10] Goods imported by Japan from Southeast Asia included deerskins, lead, tin, camphor, sugar, and spices; shipments to Southeast Asian ports included silver, copper, iron, sulphur, small firearms, and miscellaneous items. Through the European traders (transferring goods at Southeast Asian ports or delivering directly at Japanese ports) Japan imported arquebuses, cannon, gunpowder, lead for both cannon and arquebus shot, mercury needed in metallurgical processes, woolen goods, glass products, and various luxury items; and exported gold, silver, swords, handicraft and art objects. For details on exports and imports, see Iwao (2), pp. 66–69.

tion of the isolation policy, though, continued activity would probably in due time have yielded a competitive competence in shipbuilding, navigation, and trading strategy among the Japanese.

It was the more complex Sino-Japanese trade that facilitated the rise of Chinese Wave II. It was more complex because, while both governments authorized the Tally Trade and neither condoned the *wakō*, the Red Seal Ships were accredited by Japan but were not recognized by the Ming court—and hence were not permitted direct entry to Chinese ports. For over two centuries these trade institutions provided the contacts through which Buddhist priests and members of Japan's rising merchant class importing books and drugs became the agents of renewed introduction of Chinese culture (some influence stemmed also, as we shall see later, from the Japanese forays into Korea between 1592 and 1598). The advent of Chinese Wave II was marked neither by the overall enthusiasm for Chinese culture nor by the attempts to imitate its forms directly that were seen in Chinese Wave I. It was rather a much slower and longer process, omitting many Chinese cultural systems but embracing both high culture and practical techniques. The learning and techniques reached deeply into Japanese society and provided the solid base on which, in Late Chinese Wave II, the seventeenth-century intellectual outburst occurred (as we shall see in Chapter 4). And this activity (as Chapter 5 shows) provided some of the most valuable elements for the modernization processes of the Meiji era.

Western Cultural Wave I: 1543–1639

Europe's own transition to modernity was the sociocultural background of Western Wave I. Beginning with the Renaissance developments of the fourteenth and fifteenth centuries, and the Protestant Reformation followed by the Catholic Counter-Reformation in the sixteenth century, the Western transition also included the emergence of nation states, expansion overseas through trade and colonization, and later the rise of modern science and technology. Of these, it was the mercantile expansion that first had an impact on Asia, under such political figures as the Portuguese prince Henry the Navigator and others. In this great age of exploration Europeans rounded the Cape of Good Hope to reach India in 1498, leading to the global activities of Spain and Portugal and subsequently to the founding of the East India Companies of England (1600), Holland (1602), and France (1604). The advancing frontier of Western expansion into East Asia touched Japan's shores in 1543 with the arrival of some Portuguese traders on the small island of Tanegashima, south of Kyushu. The initial wave of Western cultural influx that ensued had its sources in various institutions in the West; but in Japan the forces of Catholicism, Western learning, sciences, techniques, and trading constituted, for all practical purposes, a single, overall cultural impact. This wave greatly stimulated the atmosphere of fluidity surrounding the growing tendencies toward modernization in a time of shifting political structures and struggles to reshape Japan's domestic course.

Catholicism appeared in sixteenth-century Japan much as Buddhism had in the eighth century—as a many-faceted cultural complex. The initial and primary Catholic agents were the Jesuits, who first landed at Kagoshima in 1549. The Society of Jesus had been established only shortly before, in 1540, and the Jesuit missionaries were full of the missionary zeal of the Counter-Reformation (Council of Trent, 1545–1563). They had the strong backing of the Portuguese government. Soon, Spanish power and influences were added with the coming of Franciscan and Dominican missionaries, and a handful of Augustinians, from the Spanish colony in the Philippines. The Protestant Dutch and English traders who began arriving in 1600 and 1613, respectively, had little interest in, and no connection with, Christian missions (which, for Protestants, had not yet begun on a global scale).

JESUIT ACTIVITIES

By the time of the Jesuits' arrival neither the Old Buddhism of Chinese Wave I nor the New Buddhism of the Semiseclusion

Fig. 21. Stylized depiction of a Portuguese ship; from one of the many *Nanban byōbu* ("Southern barbarians" screens) showing various aspects of Western culture as perceived by Japanese of the late sixteenth and early seventeenth centuries. (Imperial Household Agency)

Era retained their earlier levels of spiritual influence among the general populace. Their intellectual leadership was greatly diminished (and soon to be eclipsed by Confucian scholarship). Some larger temples still possessed enough secular power that they were regarded, particularly by Nobunaga, as obstacles to national reunification. Confucianism was not yet well known except among some of the court nobility and Zen priests; and Shinto at this stage remained largely an unsystematized complex of cultic sentiments and customs. These three traditional belief systems of Japan were each in such flux that they were ill prepared to directly challenge, spiritually or ideologically, the Jesuit propagation of Catholicism.

In fact, early Jesuit arrivals enjoyed an unexpected advantage: some Japanese mistook Catholicism for another sect of Buddhism (and a few missionaries used this advantage to gain converts). In any case, Japan had, without undue conflict in the past, assimilated many new Buddhist sects from the continent; but Christianity proved a bit more difficult. Its monotheism contrasted sharply with the popular pantheon of Buddhism, and its strict morality ran counter to the somewhat brutish manners of a society buffeted by the dangers and uncertainties of a century of civil wars (e.g., seniors were frequently overthrown by loyalty-flouting juniors, and among the upper classes the keeping of many concubines was the rule rather than the exception). Nevertheless, the personal excellence and zeal of many missionaries; the high ethical code kept by them and demanded of their followers; and their humanitarianism expressed in relief work for the poor and orphaned, and in care for the afflicted in hospitals and leprosaria—all had a powerful attraction for many Japanese people, despite the inevitably strange manners, ceremonies, and beliefs of the "Southern Barbarians" (*Nanbanjin*).

Along with religious work, the Jesuits attempted to implant in Japanese soil the Western system of learning. While the intellectual impact of Jesuit teaching is reviewed later, here it can be noted that the scholarly atmosphere, the systematic approach to learning, and the various educational institutions of the Jesuits made quite an impression upon learned Japanese, whose numbers happened to be on the increase in this period. That many were already looking away from Buddhism created a further openness to new ideas. Of special interest was the Jesuits' natural theology, in which naturalistic, logical explanations of phenomena in the heavens and on earth were not only often attractive but, to some, convincing.

In the last analysis, though, the impact of the Jesuits' religious and intellectual challenges was cut short, after slightly less than a century, without having substantially altered either the Japanese belief systems or the content and methods of Japanese learning. Before the Catholic religion could be fully engaged as religion, or Western learning as learning, both were expelled from Japan by the purely political means of suppression and isolation.

POLITICAL TREATMENT OF THE JESUITS

The Jesuits appeared in western Japan in concert with Portuguese traders at the height of the age of civil wars. Because of the anticipated benefits of trading with the Portuguese and of the acquisition of small arms and gunpowder, the leaders of various military factions in the western parts of the country competed with each other in granting propagation rights to the Catholics and otherwise aiding their efforts. A few local lords led their entire populations into wholesale conversion, notably the lords of the Ōtomo, Ōmura, and Arima fiefs in Kyushu. (Not infrequently conflicts with Buddhist and Shinto groups led to the precipitate burning of temples and shrines by overly zealous converts.) The new Catholic movement was encouraged also by Nobunaga as a means of rooting out Buddhist secular power, which he saw as a serious obstacle to national unity. Christianity found the situation most favorable under these receptive conditions.

Under Nobunaga the force of Buddhist secular power was blunted; and as it was further domesticated under Hideyoshi, the political utility of the Catholic missions as a ramrod against the indigenous religion disappeared. Moreover, Christianity not only made some religious and intellectual gains but also began to show signs of gaining social and political momentum. Hideyoshi came to fear the successful spread of Catholic influence—whether as a spearhead in the possible colonization of Japan, or as a new focus of allegiance to be used by provincial lords and warriors for revolt against the central government.[11] Either possibility was a serious threat to the newly won national unity. Furthermore, the Christian refusal to compromise with the traditional religions could in time reduce their value for maintaining social stability. To avoid any one or all of these possibilities, Hideyoshi adopted the policy of eliminating Catholic missions while retaining trade relations with the Portuguese—a policy continued by Ieyasu. Although such a halfway policy in time proved ineffective, it was administered with discretion for fear of losing the Portuguese trade. But when the Dutch and English began coming to Japan, they made no moves toward initiating Protestant missions. The Tokugawa regime, therefore, shifted its policy to favor trade with the

---

[11] Watsuji, p. 381.

Dutch and English, issued edicts against Christianity in 1612 and 1613, and from 1617 strengthened its suppression of the Catholics (following Ieyasu's death in 1616). A 1630 ban on importing books related to Christianity sought to block even indirect Catholic influences.

Among the Japanese were many Catholic believers and sympathizers who not only aided and abetted mission activities inside the country but also assisted Jesuit missionaries making clandestine entries into Japan, thus hindering effective implementation of the prohibitions on Christian activities. When a large-scale revolt, rallying around the Christian cause, broke out in 1637–1638 among peasants near Nagasaki on the Shimabara peninsula and the Amakusa islands, the worst anti-Catholic suspicions seemed confirmed. The Tokugawa government decided in 1639 to thoroughly suppress Catholicism and expel all foreigners except the Dutch and Chinese.

Transitions in Japanese Culture

The picture thus far is one of active Japanese engagement overseas—in the Tally Trade, the later *wakō*, the Red Seal Ships, and the Japanese towns in Southeast Asia—with a renewed openness to Chinese culture and an initial receptivity to Western activity that ended in a rigid rejection of it. The domestic scene was no less fluid; indeed, various internal pressures were a major reason for the new openness to the outside and the eventual closure of the country.

SOCIOPOLITICAL SCENE

From the outset the Ashikaga shogunate based in Muromachi was founded on an uneasy balance of power among local lords, who steadily became independent of central controls. In the middle of the fifteenth century internecine wars broke out all over the country and continued to the end of the sixteenth century when Oda Nobunaga and his successor Hideyoshi gave the land much-needed order and unification. In the interim before unification, known as the "Warring States" era (Sengoku jidai; 1467–1603), each of the lords (Sengoku daimyo) sought to increase the wealth and military strength of his own

Table 16. Steps Taken toward Isolation

Regime	Measures against Catholics	Related events
Oda		1582 Oda Nobunaga killed
Toyotomi		1585 Toyotomi Hideyoshi formally designated *Kanpaku* (ruler "on behalf of" the emperor)
	1587 Hideyoshi issued edict for expulsion of Catholic missionaries.	
		1592 "Red Seal Ships" (*Shuinsen*) system began.
	1596 Hideyoshi arrested and crucified 26 Catholics (Japanese and foreigners).	
Tokugawa		1603 Tokugawa Ieyasu became shogun. 1605 Ieyasu succeeded by his son Hidetada as shogun
	1612 Prohibition against Christianity; many Christians arrested and expelled to Manila and Macao 1614 Japanese and foreign Catholics arrested and expelled	
		1616 Ieyasu died.
	1617 Execution of Christians resumed 1622 Martyrdom of the Genna era (1615–1624); included Japanese and foreigners	
		1623 Hidetada (d. 1632) succeeded by his son Iemitsu
	1630 Ban on import of books related to Christianity	

Table 16 (*continued*)

Regime	Measures against Catholics	Related events
		1635 Ban on Japanese overseas shipping and travel
		1637–1638 Shimabara Revolt (*Shimabara no ran*)
	1638 Christianity absolutely prohibited	
		1639 Isolation policy made complete by expulsion of Portuguese

domain. Enlargement of cultivated land areas through reclamation of wastelands and opening up of new fields was encouraged, as were mining, crafts production, commodity exchange, and money economies. The most important shift in social structures derived from the daimyo decisions to move local military bosses and their troops from their bases in villages to the castle towns in which the daimyo themselves were based.

When the samurai were brought into the castle towns, merchants and craftsmen followed, leaving farmers on the land to provide raw materials and foodstuffs. The differentiation of urban craftsmen from rural peasants actually facilitated phenomenal progress in both sectors of the economy. Cultivated land area expanded from 950,000 *chō* around the end of the Semiseclusion Era to 1,500,000 *chō* in the Keichō era (1596–1615) toward the end of Early Chinese Wave II.[12] Gradual expansion of commodity production and exchange was evidenced partly in the overseas trade developments outlined earlier, yet even more in the emergence of fairly large commercial towns like Sakai, Hakata, and others. These port cities enjoyed more autonomy than the castle towns in the Sengoku period; but in both castle and commercial centers there developed a civil culture in which innovation in techniques and, after isolation, in the sciences, took place.

[12] Toyoda, p. 302, and Kodama (1), p. 28.

All sectors were threatened, however, as long as the country remained divided and at war. In Nobunaga's drive to overcome local rivalries, salvage the economy, and forestall destruction of the cities, provincial lords and wealthy merchants were forced to cooperate or capitulate. Hideyoshi further established control by the military class over all lands and farmers through a tax system based on extensive land surveys and through introduction of a hereditary status system to complete the separation between farmer and soldier. He sought additional revenue by promoting mining and by pursuing an active foreign policy—largely bluff—which included the 1592–1598 assaults on Korea. Both the mining and the overseas ventures had important effects on learning, science, and techniques.

Ieyasu undertook an ambitious policy to complete the system of centralized controls initiated by his predecessors. Appointed "Barbarian-quelling Generalissimo" (*Seii tai shōgun*) in 1603 to protect the country "on behalf of the emperor," he separated all other functions of government from the imperial office by establishing his military government (*bakufu*) in Edo in the center of the Kantō plain where his own power was strongest. He undercut the power of many daimyo by shifting some of them around to different fiefs, a policy that had the side effect of promoting further diffusion of various skills throughout the nation, as many merchants and artisans followed their masters to new locations. Interdomain exchange of commodities and know-how was most encouraged by the integration of local economies into a single national economy as a result of complete unification and, of course, by Ieyasu's improvement of the waterways and land transportation routes over which tax payments and consumer goods flowed to the new capital. Edo rapidly became the nation's largest city, partly because of the required residence in alternate years of every feudal lord, accompanied by his retainers numbering in the hundreds and even thousands. Ieyasu greatly increased the central treasury by employing Hideyoshi's device of taxation based on an exact land survey and by control of commerce, currency, mining, and the prosperous Red Seal Ships' trade—all of which were encouraged now that they were fully under control.

Ieyasu was somewhat more indifferent to the suspected threat of the Catholics and foreign traders than Hideyoshi had been; there was no bloodshed in his time. His successor, Hidetada (1579–1632), came to hold grievous fears of Western influence, and in the year following his father Ieyasu's death (1616), mounted severe persecutions against both Japanese and non-Japanese Catholics. Tokugawa Iemitsu (1604–1651) succeeded him in 1623 and not only continued the persecution of Catholics but also issued the 1630 book ban and clamped restrictions on all overseas trade and travel by Japanese thus stranding abroad forever the Japanese residents of the "Japanese towns" in Southeast Asia. Foreign trade in time declined drastically, of course, but greater control was won over domestic merchants, markets, and cities. Peasant revolts were not previously unknown in Japan, but the Shimabara Revolt (1637–1638) seemed a clear signal of the dangers of foreign influence among peasants. Iemitsu issued new laws governing peasants, including prohibition of land sales. In addition, the social and occupational mobility of all citizens was severely restricted by the minute definition of all roles and responsibilities, and this included a revision of the military code governing samurai. Thus, the centrally integrated feudal government dominated by the shogunate and served locally by local feudal lords was brought to completion and then sealed off from further external threat by the isolation policy of 1639.

CULTURAL TRANSITIONS

The entire period of Early Chinese Wave II was culturally and intellectually one of the most fluid and open in all Japanese history. Fluidity and openness did not, however, spell total discontinuity with the past. One of the three main cultural streams that ran through the Sengoku and into the Edo era was what might be called the "Muromachi legacy," or the preservation of the arts and letters and the aristocratic conventions and styles of the nobility, cultured priests, and the older military class based in Kyoto. Preservation was less a confident advocacy than a defensive action. By the early decades of the Edo period, much of the learning, arts, and skills of late medieval culture had become trapped in secretive schools (*iemoto seido*) reminiscent of the guarded transmission (*hiden*) once promoted officially in the House Learning system.

The greatest departure from traditional ways, and the primary generator of possibilities in Chinese Wave II, was the civil culture developed by wealthier townspeople who could afford to introduce new methods in business and had the leisure to pursue new interests. Advancing rapidly into social and economic prominence, they tended to disregard the subdued, refined sensibilities preserved in the Muromachi legacy, enjoying rather the flamboyant display of their rising prosperity. A less inhibited and more colorful art and social atmosphere began to displace the restrained and sometimes haughty demeanor of late medieval times. The freer, pragmatic townsmen's culture might have eventually come to dominate Japanese society had it been able to secure an independent base. But urban merchants and artisans could never have won and retained any social and economic, much less cultural, prominence without the encouragement and protection of the Sengoku daimyo. The dependency of townsmen's culture was increased under the political processes through which Nobunaga, Hideyoshi, and Ieyasu subordinated all subcultures to the central military regime. The relative freedoms of the Sengoku period were sacrificed to the Tokugawa demands for social stability.

The third and eventually most dominant cultural stream in Early Chinese Wave II was that which represented the values, skills, and ways of thinking of the new military class, that is, the Sengoku daimyo and their followers and later the Tokugawa shogunate and its retainers throughout the country. The new samurai culture was oriented partly to the ideals of the Muromachi legacy and partly to the pragmatic necessities of cities, farms, mines, and trading ports, though most of all to the preservation of its own power through the cultivation of discipline, loyalty, and administrative competence. Hence, early in the Edo era there was a certain revival of classical learning and arts, and the education of samurai was encouraged. On the other hand, the late sixteenth and early seventeenth centuries saw a certain break with conventions and more open debate on current issues, a favorable condition for the Catholic missions while it lasted. By the end of Ieyasu's time, though, "new doctrines" and especially "heresy" were suppressed. The shogunate encouraged cultural forces which strengthened its own authority and power; the promotion of Chu Hsi's Neo-Confucianism (*Shushigaku*) in its absolutist Ming interpretation is the most important example of intellectual measures taken to support the shogunate, while the edicts against Christianity represent the most outstanding case of political suppression of heresy for the same purpose.

More directly related to the revival of learning and science were two other trends: secularization and specialization. The secularization of Japanese culture proceeded on two fronts in Early Chinese Wave II. One of these followed from the turning of learned men away from Buddhism, particularly as it was domesticated by Nobunaga and Hideyoshi. The other derived from the desacralization of the traditional forms of political authority centered in Kyoto, ancient seat of the divinely sanctioned imperial line, a cause aided by the Tokugawa promotion of Confucian learning to rationalize the loyalties required of the samurai class by the new political order.

Differentiation and specialization of socioeconomic roles were not simply manifestations of the modernizing tendencies of this age: they were not always carried out to meet new necessities more rationally and efficiently. In many cases they were the means by which rights and privileges were monopolized, as the military rulers intended. By protecting occupational groups from crippling competition, their loyalty and contribution to the prosperity of the domains and shogunate were assured. Here more than anywhere else, perhaps, was seen the persistence of traditions within radical change. The hereditary succession to roles, ordered hierarchically according to their value to the ruling elite, characterized every sector of public life—the economy, politics, social structures, the arts and learning—throughout Early Chinese Wave II and, of course, long after.

In perspective, Early Chinese Wave II was an age in which

only slight shifts of the rudder in the ship of state were required to steer Japan gradually away from its traditional moorings. For a time its cultural agents were blessed with certain freedoms and new possibilities, and the culture of their time constituted the arena for a new day in learning, science, and techniques, old and new, indigenous and foreign. Nonetheless, their culture lacked depth compared with that of the period of intellectual outburst that succeeded it—in a time, ironically, of global isolation and the most rigid social controls.

Techniques Strategic to Modernizing Processes

The most important single national factor behind the diffusion of traditional techniques in Early Chinese Wave II, and behind the search for new skills as well, was the desire of the Sengoku daimyo to increase the wealth and military might of their particular domains. This impetus was strengthened by national reunification and the consequent integration of economic and transportation infrastructures, and by expansion of the commodity market.

Swords, armor, and eventually firearms were in great demand while wars raged during the Sengoku era and, to a lesser extent, afterward when peace prevailed. But farming, fishing, mining, and domestic crafts also made steady progress throughout Early Chinese Wave II—especially under the impulse of the Tally Trade, the "later *wakō*" activity, and the trading ventures serviced by the Red Seal Ships and European vessels. Developed capabilities appeared also in architecture, in castle construction, and in civil engineering, particularly where traditional techniques no longer sufficed, as, for example, in irrigation and river control, and in the excavation required for large-scale mining. Certain innovations in these fields were based on new knowledge made available in Chinese books imported through the reopened contacts of Chinese Wave II.

The most significant developments of this era took place, however, in the realm of techniques which played strategic roles in the Western transition to modern society. Some of the crucial skills actually came from China, certain of which had earlier traveled to Europe. The Chinese skills helped make possible adoption of the new Western techniques that entered Japan before her rulers grew apprehensive about Western influence.

CONTINENTAL TECHNICAL IMPORTS

Introduction of Chinese technical skills in the fifteenth and sixteenth centuries occurred mainly through the agency of the merchant class, though often with the backing of local feudal lords. The techniques of chief importance were improved silk weaving, smelting and forging, papermaking and printing, shipbuilding and use of the magnetic compass in navigation.[13] Many books on specific techniques were imported, and they could be consulted again and again for adaptation on a pragmatic trial-and-error basis. Such knowledge was sought out of a clear sense of practical need and thus was rather quickly assimilated into the mainstream of Japanese life. Also, in the military expeditions to Korea (1592-1598) a large quantity of novelties were brought home by Japanese generals, and in some cases certain technicians were also taken back to Japan. Books and movable type were most notable among new items, porcelain potters and printers among the technicians. The importance of these continental skills to change in Japan can be more easily appreciated, however, by considering them in relation to the potentially strategic techniques that came from the West.

WESTERN TECHNICAL INFLUX

The specific techniques of strategic significance to the modernizing tendencies evident in Japan in the late sixteenth and early seventeenth centuries were small firearms and cannon, shipbuilding and navigation, mining and metallurgy, and printing and paper. They were received in Japan with relative ease and in a short time began, as in the West, to add weight to the pressures for change. The specific criteria by which they are

[13] Needham, 4/1: 239f.

here considered "strategic for modernizing processes" are the following:

1. *Use* of the techniques encouraged forms of social change beyond the immediate range of the technicians, their tools, or their materials.
2. The *agents* in the introduction and use of the techniques were largely motivated by purely secular concerns (military use by samurai and commercial gain for merchants), and they were not directly aligned with the major power groups seeking to preserve traditional society, namely, the court aristocracy, the priestly class, and the older military elite.
3. Most important, the techniques could be *directly related to the basic exact sciences* and hence had some potential for hastening the early emergence of new forms of science in Japan—or at least an interest in studying Western science.

Adopting Western techniques was not so simple as importing ready-made goods. Certain elements in the performance of these skills were more or less different from the technical functions learned through use of Chinese-style techniques, and the new methods, such as use of Western nautical instruments, had to be learned the hard way through observation and trial and error. There were no organized agencies of instruction, and the language barrier was formidable. Nevertheless, the times were propitious, and in a relatively fluid and open society with an aggressive, adventurous spirit, not a few of these obstacles were successfully overcome. Compared with the difficulties of assimilating the strange forms of Western science and learning, the task of adopting new techniques was far simpler, and so more successful.

FIREARMS AND CANNON

The first weapon to use gunpowder appeared in China in the Sung period.[14] The technique was transmitted to Europe by Arab merchants whose vessels frequented Chinese ports during the Sung and Yüan eras.[15] From the fourteenth century the

---

[14] Arima (3), pp. 25–26.
[15] Ibid., p. 331, supposes the technique was also transmitted to Europe through Mongol expeditions to Russia in the early thirteenth century.

use of gunpowder became widespread in Europe, and in the middle of the fifteenth century there appeared a firearm that would have an important impact upon Japan—the arquebus. Sometimes spelled "harquebus" ("gun with a hook"), this was a musket triggered by a matchlock and supported for firing by a staff with a **U**-shaped support.

On August 25 of the year 1543[16]—the year Copernicus died and his principal work was published—a large junk drifted ashore on the small island of Tanegashima off the southernmost tip of Kyushu, bearing, as one report has it, "more than one hundred persons quite different," though interest in most reports centers on "three Portuguese."[17] Not only were they the first Westerners reported to have landed in Japan,[18] they also brought the arquebus. The island's lord purchased two of these firearms at a high price. He then ordered a swordsmith named Yaita Kinbei (or Kiyosada) to make a reproduction of these arms, which was done in the following year. When these weapons were later exhibited to military leaders in central Japan, urgent demands for more arose immediately, as this was at the height of the Warring States era. Although arquebuses continued to be imported from the West, indigenous production developed rapidly at the two main centers, Kunitomo (Ōmi district) and Sakai (Izumi district). Other manufacturing sites arose at Sakamoto (Ōmi), Funai (Bungo), Yamaguchi (Suō), and Odawara (Sagami). According to an early Portuguese traveler named Fernão Mendes Pinto, there were thirty thousand arquebuses in the Funai district in 1556 and

---

[16] August 25 of that year was "September 23" by the Japanese *Senmyōreki* calendar. Many Western books give the year as 1542, instead of 1543. The *Teppōki* (Record of firearms) by the Zen priest Bunshi (1555–1620), which is considered the most reliable source in Japanese, gives 1543; see Iwao (4), pp. 11–13. *Teppōki* was written in 1607 at the request of Tanegashima Hisatoki, lord of Tanegashima, to commemorate the role of his grandfather, Tanegashima Tokitaka (1528–1579), in producing the first arquebus in Japan.

[17] Moriya, p. 27.

[18] In 1541 a Portuguese ship drifted to the coast of Bungo district in Kyushu, but no one is known to have gone ashore.

ten times that many throughout Japan, though Pinto's report is notorious for its exaggerations.[19] Be that as it may, this firearm became one of Japan's exports to Southeast Asia in the Red Seal Ships trade.

When the arquebus reached Japan realignment of provincial powers was well advanced and reunification of the nation was not far off. Its use accelerated these processes. In the battle of Nagashino in 1575, the forces of Nobunaga, equipped with approximately three thousand of the new weapons, administered a crushing defeat to the heretofore undefeated forces, mostly cavalry, of Takeda Katsuyori (1546–1582). From that time on, the organization, tactics, and combat equipment of armies in Japan were decisively altered. Heavily armored cavalry remained at the center of large armies, but they were now supplemented by companies of foot soldiers equipped with arquebuses, as well as by troops with the traditional long lances and light armor. The leader who possessed both the foresight to adjust quickly to the inevitable changes in warfare and the economic capacity to purchase large quantities of high-quality firearms acquired an overwhelming advantage in battle.

Japanese producers of arquebuses were all originally swordsmiths with a well-established tradition that included production of high-quality steel directly from iron ore. Their swords were valued both for performance and aesthetic qualities, and great quantities were exported to China during the Sung, Yüan, and Ming periods. As the perfected techniques of steel manufacturing and forging could be readily applied to firearms production, large quantities of good quality arquebuses could be had with relative ease.

Late in the sixteenth century cannon were also introduced into Japan. The artillery of this age, however, was poor in maneuverability and dismal in accuracy and thus did not affect methods of warfare in Japan as much as did the arquebus, nor can a similarly wide diffusion be noted. Cannon were not gen-

---

[19] Hildreth, p. 30. The Japanese version of the journal of Fernão Mendes Pinto is titled *Hōrōki*.

erally used in field operations, but in sea battles and blockades, and in sieges of castles. In an effective blockade of Osaka's Ishiyama Honganji temple in 1578, Nobunaga deployed seven large ships equipped with three cannons each and protected against arquebus fire by iron plates. His ships won a smashing victory against more than six hundred smaller ships, successfully cut off supplies to the temple by sea, and forced its eventual surrender.

The cannon were of the muzzle-loading type, made of bronze. For their production the traditional skills used in casting large Buddhist statues and bells seem to have been quite adequate. Proficiency in the use of cannon was another matter, as the expeditions to Korea revealed. In both quantity and quality, the Japanese arquebuses were superior to those of the allied Chinese and Korean forces—the latter even thought this firearm a Japanese invention—though in numbers and performance, the Japanese cannon were definitely inferior.[20]

Gunnery was of strategic importance because an interest in ballistics could have led to an exact science of dynamics, though this line of development is, to be sure, an exceedingly difficult one. Even in the West, Greek rational mechanics did not evolve past statics to solve dynamics, which had to wait for the Italian Renaissance. While social conditions in Japan for a time favored discovery of dynamics, the scholarly conditions needed to duplicate the Western approach to mechanics were lacking. What was missing, specifically, were the necessary mathematical tools of algebra (which Japan was to acquire from China by the early seventeenth century) and analytical geometry; differential and integral calculus (which were not present in the Italian Renaissance but were to develop in Europe by the seventeenth century); and precise concepts of velocity, acceleration (available in the Italian Renaissance), and force and mass (rather imprecise in Galileo's late sixteenth-century studies, but perfected by Newton in the late seventeenth century). These mathematical and conceptual requirements for the for-

---

[20] On the possibility that this was a kind of mortar, see Jeon, p. 202.

mulation of a theory of dynamics were, at this time, not yet fulfilled in Europe; but Japan, by comparison, had a much longer way to go before these requirements could have been indigenously met. Furthermore, the Japanese thus far had only primitive experimental methods for studying trajectory problems.[21]

Given sustained interest, sufficient time, and continued access to developments in the West, it is not unreasonable to assume that the Japanese could have appropriated techniques based on Western dynamics. The later course of history proved that Japanese scientists could have understood concepts of dynamics and thereby reached an appreciation of Western mathematics and natural science in general much earlier. Decisions affecting these possibilities were in the hands of politicians, not scientists; and once peace was secure at home and all foreigners except the Dutch and Chinese were expelled, such questions were dropped. Japanese scholars remained quite unaware of dynamics as an exact science until a few decades before the Meiji Reforms. The shogunate did ask the Dutch to demonstrate their cannon in the early part of the eighteenth century, but the interest was purely military, not scholarly, so that the Japanese government circles were left with a strong memory of the powerful Western guns. Throughout most of the Tokugawa era both the shogunate and domains did, it is true, produce and equip small platoons with cannon and small arms, but these had no significant place in Tokugawa military organization, which stressed traditional martial arts of swordsmanship, lancing, and so on. Arms production made no progress beyond the models received early in Western Wave I, putting Japan at a serious military disadvantage when challenged by Western powers in the middle of the nineteenth century.

SHIPBUILDING AND NAVIGATION

Most of the Japanese ships used in the Tally Trade of the fif-

[21] For one of the very few studies available in English, see Itakura Kiyonobu and Itakura Reiko, "Studies of trajectory in Japan before the days of Dutch learning," *Japanese Studies in the History of Science*, No. 1 (1962), 83–93.

teenth century were of the 100-gross-ton class.²² They were adaptations of certain Chinese styles, though inferior to them. Compared to skills known to the *kentōshi* sailors of Chinese Wave I, navigational techniques by this time had made marked progress. The Tally Trade mariners knew how to take advantage of seasonal winds and used the magnetic compass acquired from the Chinese.²³ Disasters at sea were appreciably fewer than in Chinese Wave I times, even though ships now plied a direct course across the East China Sea. However, adoption of newer designs was not universal among the Japanese; most of the "later *wakō*" who took over the sea lanes following the decline of the Tally Trade used the more primitive structures of former days—keelless, flat-bottomed, one-masted craft with quite simple rigging, poor maneuverability, and low durability in rough seas.

The Mediterranean Sea served as a natural arena of interchange of shipbuilding and sailing techniques from the times of the ancient Phoenicians, Egyptians, and Greeks. Arab merchants, beneficiaries of this heritage, served as trade intermediaries between China and the Western world during Sung, Yüan, and Ming times and, along with gunpowder, brought the magnetic compass to the West around 1200. In the fourteenth century the gimbals (a set of two concentric half-rings mounted on axles perpendicular to each other to allow the compass to remain horizontal as the ship heeled) were developed in Italy. The Crusades brought Western Europe into contacts with the Arabs that stimulated improvements in ship construction, and larger ships appeared in the West from the

---

²²Tanaka (2), pp. 116–117.
²³Most probably a floating-needle or water compass. Both the dry-pivoted and floating-needle compasses were Chinese inventions; they were transmitted during the twelfth century to Europe, which not only favored the dry-pivoted compass but combined it with the Italian compass card to make it possible to get compass bearings without having the bow of a vessel pointed north, as was necessary with the water compass. Brought to Japan in the late sixteenth century by Portuguese traders, the dry-pivoted compass with compass card was taken to China by "Japanese pirate-traders" (*wakō*). Needham, 4/1: 289–291.

thirteenth century. In the fourteenth and fifteenth centuries developments in astronomy and spherical geometry were applied to improvements in the making of maps and navigational charts.[24] Research and exploration encouraged by Western states in time yielded a number of important new skills and instruments. Some of them would not have challenged East Asian science proper but would have been quite useful within the framework of traditional sailing techniques; for example, the plumb line to check ocean depths and bottom conditions, or the use of a ship's log and hourglass to determine its speed. Others, such as navigational charts showing depths and navigable routes would have added to, and possibly improved upon, the information already recorded by Chinese and Japanese seamen. Some skills, such as the ability to sail into the wind by tacking, were cause for wonder (though not for the Chinese, long acquainted with it). But the skills of potential strategic importance to Japanese science proper were those based in varying degrees on Western mathematics and astronomy, such as the quadrant and astrolabe, and declination tables of the sun (for each day in a four-year period) to determine the latitude by observing the altitude of the sun at noon. Logarithmic and trigonometric tables appeared in the West just before isolation intervened, and afterward other instruments were invented, such as the octant with a reflecting mirror (1731), the sextant (1757), and the first chronometer to give exact time (1735) for determining, with the nautical calendar, exact longitude.

The Japanese recognized the superior strength, durability, and general seaworthiness of the Western ships soon after being exposed to them from the middle of the sixteenth century and moved rather quickly to adopt many of the Westerners' techniques. Japanese ships built around 1610, or in the latter half of the Red Seal Ships period, were only slightly smaller than and inferior in construction to Western ships. The largest was

[24] World maps based on a spherical theory of the earth appeared as early as 1474, and in the following year Regiomontanus of Nürnberg produced a nautical calendar for the period 1475–1506.

in the 700–800 ton class, had three masts, and carried around 300–400 persons. Most were of the 200–300-ton class, with two masts, and could take on board about 200 persons.[25] They blended Japanese, Chinese, and Western styles. In the military campaigns to Korea (1592–1598) some 200,000 troops were transported across the Korea Strait. Various types of vessels, some specially built for troops, horses, and supplies, included 700 larger vessels and several thousands of smaller craft. Hideyoshi ordered lords whose domains faced the seas to erect construction sites, stimulating a spurt of technical progress in the domestic shipbuilding industry.

The first purely Western-style ships built in Japan were constructed under the supervision, mandated by the shogunate, of an Englishman named William (or Will) Adams (1564–1620), who had arrived in Japan in 1600 on the Dutch ship *Liefde* as its chief navigator and had offered his services to Tokugawa Ieyasu as an adviser in diplomacy and trade. Two two-masted ships were built, one 80 tons and the other 120 tons, at Itō on the Izu peninsula.[26] In 1610 the larger one was loaned to the former governor-general of the Philippines, Don Rodrigo Vivero y Velasco (?–1636), who had shipwrecked in 1609 east of Edo on the coasts of the Bōsō peninsula (Chiba prefecture) and wished to continue his voyage from Manila to Mexico. This ship, named the *Santa Buenaventura* and manned by a Japanese crew, made the trans Pacific trip in 1610.

The feudal lord Date Masamune (1567–1636) of Sendai sent his retainer Hasekura Tsunenaga (1571–1622) as an envoy to Rome via Mexico in 1613 and for that purpose had a 500-ton, Western-style ship constructed in his domain, under the supervision of a Spaniard, by the Japanese who had worked under William Adams in Itō. This ship carried Hasekura as far as Mexico; in 1616 it crossed the Pacific again directly to the Philippines where it was purchased by the governor there and made a part of the Spanish fleet (Hasekura taking other passage

---

[25] Iwao (2), pp. 61–62.
[26] Sudō, pp. 115–118.

back to Japan).²⁷ The governor ordered another ship to be built in Satsuma (part of present-day Kagoshima prefecture) for use on trans-Pacific voyages and received it in 1623.²⁸ The Dutch are known to have commented on the low cost, good workmanship, and fine timber of the Japanese-made vessels.

Early in Western Wave I the Japanese navigational arts, like their ships, were unquestionably inferior to Western ones. Especially did the Japanese mariners lack tacking skills and methods for determining location by shooting the sun. They were unaware of the earth's sphericity, and their geographical knowledge was limited to East and Southeast Asia. The *wakō* pirate-traders, for example, depended upon the simplest ships and navigational knowledge; their exploits relied more on an adventurous spirit than on advanced skills. Some of them were quick to acknowledge the superiority of both Western ships and sailing methods, however, and worked diligently to master Western techniques of shipbuilding and navigation. Adoption of new ways and means was also evident in the Red Seal Ships' operations. The earlier ones employed various Western (Portuguese, Spanish, English, and Dutch) as well as Chinese pilots. But the Japanese mariners learnt well from their Western employees, and by late in the Red Seal Ships era, almost all of their pilots were Japanese.²⁹ It is not to be expected, of course, that the Japanese pilots would have gone beyond mastery of skills as such to an understanding of the basic principles of Western astronomy, of the nautical calendar, and of the mathematics behind these skills, but neither did scholars come forth to explore these problems on their behalf.

At any rate, many Western sailing skills were rapidly incorporated into the routine operations of the Japanese merchant fleets, and in 1618 there appeared the first attempt to compile a nautical manual of Western derivation, the *Genna kōkaisho* (Navigation treatise of the Genna era [1615–1624]). Based on knowledge which its author Ikeda Koun had learned from

---

²⁷ Ibid., pp. 120–125.
²⁸ Iwao (4), p. 209.
²⁹ Iwao (1), pp. 57–60.

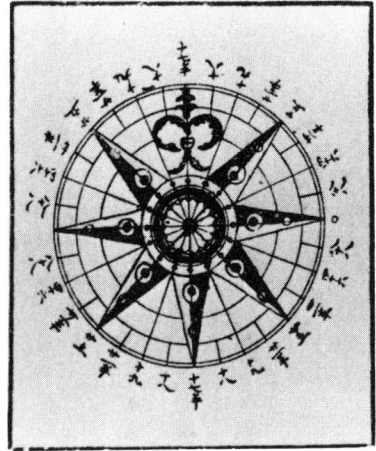

Fig. 22a. Compass drawing in the *Genna kōkaisho* (Navigation treatise of the Genna era; 1618).

a Spanish ship captain, Manuel Gonzalvez, on a voyage to Manila, its contents included:[30] explanations of the use of the quadrant, astrolabe, and other instruments for making astronomical observations; a guide to the use of solar and nautical calendars; explanations of how to determine positions at sea by the sun and stars; and a guide to the use of navigational charts of the waters between Japan and Southeast Asia. Though not printed, this manual was presumably copied and read widely. More important, until very late in the isolation period few other manuals of this kind appeared in Japan.[31] Moreover, in the time of the Red Seal Ships, Japanese vessels were constructed primarily for mercantile purposes, with no warships, in striking contrast to the European nations which, in addition to far-ranging merchant fleets, had their "invincible armadas."

Meanwhile, overseas travel to Asian markets and contacts with Western traders became matters of everyday occurrence

---

[30] For the *Genna Kōkaisho*, with annotation by Saigusa Hiroto, see Saigusa (1), 12: 1–134; see also Arima (4), pp. 118–119.
[31] Saigusa (1), 12: 14–19.

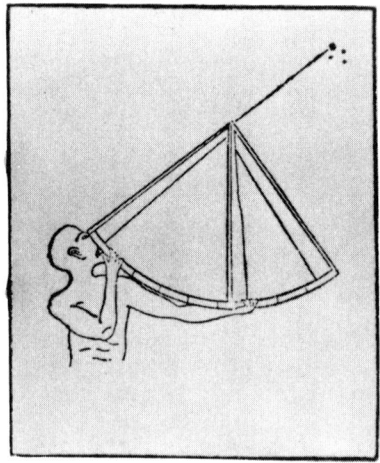

Fig. 22b. Figure of a man shooting a star; from the *Genna kōkaisho*.

for not a few Japanese citizens. Knowledge of world affairs and perspectives on the cultural, religious, and intellectual lives of other peoples were vastly expanded in the process. Unprecedented possibilities for broadening intellectual vistas beyond techniques to Western sciences were present, especially among the merchant class, in the adoption of Western shipbuilding and navigation. Beyond mastery of such techniques as the making of astronomical observations lay possibilities for comprehending Western astronomical theory and, in the process, Western mathematics. Or, use of the Western nautical calendar afforded an opportunity for gaining familiarity with Western astronomy. In nautical operations learned from Western sailors were certain mathematical methods that could have led to some very practical improvements over traditional mathematics, namely, written calculations using Arabic numerals (though the speed and efficiency of the abacus for many purposes need not be discounted). At least, familiarity with written calculations would have made possible examination of Western algebra, trigonometry, and logarithms.

In addition to the quadrant and astrolabe, contact with Western navigational operations introduced to the Japanese two other useful instruments, the telescope and mechanical watch. Continued use of these for shipping and other practical purposes could have stimulated their widespread use in Japan. Actually, the Japanese did produce the magnetic compass, quadrant, astrolabe, and mechanical watch. Watchmaking was taught them by the Jesuits, though like the oldest extant watch, made in Spain in 1583, Japanese products had no glass covers; there was no ordinary use of glass in Japan in the premodern period. But far from penetrating the maritime and scientific communities of Japan as tools for serious work, the telescope and mechanical watch were generally looked upon, and continued to be imported, as luxury goods, as objects of curiosity.

MINING AND METALLURGY

Mining was one means established by local lords in the Warring States era to increase their wealth; and gold, silver, copper, and sulphur figured large in the successive forms of Early Chinese Wave II trade. Gold and silver production experienced rapid growth from the middle of the sixteenth century but reached their peak in the early part of the seventeenth century, after which both declined; though copper production rose steadily from the middle of the seventeenth century.[32] To help strengthen central controls, Nobunaga, Hideyoshi, and Ieyasu all adopted the policy of placing the main gold and silver mines of Japan under their direct control, a policy that stimulated production. Through a Franciscan priest, Ieyasu in 1598 sent a message to the governor of the Philippines, requesting a mining expert to be sent to Japan, though in vain.[33] Much of the great volume of gold and silver mined in Japan was exported to China and the West. It is estimated that Japanese silver exports accounted for 30–40 percent of gross world production in this era.[34]

As mines opened throughout Japan, a number of technical

---

[32] Asahi, p. 398.
[33] Again, in 1609, Ieyasu requested Don Rodrigo Vivero y Velasco (Governor-general of the Philippines, shipwrecked on the Japanese coast en route to Mexico; see p. 177) to open direct trade between Japan and Mexico and to dispatch mining experts to Japan; neither request was realized. Cf. Tsuji Tatsuya, pp. 258–260.
[34] Iwao (4), p. 223.

182   Pressures toward Modern Society

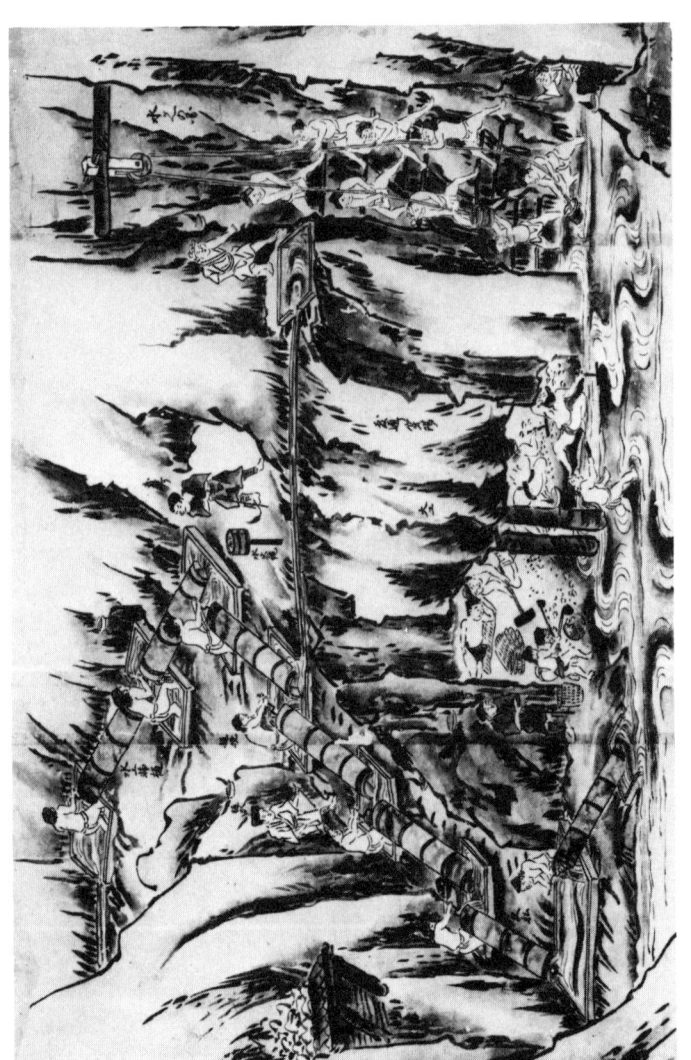

Fig. 23. Portion of a scroll showing a gold and silver mine on Sado Island during the Tokugawa era. Here are seen workers digging a vein (lower center) and drainage by pulley (right) and tubes with a kind of Archimedean screw (left). From the scroll "Sado kinzan kinbori no zu" (Depiction of gold mining in Sado Gold Mine). (The National Archives, Japan)

innovations were made. New methods were found for tunneling through hard rock formations, and deeper tunnels, sometimes several hundred meters deep, became possible. Water drainage consequently became a major problem, for which long horizontal drainage tunnels were the main answer, though a wooden pump with an Archimedean screw was also developed. Tunneling and drainage problems were a great stimulus to improved land-surveying skills.

It was in the refining techniques that foreign methods were adopted. The cupellation process was introduced from China via Korea in 1533; it was employed in the Ikuno silver mine (opened in 1542 in the Tajima district) and then spread to other mines. The method was to crush silver ore, add lead, heat, and fuse. It was then cupeled in a crucible filled with ashes, which absorbed the lead and left the silver on the ashes, which were then blown away. This method was also used to refine gold.

A Western method for separation of silver and copper, known in the West since the eighth century, came to Japan late in the sixteenth century. Called by the Japanese the "Southern Barbarians' smelting" (*Nanbanbuki*), this technique first fused lead with copper-bearing silver to separate the copper from the fused silver and lead and then cupeled the latter to get the silver. A method which extracted silver from ore as a mercury amalgam was also introduced from the West sometime in the Keichō era (1596–1615) by a Spaniard.[35] The mercury had to be imported, and when the supply was cut by isolation, this method vanished from Japan.

Mining and metallurgy had strategic significance as technical infrastructures for gunnery, ship's fittings, and production of scientific instruments. Surveying for mines strengthened the role of mathematics in society, while metallurgy might eventually have kindled an interest in chemistry, as the latter gradually came into being in Europe, except for isolation. Even after isolation, mining and metallurgy retained considerable economic importance.

[35] Asahi, p. 407.

PRINTING

Woodblock printing of full-page texts and pictures in China dates back to the seventh century, and printing of entire texts of Buddhist scriptures was done in the ninth century. The technique was developed in Japan in the eighth century and became widespread in the Semiseclusion Era under Sung influence.[36] Printing of Buddhist texts became most popular among the larger temples, the oldest existing printed book in Japan being a Lotus sutra dated 1080. The *gozan* temples initially placed equal emphasis on the printing of Buddhist sutras and Chinese classics (often reprints of Sung and Yüan texts), though the latter became the main object of later *gozan* printing.[37] When Kyoto was made a continuous battlefield in the Warring States era, many noblemen and Buddhist priests escaped and found asylum in the local towns of powerful lords. Under their influence, local cultural life rapidly improved, and the printing of Buddhist-related texts and Chinese classics was started in several provincial centers, such as Kagoshima (Shimazu fief), Yamaguchi (Ōuchi fief), Odawara (Hōjō fief), and in the commercial city of Sakai.[38]

Two new developments appeared in Japan in the late sixteenth century. One was the technique of printing with movable type; the other was the printing of secular materials by secular agents for commercial purposes. Movable type made of wood, porcelain, and copper appeared in China around 1030, and this method was widely utilized by the Koreans during the first half of the fifteenth century, when the Koreans also invented bronze movable type.[39] This was associated with

[36] In 764 Emperor Shōtoku (reigned 764–770) ordered the manufacture of one million miniature three-storied wooden pagodas which were distributed to the ten leading temples in equal number. In each pagoda was placed a sheet of paper on which a charm (*dharani* in Sanskrit; *darani* in Japanese) was printed. Some of these papers are still preserved in the Hōryūji temple and were regarded as the oldest extant printed matter in the world until the 1966 discovery of similar printed *dharani* in the Sŏkkat'ap pagoda of Pulguksa temple in Korea, which were printed between 704 and 751. Jeon, pp. 167–168; Asahi, pp. 350–351. See Figs. 2 and 3.
[37] Asahi, pp. 359–360.
[38] Ibid., p. 363.
[39] Jeon, pp. 175–184; also Nathan Sivin's preface to Jeon, pp. xvi–xvii.

the Yi dynasty's aggressive efforts to found its policies on Confucianism. The transformation of scholarship by the Chinese inventions of printing and papermaking occurred much earlier in the East than in the West, where printing with movable type (using a press for the first time) did not appear until the middle of the fifteenth century. The late sixteenth-century invasions of Korea provided the occasion for introducing movable type into Japan, but it was used chiefly by the established groups, such as the imperial court, the governments of Hideyoshi and Ieyasu, and some Buddhist temples, and mostly for printing Chinese classics and Buddhist texts. Commercial printing continued to utilize traditional woodblock methods, as it was cheaper and frequent reprints were possible without resetting type. Here the technique was less important than the fact that in this age of realignment of political and cultural values, books of solely secular interest were printed for commercial distribution. Prior to this, printing had been almost exclusively confined to religious and classical literature under the auspices of established religious and political groups for their own purposes.

In 1590 the Jesuits brought to Japan a printing press that used movable lead-alloy type. In time a number of quite acceptable Japanese translations of Catholic doctrinal and inspirational works (e.g., Thomas a Kempis's *Imitation of Christ*) were printed, as were also some selections of general Western and Japanese literature (e.g., *Aesop's Fables* [in Japanese, *Isoho monogatari*] and the *Heike monogatari* [Tale of the House of Taira]). Actually the press was first put to use in Macao, while waiting two years for shipment to Japan, to produce two Latin imprints. Set up first in Japan at Katsusa in Kyushu, a *Life of the Saints* was printed in Japanese in 1591, but with Latin letters (the first use of romanization in printing in Japan). Latin type had been brought from Europe, though various kinds of materials were used by Japanese workmen to produce type domestically. The first Japanese type was cut from wood, though in time metal type was cast and used exclusively, at the latest by 1598.

The Katsusa site was chosen for the printing operation because the Jesuits' college was located there. As the press, along with the college, was moved several times (see p. 189), production was limited until it was finally removed to Macao in 1614 when the college was closed. Not counting the works printed in Macao, twenty-seven of the works printed in Japan are extant,[40] and about the same number are said to be referred to in other sources.[41] The latest known print is dated 1611.

Afterward, and until the Meiji era, printing in Japan was done chiefly with woodblocks. The custom of copyright protection for publications also evolved, though it was not applied to printing with movable type, which was disassembled after use. Thus reprint rights were not regarded as permanent and needing protection. Carved woodblocks could be stored for repeated use, allowing small but frequent printings (as in China), and thus were considered worthy of copyright protection by the shogunate.[42]

Little trace of any of the Jesuit printings can be found, as all forms of Catholic influence were systematically suppressed for more than two hundred years. But neither the removal of the Western press and movable type, nor the fact that Korean movable type was not extensively used, seem to have affected the continuing commercial diffusion of many kinds of literature or otherwise deterred the Japanese cultural and intellectual processes. The outburst of intellectual activity in the seventeenth century and later scholarly developments as well witness to the adequacy of traditional woodblock printing in the nurture and diffusion of knowledge at this stage.

Learning in the Transitional Period

CHINESE LEARNING

Two major deviations from past patterns occurred in Early Chinese Wave II, both of which helped lay the foundations

[40] Asahi, pp. 365–366.
[41] Laures, p. 140.
[42] Asahi, pp. 379–380.

for the more sophisticated developments of the seventeenth century. They concern the focus and agency of intellectual influence from China.

Zen priests, it will be recalled, studied Buddhism and Confucianism together in the Semiseclusion Era. In Early Chinese Wave II there appeared, though very gradually, a tendency to consider Confucian thought by itself. It became for some intellectuals a secular alternative to the religious outlook of Buddhism and Shinto and gave learning a new lease on life. From the Sung period Buddhism in China had receded into the background as Neo-Confucianism assumed a central position. A truly corresponding situation in Japan did not arise, however, until the great intellectual revival in the seventeenth century.

While Buddhist priests initiated new studies of Chinese thought in early Chinese Wave II, the key roles in introducing and using Chinese knowledge in time shifted to private citizens who were more sensitive to contemporary social change. These men found a new ground for learning in the more secular outlook of Chinese rationalism—as expressed in Confucianism or Chinese science—and in a rekindled interest in practical, technical matters. For the first time in Japan social conditions favored a relatively autonomous merchant class motivated by local and private concerns rather than national policies or religious purposes. But neither the merchants nor the local military leaders who sometimes supported them were interested in duplicating the educational or research institutions of contemporary China. There was no significant establishment of an institutional basis, adopted or indigenous, for traditional learning and science in the early (preisolation) phase of Chinese Wave II. And though many printed books and some technical items were acquired in the late sixteenth-century Korean campaigns, their intellectual importance does not properly belong to this period. The gradual awakening of scholarly concerns in Early Chinese Wave II came to fruition, then, only in the late (isolation) phase of Chinese Wave II.

WESTERN LEARNING

The Jesuit missionaries were, of course, the main agents for introducing Western learning. As an integral part of their missionary work they erected a variety of institutions for education and research, from schools for primary education and catechetical instruction to higher-level schools for training an indigenous priesthood. One of the purposes of the primary schools was to overcome Buddhist influence implanted in the minds of pupils, but the Jesuits intended also to provide a general education for those in their charge. The first of such schools was founded in Funai of Bungo district in northeast Kyushu. One estimate, based on records of 1583, indicates that there were around two hundred primary schools in Kyushu and Honshu.[43] Instruction in these schools included: basic Christian beliefs; reading and writing Japanese; composition; arithmetic; music, both vocal and instrumental; and painting.[44] These schools were also open to girls.

In 1580 the Jesuits initiated a program for training Japanese priests. The basic institution for this purpose was a college (*collegio*), though students were generally expected first to complete work in one of two seminaries (*seminario*; three were originally planned), which were open as well to sons of ranking samurai, who had no plans to become priests. The first seminary was opened with twenty students in the Arima fief of Kyushu in 1580. By 1590 its enrollment had swelled to one hundred students. Due to varying treatment of the Jesuits by local and central governments, this seminary and the college were forced to move from place to place. The seminary founded in Nobunaga's base town of Azuchi in 1581 was burned (along with Azuchi castle) soon after Nobunaga met his end during a surprise attack on Kyoto in 1582. Relocated first in Kyoto at a local church found to be too small, it was moved again to Takatsuki (near Osaka) in 1585, under the sponsorship of Takayama Ukon (1552–1614), a

---

[43] Ebisawa (3), p. 61.
[44] Ibid., pp. 60–61.

Christian and a trusted vassal of Hideyoshi. Growing suppression in the capital region led to its being combined with the seminary in Arima in 1587, when Hideyoshi shifted from tolerance to an overtly anti-Christian policy.

The core curriculum of the seminaries consisted of theology and philosophy; supplementary instruction included Japanese language and literature, Latin, Portuguese, history, mathematics, music, art, and for some, instruction in copperplate printing. Except for Japanese and literature, teaching was done initially by Portuguese priests, though various responsibilities were later assumed by Japanese teachers. Of the four young Japanese Christians who visited the Vatican as envoys of three Christian daimyo in Kyushu and returned in 1590 (bringing the Jesuits' printing press with them), one joined the staff of the Arima seminary.[45]

The Jesuit college, besides training future Japanese priests (the first were ordained in 1601[46]), also served as the central administrative agency of the Jesuit mission in Japan. It was the base for printing operations, and from 1580 the missionaries undertook systematic studies of Japanese culture and language there. Established originally in Funai, Bungo district, in 1580, it was later compelled to relocate repeatedly due to the intensified persecutions of the Catholics. The invasion of Funai by forces of the Shimazu fief precipitated the first move in 1587 to Chijiwa in the Arima fief. Later the college was moved to Katsusa in 1590, to Amakusa in 1592, and finally to Nagasaki in 1598, where the missionaries were allowed to stay, ostensibly to minister to the Portuguese trading community. The college was finally closed in 1614, after persecution gained added intensity under Hidetada. In the short span of thirty-four years it had been located in five places.

---

[45] Matsuda (3), p. 204.
[46] They were Luis Naibara and Sebastio Kimura, who were among a number of Japanese who, because of worsened political conditions in Japan, went to the Jesuits' theological college in Macao before coming back to Japan and being ordained (not a few became martyrs). Cf. Cieslik, pp. 7–8; Jennes, p. 114. See also n. 48.

Fig. 24. Sketches of Jesuit academic institutions in Japan during Western Wave I, used to decorate book pages. (a) Novitiate in Usuki, Bungo district, Kyushu; (b) Seminary in Arima, Hizen district, Kyushu; (c) Seminary in Azuchi, near Kyoto; (d) College in Funai, Bungo district, Kyushu. (Courtesy of Michael Cooper, S.J.)

Learning in the Transitional Period

Seminar. Arimaense in Iapone

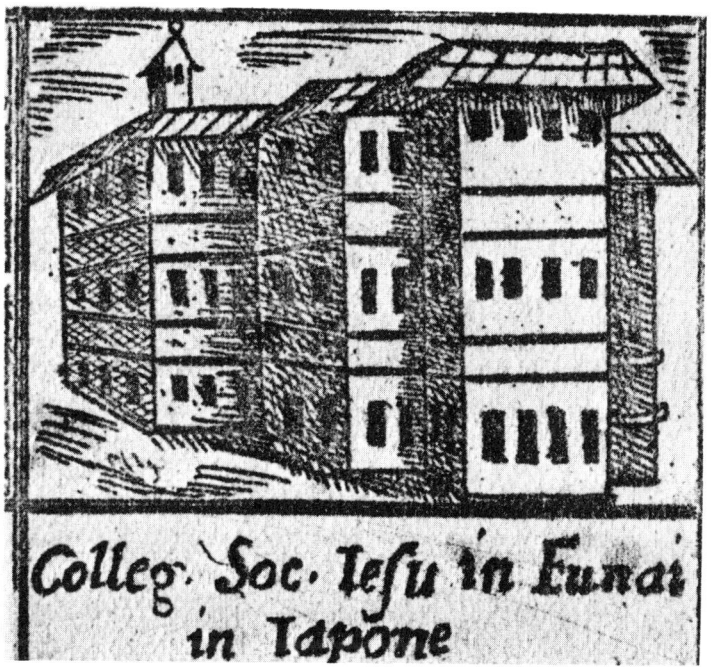

The college's curriculum reflected the Aristotelian norms of late medieval Scholasticism, but also, due to Renaissance influences, Platonic norms. Moreover, as an institution devoted to instilling Jesuit discipline, the curriculum was not merely systematic, but rigid; training was more indoctrination than education, and by definition restrictive. Nonetheless, the curriculum was quite broad, and included theology, philosophy, canon law, jurisprudence, logic, natural science, rhetoric, and music.[47] The curriculum and the institution established to teach it amounted to an epochal event—the first organized agency for higher learning with systematic coverage of a whole body of knowledge to appear since the extinction of the *ritsuryō* academic system founded in Chinese Wave I.[48]

It was of equally epochal significance that the natural sciences (albeit of a kind soon to become outdated in Europe) and mathematics were part of the seminary and college curricula. Though their relative status in the curricular scale may have been low, the important point is that they had any clear and definite place at all; that they were regarded as indispensable to education, especially for men entering religious orders, and as useful tools for propagation of the faith as well as integral to a systematic (Christian) world view.

The studies in Japanese language and culture at the college produced, for the short time involved, a number of remarkable results. Sometime in the decade after 1550 the able lay brother Duarte da Silva produced a grammar of the Japanese language and a Japanese-Portuguese dictionary, and later a Latin-Portuguese-Japanese dictionary. Again, in 1603, another Japanese-Portuguese dictionary appeared, possibly by the lay brother Juan Fernandes (1525/6–1587), who was also known for his linguistic ability. One of the Jesuit fathers, João Rodri-

---

[47] On the Jesuit college, see Ebisawa (3), pp. 65–69.
[48] Even the Jesuit college in Japan never reached the lofty goal envisioned for it: a base to train a Christian elite and to confer academic degrees. Because of persecutions in Japan, this goal had to be carried out instead in Macao, where a number of escaped or exiled Japanese Catholics were trained, many of whom returned (usually secretly), and some were martyred.

gues Tcuzzu (1561?–1633), finished another grammar of the Japanese language in 1604.[49] Prior to the appearance of these grammars the Japanese language had never been subjected to systematic analysis. Presumably these linguistic tools, apart from use by missionaries for learning the language, were helpful in producing the printed translations of many doctrinal and inspirational materials for evangelistic use and the selections of Western and Japanese literature for educational use. It is said that among the translators and editors of this body of Christian and general literature there were a number of former Buddhist (mostly Zen) priests. Most of the translations were in plain, unstilted language, and their literary quality is considered high.[50]

The Jesuits' attempt to introduce the basic framework, though of course not the whole body, of late scholastic learning to the Christian community in Japan, and through it to the country as a whole, was not simply disinterested educational service, but one means for winning Japanese converts. That intellectual framework was essential to their natural theology, which the Jesuits hoped would appeal to learned men in the samurai and merchant classes, as well as to Buddhist clerics. Interpretations of the mutual relevance of divine and natural knowledge were something that many Japanese intellectuals were able and eager to understand, and not a few accepted with conviction. Many of the larger churches in Japan at the time had terrestrial globes and even instruments for making astronomical observations, and such equipment often attracted inquisitive minds.[51]

After the Jesuit academic program collapsed under persecutions, a systematic effort to encompass a whole body of learning and science was never again projected by any agency in Japan

---

[49] Cooper, pp. 224–236, gives an excellent discussion of this grammar, *Arte da Lingoa de Iapan*, including problems of the date of publication (1604–1608), its limitations as a grammar, and some interesting nongrammatical content; reference is also made to a second, shorter, but superior grammar, *Arte Breve da Lingoa Iapoa* (Macao, 1620). See also Ebisawa (3), pp. 78–81.
[50] Ebisawa (3), pp. 88–91, 95.
[51] Ibid., p. 149.

until the sweeping changes wrought by the Meiji Reforms. This estimate does not discount the value of the various shogunate and domain schools that appeared late in the Tokugawa period; while they did, in fact, add to traditional learning some selected courses in sciences, traditional and even Western, they were not comparable in scope either to the preceding Jesuits' program or to the succeeding Meiji institutions.

Still, the Jesuits' impact on the Japanese intelligentsia should not be overestimated. Nor should it be forgotten that they were quite removed from the main center of intellectual life in Kyoto.[52] In China the Jesuits made direct contact with scholars in the capital and other intellectual centers, produced many important translations, including scientific works, and a few were even appointed to posts in the central administration. No such claim can be made for the Jesuit enterprise in Japan in the late sixteenth and early seventeenth centuries. Rather, Western Wave I ended with the uncompromising expulsion of non-Japanese Catholics, suppression of their native brethren, and closure of the nation to most international intercourse. There were, of course, tertiary effects that lingered, but one searches in vain for evidence of any fundamental alteration of the form, content, or methods of traditional learning and science.

The Sciences in Transition

Science shared in the transitional, fluid character of Early Chinese Wave II, pressing toward new forms. The renewal of activity in the Chinese-style sciences was based not on revival of earlier Chinese Wave I forms but on the introduction of contemporary Chinese science from the continent. This was done not by government, nor even principally by Buddhist agents, but mainly by members of the merchant class and therefore for practical purposes. The practical knowledge they imported was diffused and fixed firmly in many sectors of society.

---

[52] A "university" (*academia*) was once planned for Kyoto but was never realized. Cf. Ebisawa (3), p. 70.

Although the level of science imported in Early Chinese Wave II was largely elementary, it served as the foundation for the upsurge of scientific and other learned activity that followed closure of the country.

Western scientific ideas and techniques did not fare so well. The Western sciences were, to be sure, quite different in some basic respects from the traditional Chinese-style sciences of Japan, but even in the brief period of exposure they began to take root in Japanese soil and could have borne rich fruit. The pressures for change were strong, the agents of change were forward-looking, and the desire for new knowledge was intimately related to concrete social and economic concerns of the times. The possibilities for effective, if not revolutionary, adaptation of the Western sciences and techniques were far from small—and their loss through national isolation exceedingly great.

The failure of Western science to make any permanent impact on its Japanese counterpart can be explained, in large part, by fundamental differences that made it difficult to combine or accommodate Western scientific thought within the traditional Chinese-style framework. Also, the Jesuits, though learned, were not specialists and had only limited aims in teaching science. The European traders and mariners, though skilled, were not educators. Linguistic barriers and official duties prevented either group from delving deeply into the complex background of Japanese science. The foreigners were too distant from Kyoto and had little real encounter with specialists in the mainstream of Japanese scientific or, for that matter, scholarly life. Frequent relocations of their educational institutions compounded the Jesuits' difficulties in trying to transmit Western science and learning. Finally, the increasingly harsh persecutions hampered their work. Taken together, the barriers faced were of such magnitude that it is too much to expect that the Jesuit labors should have left a permanent imprint on Japanese science.

Some conditions did favor the Jesuits' near-century in Japan. Their own efforts were industrious and able, and the close cooperation of Japanese associates and the warm reception of their teaching by some intellectuals augured well for the future. Compared to the rather speedy assimilation of Western techniques, though, the reception of Western sciences in Japan was, on balance, quite poor. As in early contacts with Chinese culture, it would seem that more time is needed for the effective transmission of science proper than for the transfer of techniques.

ASTROLOGY AND CALENDRICAL ASTRONOMY

No significant influx of Chinese astrology or calendrical astronomy occurred during Early Chinese Wave II. It was a time rather of the spread of traditional calendrical techniques into some localities outside Kyoto, continued interest in prognostications based on the calendar-almanac, and some minor changes in authority over the calendrical office. That a certain momentum was built up during the period is evident from the outbreak of activity that appeared soon after isolation began and was continuous with the traditional concerns of Early Chinese Wave II. The Western astronomical and calendrical sciences as such had little direct impact on Japanese science during Western Wave I, though their potential influence was not meager.

CHINESE-STYLE ASTROLOGY AND ASTRONOMY

Shogun Yoshimitsu, who opened the negotiations that led to the Tally Trade, also proposed adoption of the contemporary Chinese calendar; but he was defeated by opposition from his own retainers. There is no known instance of any Japanese specialists going to China to acquire current knowledge of astrology or calendrical astronomy. That interest in Chinese thought in these fields was not entirely lacking is suggested by the appearance in 1414 of the *Rekirin mondōshū* (Collection of dialogues on the calendar), by the court astronomer Kamo no Arikata. Devoted largely to prognostication in the *onmyōdō* tradition, the opening section, which makes brief references to Chinese cosmology, exhibits a strong Neo-Confucian flavor—

presumably transmitted through the learning of Buddhist priests in the late phase of the Semiseclusion Era.[53]

During the confusion of the Warring States era, calendars came to be printed and, in some cases, calculated independently of Kyoto. Independent calculation seems to have been the case in Mishima, which had the longest history of calendar issuance outside Kyoto, though in at least three years—1374, 1582, and 1583—its calculations are known to have differed from the official Kyoto calendar (as did, at times, those printed elsewhere, e.g., Osaka, Nara, Ōmiya, and the Ise and Aizu districts).

The emergence of specialists outside Kyoto was indicative of the declining competence and prestige of the calendrical office in Kyoto's Institute of Divination. The Kamo family in charge of calendar production ended with Kageyukōji Aritomi around 1560. Thereafter, the Tsuchimikado family (descendants of the Abe line) assumed responsibility for both astrology and calendrical astronomy. Differences between the local and Kyoto calendars cast doubt on the level of competence, as they all relied on the astronomical tables of the *Senmyōreki* (Japanese for the Chinese *Hsüan-ming li*) used in Japan since 862. In Nobunaga's time one of the local specialists charged the Tsuchimikado house with mistakes in placing the intercalary months, and the latter was rebuked by Nobunaga. The court astronomers were beginning to lose their traditional monopoly over the "secret learning" of astrology and astronomy.

Serious work in these fields did not occur, however, prior to isolation. The spread of interest in calendars during Early Chinese Wave II was largely connected with the widespread popularity of prognosticatory annotations, so appealing in times of peril and strife. A treatise entitled *Hoki naiden*,[54] which dealt with such fortune-telling, was printed around 1612, and

---

[53] Nakayama (6), p. 43.
[54] This is regarded as the first printed book on astrology in Japan; it treats mainly calendrical annotations and prognostications. It is traditionally attributed to Abe no Seimei, tenth-century court astronomer (see p. 123).

was reprinted many times prior to isolation. Calendars and the astronomical treatises used in their computation remained unrevised. Moreover, the cosmic perspective most common among Japanese intellectuals continued to reflect the traditional view of the Chinese literati that heaven is round and the earth flat.

WESTERN INFLUENCE

The Jesuits' teaching of astronomy as such was straightforward, though oriented to their natural theology as a means of propagation. That is, they appealed to concepts of the structure of the universe, the courses of the heavenly bodies, and even to solar and lunar eclipses as manifestations of natural law and part of a created order. Rather than astronomy proper, this was natural philosophy, argued in part from astronomical sources, often in the context of controversy and thus from theological grounds as well. But the Jesuits did believe there was a valid intellectual avenue to conversion. In a letter written January 29, 1552, from Goa to Ignatius Loyola, Francis Xavier (1506–1552) urged that personnel recruited for the Japan mission be not only "philosophers ... well trained in dialectic" but also "well acquainted with cosmic phenomena, because the Japanese are enthusiastic about listening to explanations of planetary motions, solar eclipses, and the waxing and waning of the moon," for "all explanations of natural philosophy greatly engage that people's minds."[55]

The Jesuits' views proved shocking to some, especially learned men who understood their claims. But most people knew nothing of a field heretofore monopolized by House Learning. The Jesuits could present their geometric models of orbits in the heavens and their coordinated explanations of strange happenings such as eclipses, earthquakes, lightning, and floods; they could insist that the earth is round and extends far beyond Asia. But to nonspecialists their arguments, however rational on Western premises, were unlikely to appear radically different from customary images rooted in native astrology and divination.

[55] Quoted in Nakayama (6), p. 81.

Jesuit teaching did not, in any event, represent the latest European developments in astronomy. Copernicus' theory, marking the shift from the Ptolemaic geocentric system to a heliocentric system, was published in 1543, the year the first Portuguese landed at Tanegashima. Only later, in the first two decades of the seventeenth century, did Kepler use the observational data of Tycho Brahe to extend elliptic orbits to planetary movements. Moreover, Galileo's series of revolutionary discoveries in astronomy through use of improved telescopes came early in the seventeenth century, while his work in mathematical dynamics reached a peak only in the middle of that century—all too late to have been a part of the Jesuit teaching in Japan. Instead, it was the Ptolemaic geocentricism accommodated to Aristotelian philosophy which the Jesuits presented in Japan.

Even so, some Japanese intellectuals were rather impressed by theories which to them were new. Not a few discovered in debates with the Jesuits that appeal to tradition did not necessarily constitute a successful counterattack, and the moment of realization became an occasion of conversion to Catholicism for some of them. Most intellectual encounters probably ended in the kind of stand-off that characterized the famous dispute in 1606 in Kyoto between the famed Neo-Confucianist Hayashi Razan (1583–1657) and the Jesuits' ablest Japanese spokesman, Fucan Fabian. A former Zen priest, Fabian in 1605 wrote a tract entitled *Myōtei mondō* (A dialogue between Myōshū and Yūtei), which included an affirmation of the earth's sphericity. In the 1606 dispute, Fabian claimed that the earth is round, with heavens above and below, and that if one sailed far enough to the west, he would return to his starting point. Razan opposed these claims by asserting that all things are ordered in proper above-below relations, and that heaven is above and earth below; so, the earth cannot be spherical.[56] In any case, neither the astronomical concepts of the Jesuits nor their related cosmic views penetrated deeply enough into the Japanese mind to survive persecution and

[56] Nihon Gakushiin (5), pp. 129–130.

isolation. Moreover, Japanese specialists at the time were more interested in calendars, while the Jesuits considered cosmology the weightier subject. Even if their interests had coincided more perfectly and the teaching of the Jesuits had survived, the actual work of calendrical production could not have been restructured on the basis of that teaching alone.[57] The significance of Jesuit teaching lay in its potential for stimulating new perspectives.

Among the Jesuits in China at the time there were qualified specialists who were in direct contact with able Chinese scientists in the capital of Peking. In the early seventeenth century the key figure among Jesuit scholars, Matteo Ricci (1552–1610), and his colleagues, including two Chinese scholars, produced a series of works that dealt with such subjects as cosmology, astrology, astronomy, observational instruments, and mathematics—including a translation of the first six chapters of Euclid's *Elements*. Unfortunately, the works by Ricci and his followers, including various scientiflc works, were among the prime targets of the ban on importation of all books related to Christianity imposed by the Tokugawa government in 1630.[58] Even if the scientific works of the Ricci group (some of which were completed after his death) had not been prevented from entering Japan, unless other conditions —the distant remove from Kyoto, the low level of Japanese astronomy, and especially the persecutions—were not also altered, it is questionable whether the eclipse of Jesuit knowledge could have been forestalled.

The calendar used by the Portuguese mariners and traders, as well as by the Jesuits, was a purely solar calendar. Though referring to the same sun that was one of the points of reference for the lunisolar calendar used continuously in Japan since Chinese Wave I, the intellectual frame of reference was as foreign to the Japanese as were the "barbarians" who brought

---

[57] Nakayama (6), p. 88. Even works appearing after the Jesuits were expelled, but reflecting their limited influence, were inadequate for technical calendrical work; see Ibid., pp. 88–98.
[58] Ibid., p. 84.

it. The Jesuits published a liturgical calendar based on the solar calendar, which was distributed in great quantities but was merely a parish guide to days for religious observances (it remained in clandestine use throughout isolation[59]). Of much greater importance was the Westerners' nautical calendar with sun-declination tables, which Japanese navigators in time learned to use; and mention has been made of the Japanese-authored nautical manual, *Genna kōkaisho* (p. 178), which included explanations of both solar and lunisolar calendars; use of the astrolabe, quadrant, and navigational charts; and how to determine positions at sea by the sun and stars.

However interesting the historical details relative to Western calendars and astronomy in Japan during Western Wave I, or however crucial their scientific and technical potential may have been, the fact remains that they were excluded from the Japanese intellectual scene along with those who brought them, without having been seriously or sufficiently compared with their traditional Japanese counterparts. The mainstream of astrology and astronomy in Japan remained, where it had always been, within the orthodox framework of Chinese-style science. Only a few years after isolation, though, serious work was to begin to revive and eventually reform the traditional astronomical and calendrical arts.

MATHEMATICS

TRENDS IN TRADITIONAL MATHEMATICS

The House Learning of the Ozuki and Miyoshi households had already become inactive by the end of the Semiseclusion Era, and their lineage is untraceable in Early Chinese Wave II.[60] Rudimentary arithmetic first offered in the temple schools and in *ōraimono* (p. 135) continued in use, and some of the Buddhist priests active in the Tally Trade are thought to have been competent in practical mathematics. Expanded activity

[59] Jennes, pp. 196–197, 201.
[60] Nakayama (6), p. 154.

in domestic and overseas trade together with the technical needs of navigation, new temple and castle construction, land-surveying, mining operations, and so on, contributed to a general leveling-up of practical mathematics, which later provided the stimulus for renewed introduction of contemporary Chinese mathematics.

Merchant households in touch with China introduced a number of mathematical texts into Japan in Early Chinese Wave II. One of the most important of these was the 1593 work by Ch'eng Ta-wei (1533–ca. 1592), the *Suan-fa t'ung-tsung* (A systematic treatise on arithmetic). Oriented to the practical needs of government and commerce, it dealt largely with the use of the abacus, which was widespread in Ming China.

Having reached a peak in the Yüan era, mathematics in China during the Ming period was marked by general decline. Most mathematicians could no longer treat the more difficult problems of former times, especially the "heaven-origin" algebra. In fact, the two key texts on heaven-origin algebra, Chu Shih-chieh's *Suan-hsüeh ch'i-meng* (1299) and *Ssu-yüan yü-chien* (1303), had been lost in China. But the former was among books brought home from the Korean expeditions of the 1590s, and it was the most influential in lifting higher mathematics in Japan from the low level to which it had sunk in the Semiseclusion Era. (In 1658 Hisada Gentetsu made a reprint of it from a Korean version preserved in Tōfukuji temple in Kyoto.[61])

Late in Early Chinese Wave II, from the 1620s, interest in mathematics suddenly mushroomed, as seen in the rapid rise in the quantity and quality of published works. Many editions of manuals on practical mathematics for both general and specialized uses were printed and marketed for the first time in Japanese history. A review of several of these books is sufficient to indicate both the greatly intensified demand and the level of interest and ability in Early Chinese Wave II.

---

[61] *Suan-hsüeh ch'i-meng* was first printed in Korea sometime between 1419 and 1450. The first reprint of this book in China was made in 1839. Hirayama Akira (2), p. 28.

Table 17. Key Mathematical Works Published in Japan in Early Chinese Wave II (1401–1639)

Author	Title	Date	Remarks
Mōri Shigeyoshi	Warizansho	1622	Printed in Japanese, for general use; oldest extant printed mathematical book in Japan
Momokawa Chihei	Shokan bumono	1622	Handwritten, in Japanese; copied and transmitted among disciples
Yoshida Mitsuyoshi	Jinkōki	1627	Printed, in Japanese, for general use; widely reprinted and plagiarized
Imamura Tomoaki	Jugairoku	1639	Handwritten, in Chinese; copied and transmitted among disciples
Imamura Tomoaki	Inki sanka	1640	Printed, in Japanese, in verse form for memorization; companion to the Jugairoku

The *Warizansho* (Manual of division) by Mōri Shigeyoshi was in most respects a simple book, a compilation of methods known by the end of the Semiseclusion Era and used throughout Early Chinese Wave II until its appearance. It manifested no direct or recent continental influences, except that it treated explicitly—in detail and in plain language—the methods of division done on the abacus (in Japanese, *soroban*). The original title of this book is unknown, but from the problems treated and the table of contents, it is usually referred to as the "Manual of division."[62] (Addition and subtraction on the abacus are quite easy, and multiplication is about the same as on the counting board; but division is fairly difficult.) Its methods for calculation of area and volume, as well as its methods for land-surveying, were simple compared to later books. Moreover, it employed no technical terms like "sphere" but used only conventional phrases such as "ball-like round things" (*tama no gotoku marukimono*).

Momokawa Chihei's (?–1638) *Shokan bumono* (Various ap-

[62] Katō (6), 1: 82.

proaches to area and volume) originally appeared in two parts, but only the second part now exists. It treats methods of calculating area and volume for use in civil engineering and architecture. Both it and the *Warizansho* are quite elementary, typical of Japanese mathematics before the renewed introduction of Chinese sources.

A more important work, incorporating contemporary Chinese mathematics, was the *Jinkōki* (Treatise on [numbers] great and small; sometimes *Jingōki*), published in 1627. Yoshida Mitsuyoshi (1598–1672), a disciple of Mōri Shigeyoshi, wrote it after first absorbing the content of the *Suan-fa t'ung-tsung*, which he rearranged, omitting the most difficult parts.[63] It explained multiplication, division, and square and cube root extraction done on the abacus, focusing on commercial and technical problems often encountered in daily life, as in the *Warizansho*. Mathematical games were also included but no numerology such as appeared in earlier texts. Like all its Chinese predecessors, it was a collection of solutions to specific problems rather than a treatise on the principles of mathematics. It used technical terms like *kyū* for sphere, instead of the cruder "ball-like round things" found in the *Warizansho*. That there was great demand for the *Jinkōki* or works of its kind is clear from the fact that printers reprinted it without the author's permission, revising or abridging it as they pleased, under many different titles such as *Eitai (eternal) Jinkōki* or *Fūki (rich and noble) Jinkōki*. The author himself revised it three times, in 1631, 1634, and 1641. In the 1641 version, entitled *Shinpen (newly revised) Jinkōki*, are found twelve "unsolved problems" (*idai*). This challenge to the reader was the beginning of a custom that was to play a very important role in the development of mathematics in Japan after isolation (see Chapter 4). Here it suffices to point out that the *Jinkōki* and its plagiarized versions (numbering more than four hundred) became standard texts for practical mathematics until Meiji times.[64]

---

[63] Hirayama Akira (2), p. 13.
[64] Ibid., pp. 17–18. It is estimated that several thousand copies of *Jinkōki* were printed in the Kan'ei era (1624–1644) alone; Ibid., p. 16.

The *Jugairoku* (Record of a young boar) by Imamura Tomoaki, another disciple of Mōri Shigeyoshi, was not a mere enumeration of solutions for various problems, as was the *Jinkōki*, but gave rules to be followed for each problem, a more systematic approach than that found in other works. Written in Chinese, it was a kind of teacher's guide for use with its companion volume in Japanese, the *Inki sanka* (Mathematical verse for memorization), by the same author, which put the rules of the *Jugairoku* into verse form for easy memorization.

These books were all written by private citizens, primarily for use by town-based merchants and craftsmen, though they were equally useful to lower-ranking samurai responsible for tax collection and other fiscal matters, logistics of military maneuvers, and construction of castles, bridges, dikes, and the like. Their extensive use fixed the public image of mathematics as a tool for utilitarian ends, whether commercial, technical, or military. For the more abstract Japanese mathematics known as *wasan* to be born through the agency of samurai scholars later in the seventeenth century, that image had to be recast as more than utilitarian, if not, indeed, nonutilitarian. Mathematicians of the transitional period suffered no feelings of inferiority from being engaged in practical problems, but were consciously responding to contemporary social and technical needs. It was a distinct loss that the more confident, pragmatic attitude of Early Chinese Wave II did not carry over into the late (isolation) phase of that wave, when the mathematics called *wasan* became, instead, a leisured pursuit that boasted of its separation from useful activity. The tradition of the Nine Chapters did, however, enter the *Jinkōki* through the *Suan-fa t'ung tsung* and greatly conditioned the character of mathematics in Japan up to the Meiji period. Finally, as publication of mathematical works started two decades before printing of astrological and astronomical texts was begun, it is not surprising to find mathematicians first among those to make commentaries on calendrics.

WESTERN INFLUENCES

Though the means for introducing Western mathematics into Japan in Western Wave I were as good as could be devised under the circumstances, they were not good enough. The small tributary of Western mathematics never flowed into the mainstream of traditional Chinese-style mathematics. The historical record suggests little or nothing in the way of achievements. The possibility of getting further into the pure mathematics of the West, and thus, into related fields of Western science, went unfulfilled. Of the known disputations held between Jesuit and Japanese scholars, mathematical questions do not appear to have been touched upon.

From the time he went to Peking in 1601 until his death in 1610, Matteo Ricci worked diligently with fellow Jesuits and Chinese colleagues to complete a number of treatises on Western mathematics, and several works were completed by Ricci's associates after he died. These works covered the four basic arithmetical operations, arithmetical and geometric progressions, square and cube root extractions, and also the first six chapters of Euclid's *Elements* (completed in 1607). Written in Chinese, these works could have greatly enhanced the educational efforts of the Jesuits in Japan, but, like other books of the Ricci group, they were banned from Japan by the Tokugawa regime.

Much greater potential seems to have been inherent in the adoption of Western navigational techniques, discussed earlier in this chapter. For the performance of traditional Chinese-style mathematics represented in the Japanese publications noted above, the abacus was not only convenient but essential. There was no reason why the use of Arabic numerals in written calculations was needed in Japan at the time. This does not mean that they would not have been helpful. Western physical science could not be fully grasped without some proficiency in Western mathematics. Japanese mathematicians later in the seventeenth century found it necessary to invent their own symbolic representations for algebraic problems where the abacus did not suffice. But very little trace of Western mathematics is found after isolation began, and it had no essential influence on the development of *wasan*.

MEDICINE

Because the sciences other than medicine had rather scant and scattered histories until very late in the sixteenth century, toward the end of Early Chinese Wave II, the case for dating this wave as early as the fifteenth century—so far as the sciences are concerned—may not be clear. However, it was precisely in the fifteenth century that the field of traditional medicine began to develop most significantly—a development sustained for a long time thereafter. It was also in medicine that the Japanese response to Western Wave I was strongest and most enduring.

CHINESE-STYLE MEDICINE

In the Ming period (contemporary with Early Chinese Wave II), medicine, like other academic fields in China, experienced no remarkable developments. Ming medicine was dominated by the Liu and Chang schools formed in northern China during the time of Jürched rule (Chin dynasty) and the Li and Chu schools formed in southern China during Mongol rule (Yüan dynasty); for brevity the former two are often referred to together as "Chin medicine" and the latter two as "Yüan medicine." On a more popular level in both China and Japan, the less theoretical practice based on the manual of "Official Prescriptions" was widespread.

It was the two later Yüan schools that entered Japan in Early Chinese Wave II; the somewhat earlier Chin schools came to Japan in Late Chinese Wave II, after isolation. Because it was in Early Chinese Wave II that medicine—for the first time in the traditional sciences—became the means by which the overall Chinese conceptual background began to be fully understood and related to conditions in Japan, it may be helpful here to review briefly the "phase energetics" on which the Chin and Yüan schools were based.

Among the many medical classics compiled or revised and reprinted by imperial order in the Sung era, the most widely read and influential were the Yellow Emperor's Inner Classic (*Huang-ti nei-ching*), which was mostly medical theory (physiology and pathology) with little on therapy and no prescriptions, and the equally theoretical but more clinical work on therapy and prescriptions, entitled On Cold Damage Disorders (*Shang-han tsa-ping lun*). Most other medical texts were in general consistent in approach with these two books. The Official Prescriptions, because of its bureaucratic orientation, took a much simpler approach to prescriptions and contained virtually nothing on theory.

Although the Yellow Emperor's Inner Classic set out the theoretical foundations of classical medicine in terms of *yin-yang* and Five Phases concepts, many of the implications of the theoretical assumption of macrocosmic-microcosmic correspondence were not fully worked out.[65] It was Wang Ping who, in his 762 edition of the *Nei-ching*, first gave definite form to the theory of phase energetics to clarify and systematize the various macrocosmic, and thus the physiological and pathological, implications of this classic (see p. 86). One of the Sung forerunners of Chin and Yüan medicine, Liu Wen-shu, in 1099 wrote a commentary, *Su-wen ju-shih yün-ch'i lun-ao*, on the first part (*Su-wen*) of the *Nei-ching*; this commentary included seven sections that explained in further detail the theory of phase energetics. Only from the Sung era was the *Nei-ching* interpreted in a more or less homogenous way, and it was much later, in the Ming and Ch'ing eras, that reconciliation with the *Shang-han lun* (the original text of which did not use the Five Phases concept) was undertaken.[66] Any summary of phase energetics is necessarily inadequate, partly because the theory itself is extremely complex, but also because of the different uses made of the *yin-yang* and Five Phases cycles in describing both macrocosmic and microcosmic dimensions.

Already central to a developed philosophical tradition, the *yin-yang* concept was not explicitly represented by a cyclic alternation until the T'ang and Sung periods.[67] To the usual notion of increase and decrease of *yin* and *yang* was added an additional phase of "fullness" to make a six-sequence cycle of

[65] Porkert, p. 42.
[66] Ibid., p. 43; Ōtsuka Keisetsu, pp. 21–28, 49–52.
[67] Porkert, p. 43.

increase-fullness-decrease where *yin* and *yang* alternated. This six-sequence cycle became, in phase energetics theory, the basis of Six Energetic Configurations (*liu-ch'i*), which constitute six climatic or immunological situations of different quality. The Five Phases were made the basis of Five Circuit Phases (*wu-yün*), each also of different quality. The variable qualities of the Six Energetic Configurations are specified by reference to the six-sequence *yin-yang* cycle and the Five Circuit Phases. Among the latter the phase associated with Fire is split into two to permit a six-part correlation with the Six Energetic Configurations.

Since the point of phase energetics is to describe the influences of temporally variable meteorological, climatic, and immunological conditions on the physiological, pathological, and therapeutic functions, the circuit phases and energetic configurations were further correlated with the already developed calendrical theories of macrocosmic variations. What phase energetics does is to make specific links between meteorological and climatic conditions.[68]

The first of the Chin schools grew out of the search by Liu Wan-su (1110–1200)[69] for a theoretical foundation for the prescriptions of the On Cold Damage Disorders. On the premise that diseases come from an energetic imbalance, he held that *yang* activity characterized by the phase Fire is the

---

[68] Porkert's analysis of phase energetics is a remarkable intellectual feat. He acknowledges that phase energetics "is the weakest link within the theoretical system of China's scientific medicine" and that "from the outset speculative combinations have held sway over empirical data." Its two principal defects, he notes, are that (1) "it is coupled only to the rhythms of sun and moon and does not permit precise statements on the qualities of events whose temporal dimension exceed 60 years," and (2) "it completely disregards geographical coordinates" and can therefore say nothing about meteorological and climatic conditions at specific places. Porkert believes, however, that the necessary research tools—"adequate astronomical instruments, a far-flung network of precisely coordinated observations, uniform and exact measurements of time, and calibrated instruments"—are now at last available for use in exploring the worth of phase energetics as a theoretical model (though not necessarily all its details) for modern therapy. Porkert, pp. 59, 105–106.

[69] Like countless others in China and Japan, Liu Wan-su and the three men subsequently discussed had, besides their given names, names taken upon

most frequent cause of illness. As Fire is correlated with cardiac functions in the Five Phases theory, the usual approach was to increase the renal function, which is correlated to its conquering phase, Water, and thereby restore the balance of somatic activity. Water (the most intense phase of *yin* energy) is correlated to cooling drugs, and hence Liu's school came to be known as the "cooling medication school" (*Han-liang p'ai*; *kanryōha* in Japanese).

Chang Ts'ung-cheng (1156–1228), also of the Chin line, followed Liu's ideas for the most part but advocated the view that for new diseases a new therapy was needed. His emphasis was to eliminate the cause of exogenous diseases by the use of sudorific, emetic, and cathartic drugs to induce, respectively, sweating, vomiting, and excretion. This was not really a new approach but merely a move toward the more general use as early as possible of strong drugs—which, in the hands of unskilled practitioners, sometimes killed the patients. Chang's school was known as the "cathartic school" (*Kung-hsia p'ai*; *kōgeha* in Japanese).

Early in the Yüan period Li Ai (1180–1252) criticized the theories of Liu and Chang, claiming that external pathological agents can enter the body only when it is weak. What is needed, he reasoned, is not so much to attack invasions from the outside with strong drugs, as to strengthen the body's resistance to them. What Liu and Chang regarded as exogenous pathological forces is better understood in terms of the decline of health-maintaining *ch'i* within the body. All this was a reversion to a more classical approach. What was new was his broad application of preventive strengthening of the splenetic and stomach functions, which are correlated to Earth;

reaching adulthood, as well as literary names. Because of the importance of these four men, and because they are variously cited in the literature, we give all three names of each below:

Surname	Given name	Adult name	Literary name
Liu (1110–1200)	Wan-su	Shou-chen	Ho-chien
Chang (1156–1228)	Ts'ung-cheng	Tzu-ho	Tai-jen
Li (1180–1252)	Ai	Ming-chih	Tung-yüan
Chu (1281–1358)	Cheng-heng	Yüan-hsiu	Tan-chi

thus his school was also known as the "Earth-supplementing school" (*Pu-t'u p'ai*; *hodoha* in Japanese).

Following Li, Chu Chen-heng (1281–1358) insisted that *yang* energy when disordered tends toward overabundance in the body, but *yin* tends to be deficient. His therapy, then, sought to raise the level of *yin* energy, especially its role in the splenetic and stomach functions, by use of mild drugs. His approach became known as the "*yin*-nourishing school" (*Yang-yin p'ai*; *yōinha* in Japanese). Chu also wrote a book, *Chü-fang fa-hui* (Elucidation of the Official Prescriptions), attacking the mechanical use in his native south China of drying drugs as advocated in the manual of Official Prescriptions.

Japanese medicine during Early Chinese Wave II can best be described by dividing the period into two parts. In the century and a half from 1401 to 1550 there was an acceleration in the transfer of Chinese medical knowledge to Japan, mainly through Buddhist priests who went to Ming China exclusively for the study of medicine. While Chin and Yüan medicine did not yet form the mainstream of Japanese medicine, many medical books were written, and the specialization already seen late in the Semiseclusion Era now increased. In the century before isolation, from 1550 to 1639, Li and Chu medicine (Yüan medicine) became the mainstream, specialization became even more prominent, secularizing tendencies were quite strong, and many medical books were printed.

Though medicine in Japan by the end of the Semiseclusion Era was dominated by Buddhist priests, they were always priests first and medical practitioners second. In the first part of Early Chinese Wave II (1401–1550) there appeared not only priests primarily concerned with the practice of medicine but also a new class of secular medical men. Some members of both groups went to China for the sole purpose of studying medicine, accelerating the influx of contemporary medicine into Japan. A forerunner of this activity late in the Semiseclusion Era was Takeda Masayoshi (1338–?), son of a court nobleman, who stayed in Ming China from 1369 to 1378, studying medicine and Confucianism; he returned to Japan to practice medicine, taking on the Buddhist tonsure. Yoshida Sōkei (?–1572) went twice to China, in 1539 and 1547, and on the second visit treated the Ming Emperor Shih-tsung. The most important figure in the 1401–1550 period was Tashiro Sanki, who stayed in Ming China for eleven years to study Li and Chu medicine. Returning to Japan, he first practiced in Kamakura, but in 1509 moved to Koga (base of the Ashikaga shogunate's commissioner for the Kantō plain), at which time he also discarded his Buddhist robe, returned to secular life, and married. Highly esteemed in the Kantō area for his skill, Sanki himself exerted no national influence; his importance lay in his teaching Li and Chu medicine to Manase Dōsan, who established for it a central place in Japanese medicine in the 1550–1639 period.

In the 1401–1550 phase of Early Chinese Wave II medical practice was essentially a continuation of the general tendency of the late Semiseclusion Era, namely, heavy reliance on the Official Prescriptions manual, with a strong undercurrent of mixed Chinese and Buddhist ideas. Chin and Yüan medical theories of phase energetics were only fragmentarily introduced and only partially understood in the 1401–1550 phase; they still lay beyond the grasp of most Japanese. But medical literature increased remarkably (approximately fifty works are known), even if most of the books were neither published nor printed. The major exception was the 1528 reprinting of the Ming period collection *I-shu ta-ch'üan* (Encyclopedia of medicine)—the first medical book printed in Japan[70]—sponsored by Asai Sōzui (?–1532), a gynecologist who practiced in the city of Sakai. Moreover, specialization spread steadily; already visible in ophthalmology and dental and oral hygiene, it now expanded to include, for example, a new group of specialists to treat combat wounds of men embroiled in civil wars. This was a departure from peacetime surgery's focus on minor lacerations and boils. Combat surgeons (*kinsōi*) had only very

---

[70] The full title of *I-shu ta-chüan* is *Hsin-kan ming-fang lei-cheng i-shu ta-ch'üan*. A Confucian scholar as well, Asai Sōzui was responsible for the reprinting of the Analects (completed the year after he died). *Nihon rekishi daijiten*, 1: 73.

primitive instruments at this stage and generally kept their skills secret in various closed schools distinguished only by minor differences in techniques. They also practiced midwifery on the not so curious logic that wounds and childbirth have in common bleeding not caused by illness.[71] These were not, of course, specializations in the modern sense; they were a step short of, not a step beyond, broad training for general practice.

In the last century of Early Chinese Wave II (1550–1639) the Li and Chu schools of medicine became fully established as a medical tradition in Japan—the first time for a medical system as a whole to be assimilated by the Japanese. As such it not only preceded the full assimilation of Chinese astrology, astronomy, and mathematics, but of Neo-Confucianism as well. This epochal event took place while all sectors of medicine were changing—through secularization and specialization, printing of medical books, and continued introduction of contemporary Chinese knowledge, especially of Yüan medicine. This was also the century of exposure to European medicine, which we evaluate later on; but the introduction of Western medicine amounted, at best, to a mere page in Japanese medical history, which was dominated in this age—and even more substantially in the next, when Western medical influence dwindled—by contemporary Chinese medicine.

Manase Dōsan (1507–1594) was born in Kyoto in 1507 and in his youth entered the Buddhist priesthood. He later attended the Ashikaga School in the northern Kantō plain to study Confucian thought and there learned of Tashiro Sanki's medical practice in nearby Koga. He then spent fourteen years (1531–1545) studying Li and Chu medicine, seven of them under Sanki (d. 1537), and returned to Kyoto in 1545 to practice and teach. Dōsan dissociated himself from Buddhist orders and became a secular practitioner, as had his teacher. (In 1584 Dōsan received Christian baptism, though this had no known effect on his medical thought and practice.) Dōsan's practice won for him a good reputation, and among his

[71] Ishihara (2), p. 68.

patients were members of the imperial family, some of the court nobility, and high-ranking samurai in the Ashikaga shogunate, as well as Hideyoshi and Ieyasu.

Had Manase Dōsan confined himself solely to his successful practice, he would have been well known in his time but little noted by historians. Turning his hand also to medical education, he became a pivotal figure in the history of Japanese medicine. In his private academy, called Keitekiin, he devoted himself to making the whole system of Li and Chu medicine intelligible to several hundred students through clear and concise explanations. He also incorporated his teaching into a comprehensive textbook of Li-Chu medicine, entitled *Keitekishū* (Collected orientations; 1574), which was copied and used by his disciples (this text was later printed in 1648).[72] Dōsan had many able followers who excelled in theory, practice, and education, a fact that contributed much to the establishment of Li and Chu medicine as the mainstream of Japanese medicine. His adopted son, Manase Gensaku (1548–1631), became the attendant physician to Tokugawa Ieyasu. In the development from Tashiro Sanki to Manase Dōsan, and in the subsequent domination of Japanese medicine by the Li and Chu traditions, there was no mixture of Buddhist medical ideas; Buddhist influence on medical theory and practice was

---

[72] In addition to *Keitekishū*, Dōsan wrote many other books; at least eighteen (including *Keitekishū*) are extant. One of these gives "Fifty-seven articles for the judicious practitioner," such as: medicine is "the manifestation of compassion" (art. 1); "diagnosis begins with determination of the nature of disease by examining the pulse" (art. 2); "treatment for the same disease differs according to youth, adults, and the aged" (art. 15), as well as locality and climate (art. 11), not to mention social and economic setting (art. 51); "drugs, like military weapons, exist only to attack evil; even though a drug may be mild and nonpoisonous, it must not be used unless there is some disease to subdue, and all the more so if the drug is strong and poisonous" (art. 23); and "the cosmic determinants based on the Five Circuit Phases and Six Energetic Configurations should be generally memorized" (art. 53). Dōsan's spirit shines through in another, "when you err in diagnosis or therapy, do not hesitate to correct your mistake" (art. 32), and one on preventive concerns, "the essence of therapy is to cure diseases before they occur, not after the patient has already fallen ill" (art. 12). Fujikawa Yū, pp. 196–203.

greatly reduced in the latter part of Early Chinese Wave II. The trend toward secularization of Japanese medicine was in fact quite remarkable. Although most medical practitioners continued to don Buddhist clerical garb and shave their heads, they openly concentrated on their medical work and were, in actuality, secular persons. Another seemingly contradictory custom paralleled the Buddhist costumes of medical men. Sometime late in the fourteenth century the priestly ranks came to be granted to priest-practitioners for their excellence in medicine. Either the court or the shogunate could recommend medical practitioners to these ranks, awarded by the court, provided they adhered to the formality of wearing clerical robes. Designation of the priestly ranks to medical men continued to the middle of the nineteenth century, despite the fact that, by that time, many of Japan's leading figures in medicine had strongly advocated throwing off the Buddhist mantle.

Meantime, specialization became more the rule than the exception in advanced medicine. In addition to specializations already noted—surgery, obstetrics, pediatrics, gynecology, ophthalmology, and dental and oral hygiene—specialists also appeared in two uniquely Chinese fields, acupuncture and moxibustion, and several schools arose in these fields. Here too foreign contact was responsible. Irie Yoriaki, after studying acupuncture in Japan, was instructed by a Chinese specialist captured in the Korean wars. Yoshida Ikyū went to Ming China around 1560, staying for seven years to study acupuncture.[73]

WESTERN MEDICAL INFLUX

The chief source for knowledge of Western medicine in this period was the Catholics' charitable work, and the most prominent medical practitioner among the Europeans was a Jesuit named Luis de Almeida (1525–1584). Coming to East Asia as a wealthy trader, Almeida was so impressed by the work of the Jesuits that he joined them in 1556. For most of

---

[73] Ibid., p. 246.

his career he was a lay brother, though he was ordained to the priesthood late in his period of service in Japan. Deeply moved by reports of children abandoned by desperate mothers in wartime circumstances, he helped establish a foundling home in 1556 at the Jesuit base in Funai, Bungo. Soon after, in 1557, a hospital was built at the same site, to which was added also a leprosarium. A second and larger hospital was built in 1559.[74] One of the more interesting sidelights of the record is that surgical operations were performed on the hospital's veranda—to have plenty of daylight, though also, it is said, to dispel a rumor that "the barbarian padres eat human flesh."

A Portuguese licensed to practice medicine before coming to Japan,[75] Almeida developed a Western-style hospital, not just a small or temporary clinic, though Japanese physicians practiced most of the internal medicine, and of course used traditional Chinese-style knowledge and drugs. The Jesuits did not advocate the use of European internal medicine in Japan. Some time was given to educating both Japanese and missionary personnel in Western medical practice, including surgery. Almeida was in general command of the hospital but seems to have devoted most of his time to surgery. It was, of course, no loss whatever that internal medicine was in Japanese hands and performed according to traditional Chinese practice, for the therapy of European medicine stemmed from the Galenic humoral pathology, which was hardly superior.[76]

Just as this base for introducing Western medicine to Japan in a more or less systematic way was being developed, it suffered a severe blow—this time not from the Japanese side, but from Jesuit headquarters in Rome. The missionaries were forbidden to practice medicine themselves, but were to give full time to the direct apostolate. The headquarters' decision was issued in 1558, though it was 1560 before word of it reached Japan. Almeida in 1561 left the hospital in the hands of the

[74] Ishihara (2), p. 89.
[75] Ibid., p. 89.
[76] Ibid., p. 90.

Japanese Christian staff, and it is said to have flourished in the 1570s.[77] The hospital was burned down in 1586, along with the church, when the Shimazu army invaded Funai, and was never restored.

The Funai hospital was only one of the several hospitals, clinics, and leprosaria built by Christian groups in Japan, though the one most widely referred to. Late in the sixteenth century other Catholic orders—Franciscans, Dominicans, and Augustinians—entered Japan for mission work. The Franciscans especially were noted for including leprosaria in the work of their mission stations in many parts of the country. To a church built in Kyoto in 1594 they added a hospital in the following year, and later a leprosarium. Other leprosaria were built by the Franciscans in Sakai in 1594 and in Nagasaki in 1595. All were destroyed in Hideyoshi's persecutions of 1596.[78] As leprosy has not been conquered even in our time, it was certainly no small service to give comfort, physical and spiritual, to the distressed, and usually ostracized, patients (though at the time many who were considered lepers suffered from only lesser skin diseases).

The European traders—Portuguese, Spanish, Dutch, and English—who came to Japan generally made provision for their own medical care, often with licensed practitioners like Almeida on board their ships. Inside Japan they maintained no medical facilities which, like Almeida's hospital, might also have had Japanese doctors. The Dutch were the only Europeans who outlasted the isolation decrees, and after isolation a physician or surgeon was stationed at their settlement in Nagasaki. In time Japanese were able to establish contacts with the resident Dutch medical officer, and this intercourse later contributed much to the development of "Dutch medicine" (*Ranpō*) in Western Wave II (see Chapter 5).

On strand of the story did begin to unwind, however, before the end of Western Wave I, namely, the practice by Japanese

---
[77] Ibid., p. 92.
[78] Ibid., pp. 93–94.

of "Southern barbarian surgery" (*Nanban geka*) learned from the Catholic foreigners before isolation, and sought afterward at the Dutch trading post in Nagasaki (but renamed *Kōmō geka*, i.e., "Red-hairs' surgery," to avoid connection with Catholicism). At least two Japanese are known to have gone overseas to study Western surgical techniques. Handa Jun'an studied surgery in Macao sometime in the Keichō and Genna eras (1596–1624, inclusive) and after returning to Japan enjoyed a good reputation as a surgeon.[79] Kurisaki Dōki (1566–1651) received training abroad, possibly in Macao, and returned to practice surgery in Nagasaki.[80] Practitioners of *Nanban geka* were favorably accepted by many Japanese clients, and the surgeons themselves felt that Western techniques were superior to traditional Chinese-style surgery—enough to continue practicing it after isolation under a different name.

Evaluation of Western medicine in Japan in the late sixteenth and early seventeenth centuries as against contemporary Chinese-style medicine in Japan is far from easy. Truly modern European medicine emerged much later, in the nineteenth century, with the discovery of bacteriology, vaccination, and serum therapy by men such as the Frenchman Louis Pasteur (1822–1895), the German Robert Koch (1843–1910), and the Englishman Edward Jenner (1749–1823). Prior to that neither European nor Chinese practice could claim any essential superiority in the cure of diseases. At the time of Western Wave I both Eastern and Western drugs could be lethal or curative according to the dose. But in surgery, a field discouraged in Chinese practice in favor of stimulating the body's own healing powers, the Western techniques introduced into Japan attracted considerable attention. This is all the more striking since European surgery itself was still far short of the early modern standards developed after the French surgeon Ambroise Paré (1517?–1590), working mainly in combat

[79] Fujikawa Yū, p. 265.
[80] Koga (1), pp. 34–42.

medicine, made some of the first improvements in the treatment of gunshot and cannonball wounds, in amputations and operations on broken bones and dislocations, and in the invention of various surgical instruments. And, of course, all surgery remained a most hazardous undertaking before Joseph Lister (1827–1912) and others made advances in sterilization, and before developments in anesthesia reduced the problem of shock, which, in most early surgery, was as much the cause of death as infection.

The evaluation of Western medicine vis-à-vis Chinese medicine in this period should not, perhaps, be made solely on the basis of what was actually introduced into Japan by missionaries and traders; account should also be taken of the potential significance of continued contact had isolation not been imposed. The 1543 date of the arrival of Westerners in Japan coincided not only with the publication of Copernicus' great work in astronomy but also with that of the epoch-making masterpiece of Andreas Vesalius (1514–1564) in anatomy, *De fabrica corporis humani*. This work was first received in Europe with widespread contempt. So it was also with William Harvey's (1578–1657) revolutionary work on blood circulation and the function of the heart, his *De motu cordis*, published in 1628. Although these innovations appeared in Europe before isolation shut Japan off from most contact with the West, too much stood in the way of their being communicated readily to the distant East. But these two achievements of the sixteenth and seventeenth centuries marked a new tendency in European medicine, which now began to dissociate itself from the Galenic doctrines so that detailed observations and careful experimentation became matters of course in medical research. Knowledge of various diseases rapidly accumulated and was applied to the practice of internal medicine and surgery. In Europe a clear line of demarcation divides medical knowledge and practice before and after these epochal events. The actual content of European medicine introduced in Western Wave I was based on knowledge antedating the sixteenth- to seventeenth-century breakthroughs; and even the surgery predated Paré's improvements, though it surpassed the traditional techniques used in Japan.

The time was ripe for continued encounter. Japanese interest and activity in medicine, already greater than in other scientific fields, would be sustained at high levels for generations, and even without Western stimulus would make several forward strides, notably in anesthetics. Continued interchange with a steadily improving European medical science could have only quickened the pace of Japanese medical advance, which eventually came about as scholars pushed traditional Chinese-style medicine to its limits and, from late in the eighteenth century, once again made Western medicine the focus of their studies.

4

The Seventeenth-Century Intellectual Outburst

National Isolation and the Peak of Chinese Cultural Wave II: 1639–1720

Seventeenth-century Japan experienced a renaissance of learning and science without parallel in any previous period, in which the whole breadth of Chinese learning and science came to be understood critically, comprehensively, and creatively. A new class of Confucian scholars was born that advocated a variety of Confucian schools, and new methods of textual criticism were developed. From the 1620s mathematical texts were printed and marketed; extensive calendrical studies mushroomed from the 1640s, leading to the first Japanese revision of the traditional system in the 1680s; different schools of Chinese medicine were introduced and explored from mid-century on; and original treatises in such fields as agriculture and materia medica appeared around the end of this landmark century that also boasted great literature by such men as Matsuo Bashō (1644–1694) the poet, Ihara Saikaku (1642–1693) the novelist, and Chikamatsu Monzaemon (1653–1724) the playwright.[1]

The isolationist measures of the 1630s, far from stopping or even hindering Chinese Wave II, actually favored its escalation until, for the first time in Japanese history, Chinese science and Confucianism were really understood as comprehensive systems. The national academic infrastructure took on new shape as the shogunate appointed its own Confucian teachers, pro-

---

[1] One of the earlier English-language studies to recognize the importance of the seventeenth century in Japanese cultural history as a time of intellectual renewal is *Renaissance in Japan: A Cultural Survey of the Seventeenth Century*, written in 1938 by Canadian diplomat and scholar Kenneth P. Kirkwood (available in a 1970 paperback reprint published by Charles E. Tuttle Co., Tokyo). At the time a "cultural survey" could still ignore the larger scope of culture, including learning and science. Kirkwood's book is centered on the three literary figures named in the text.

moted Confucian training of samurai, and sponsored libraries and printing. Many feudal domains followed suit, assigning their own Confucian teachers, founding fief schools (*hankō*) and supporting collection and publication of books. Far greater in number, however, were the hundreds of private academies that taught subjects ranging from Confucian classics, mathematics, and medicine to flower-arranging and the tea ceremony. Scholars, books, and academies were mobilized in this peak period of intellectual creativity to produce a firmly rooted research tradition and a love for learning that distinguished the Tokugawa legacy of learning.

Isolation did, however, successfully curtail Western influence. Tertiary effects of Western Wave I lingered beyond isolation, notably in cosmology and surgery, but these had little to do with the major intellectual developments of the seventeenth century. Early in the eighteenth century, as the intellectual outburst began to lose its momentum, the eighth Tokugawa shogun, Yoshimune (ruled 1716–1745), relaxed the ban on foreign books as part of his program for calendrical reform, creating a new but small opening to Western knowledge. As traditional learning and science increasingly confronted problems they could not solve, Japanese intellectuals began in earnest once again to turn to Western sources, a movement we call Western Cultural Wave II (see Chapter 5).

Historically, the development of science and learning in Japan depended, as we try to show, upon Chinese and Western influences. Yet the high-water mark of intellectual creativity in almost every scholarly field occurred precisely in the seventeenth century when isolation kept the Western input at a minimum. The direct input from China came mostly through books and a few refugee scholars; no Japanese went abroad to study.

INTERNATIONAL TRADE RELATIONS

The gateway for the now diminished foreign cultural contact was the city of Nagasaki where, in accordance with isolationist rulings, only Dutch and Chinese were allowed to trade. Strict controls were exercised over the number of ships permitted port entry, the kinds of goods exchanged, the methods of transaction, and so on. The shogunate also assumed a monopoly over trade receipts to prevent any counterforce arising in western Japan from overseas trade profits. Overall trade volume rose soon after isolation began but in time fell off considerably and lost its importance in the national economy.

Initially Chinese traders in Nagasaki were permitted to live among the Japanese. In 1689 they were confined to a single settlement (of more than a thousand),[2] but continued to enjoy more or less normal intercourse with Japanese citizens. Though no formal diplomatic relations were concluded with China, trade with the Chinese merchants was rather brisk throughout the Edo era. Exports to China were mainly silver and copper;[3] raw silk and woven silk goods were the chief imports,[4] though

Table 18. Major Isolationist Measures of the Tokugawa Regime

1616	Entry by foreign vessels was restricted to the two ports of Nagasaki and Hirado.
1633 1635 1636	Successive prohibitions were placed on overseas travel and trade by any Japanese, and those already abroad were barred from returning to their homeland, upon penalty of death.
1634	The shogunate ordered the Nagasaki merchants to build a tiny artificial island in Nagasaki harbor, and the Portuguese traders were ordered to move from Hirado to this island, called Dejima.
1635	Construction of vessels capable of navigating the high seas was prohibited, and all such vessels were destroyed.
1639	All Portuguese ships were banned from all Japanese ports; illegal attempts at entry were repelled by armed attack, and violators were put to death. (Spanish ships had already been banned in 1624; and the British, unable to compete successfully with the Dutch and Portuguese, had given up the Japan trade in 1623.)
1641	The Dutch trading company in Hirado was moved to Nagasaki and confined to Dejima.

[2] Iwao (4), pp. 413–421.
[3] Copper was needed by the Ch'ing dynasty for minting coins—a demand unmet by the relatively low level of copper production elsewhere in Asia. Ibid., p. 410.
[4] Japan imported great amounts of raw silk up to the middle of the Edo era; but when she opened her ports to world trade in the middle of the nineteenth century, silk was one of her largest export items. Ibid., p. 412.

Japanese imports of Chinese drugs and minerals remained considerable. Because of the shogunate's support of Confucianism (discussed in the next section), and also because accomplishment in the Chinese classics became an important index of cultural attainment, Chinese books—excluding any related to Christianity—were also imported in great quantities. This constituted the primary channel through which Chinese Wave II continued to flow and, of course, was a major stimulant to the seventeenth-century intellectual outburst.[5]

The Dutch traders in Nagasaki provided the single, slim channel to markets beyond China, through which Japan exported mainly gold, silver, and copper, but also camphor, sulfur, porcelain, and sundries; imports were raw silk and woven silk goods (produced in China), cotton goods, sugar, deerskins, and indigo dye from Southeast Asia, and woolen goods and miscellaneous items from Europe.[6] Unlike the Chinese, the Dutch traders could not live in Nagasaki proper but were confined to the artificial island of Dejima just offshore in Nagasaki harbor. Dejima measured about thirteen thousand square meters (almost half the thirty-one thousand square meters of the more numerous Chinese settlement in the city), and several dozen Dutch personnel were always in residence, though they were denied normal social relations with the Japanese. Exchange transactions were conducted through offi-

---

[5] The Ryukyu Islands served as a kind of provincial parallel to the place of Nagasaki as a trade outlet for the nation. The Shimazu clan of Satsuma domain had special trade relations with the Ryukyu chiefs prior to the Edo era and brought the latter into vassalage in 1609 without disturbing the Ryukyu's formal tributary relations with China. The northern islands of the Ryukyu chain were then used as an outlet for trade with China and Southeast Asia, though on a small scale. Tsushima domain on the island of that name northwest of Kyushu also served as a base for small-scale trade with Korea. Each time the Japanese shogun changed, the Yi dynasty in Korea dispatched a representative with greetings to the new shogun. Cf. Naramoto, p. 73. These minor exceptions to the dominant Chinese trade through Nagasaki in no way altered the character of Late Chinese Wave II; rather, they served as secondary channels for the continued Japanese access to Chinese science and learning.

[6] Iwao (4), pp. 390–391.

Fig. 25. Dejima island (center, right) with residences and warehouses, and three Dutch ships in the foreground. The Chinese trading settlement is in the background. From a screen painting of Nagasaki harbor (Nagasaki City Museum)

cial Japanese interpreters assigned by the shogunate to the Dutch company, and intercourse was limited purely to commerce. Because of the severity of the repression of foreign Catholics early in the Edo era, the Dutch traders studiously avoided any religious or intellectual activities.[7] The Japanese interpreters with equal caution avoided any show of interest in Dutch affairs beyond official duties, fearful of being suspected of clandestine Catholic connections. This became the generally

[7] Ibid., p. 375.

established pattern of Dutch-Japanese relations in Nagasaki in the seventeenth century. Significant exceptions to this pattern came later in the Tokugawa period with the rise of "Dutch learning," one of the themes of the next chapter. The best-educated men on Dejima were the regularly stationed medical officers, mostly surgeons. Of these, Willem ten Rhijne (1647–1700), Engelbert Kaempfer (1651–1716), Carl Pieter Thunberg (1743–1828), and Philip Franz von Siebold (1796–1866), who came to Dejima in 1674, 1690, 1775, and 1823 respectively, possessed more than ordinary academic qualifications.[8] Siebold was to be particularly important in the development of "Dutch learning" as a specific Japanese activity.

SOCIOPOLITICAL CONTEXT OF THE INTELLECTUAL OUTBURST

Coming almost exactly one millennium after the Taika transformation of Chinese Wave I, isolation wrought vast changes in the sociopolitical context of Japan. The basic conditions of Early (preisolation) Chinese Wave II were now reversed: peace, stability, and national seclusion replaced civil strife, shifting allegiances, and unsettling Western contacts. The new peaceful posture and domestic stability of Late Chinese Wave II were maintained by the Tokugawa hegemony over nearly three hundred daimyo who exercised semiautonomous rule in their respective domains. But there was no longer much need for the open use of military force, and the large samurai class—a fairly constant 5 to 6 percent of the total population throughout its rise from about eighteen million to some thirty million in the first half of the Edo era[9]—had to be remolded into civil administrators. Hence, the samurai were enjoined to master both literary and martial arts.

For only a limited number of small specialized platoons did martial arts include small arms and cannon, now that reunification of the nation and expulsion of unwanted foreigners were secured. The vast majority of the samurai mastered only medi-

[8] Bowers, pp. 32, 213.
[9] Yazaki, pp. 133, 143; Reischauer and Fairbank, pp. 606, 629.

eval soldiering skills: swordsmanship, lancing, archery, and horsemanship—martial forms retained as essential to the samurai spirit.

Their literary training was focused chiefly on the Chinese classics, and especially on Confucianism, which the shogunate chose to promote as an ideological buttress to the military's dominant role in the social hierarchy. Although the Confucian canons are anything but militarist, as interpreted in the Ming they supported an authoritarian, stability-oriented hierarchic order. Government promotion of learning, although it did not seek to reproduce Chinese academic institutions as in Chinese Wave I, was recurring for the first time since that period. Moreover, it created an atmosphere in which scientific studies —again, for the first time in a millennium—were recognized as legitimate concerns of specialists and nonspecialists alike; and some scientific functions, as we shall see, were even incorporated into government agencies. Meantime, the warrior-gentry turned martial-administrators were the chief patrons of the new class of professional Confucian scholars whose ranks swelled rapidly under Tokugawa encouragement. Learned men of the military class constituted the mainstream of intellectuals in the Edo period, and thus the image of the cultured person was of genuine competence in the Chinese, especially Confucian, classics and in related art and literary forms. The overall cultural ethos diffused under samurai leadership was a fusion of Confucian and martial ideals, and from it there permeated deeply into Japanese society a spirit of high respect for learning and a willingness to cultivate a samurai-like discipline to obtain it.

Although the "primacy of agriculture" was continually extolled by the military class and farmers were second only to the samurai in the status hierarchy, the rural sector was consistently sacrificed to the urban sector. The gains of agriculture[10] went

---

[10] Cultivated land area almost doubled from 1,600,000 *chō* in the Keichō era (1596–1615) to 3,000,000 *chō* in the Kyōhō era (1716–1736). Kodama (1), p. 28.

into financing the growth of large feudal cities[11] and improving land and coastal transportation systems that connected them. Measures were imposed to keep city merchants out of the rural sector,[12] but among the merchants and artisans who served the extravagant demands of samurai concentrated in the larger castle towns and cities, some became rather well-to-do. Wealthier merchants had, then, the leisure and money for cultural and intellectual pursuits. It was the larger cities, and especially Kyoto, Osaka, and Edo, that provided the main context of the seventeenth-century intellectual renaissance, especially as the locus for various private academies and for the printing and marketing of secular books. Most merchants of means patronized popular arts and styles. Not a few used their relative freedom to introduce up-to-date techniques, and the interests of some extended even into the precincts of science proper.

The court nobility had lost virtually all political and economic power by the Edo period, and none of them exercised any leadership in the intellectual outburst. They contented

---

[11] No more than one in seven Japanese lived in a city or town in the seventeenth century, but the actual concentrations were quite large. Townsmen in Edo numbered 501,394 in the year 1721, and as samurai are believed to have comprised sixty percent of Edo's total population, the total is judged to have reached or exceeded one million by the end of the seventeenth century (and did not vary significantly up to the end of the Edo era). Kyoto is thought to have held a rather constant population level of between 400,000 and 500,000 throughout the Edo period. Osaka jumped from 279,610 in 1625 to 389,866 by 1736. Kanazawa and Nagoya were both over 100,000; Kagoshima and Hiroshima, over 70,000; and Sendai had more than 60,000. At least fifteen castle towns had between 20,000 and 40,000 population. As late as 1696 Nagasaki had a population of 64,523, but as its trade was already declining, its population also dwindled rapidly to 41,500 by 1715. Yazaki, pp. 133, 134, 137, 256.

[12] Merchants were restricted from setting up independent residences or markets in farming villages; but to free merchant guilds (*za*) from traditional ties to temples, shrines, and the court aristocracy, and thus bind them to samurai-controlled social structures, individual households were often given monopolistic privileges in castle towns, especially in the initial stages. From the mid-seventeenth century, protective associations (*kabu nakama*) among established merchants were encouraged to prevent encroachments by newcomers that might disturb domain or national interests. The highest (samurai) and lowest (merchant) status groups thereby became increasingly interdependent. See Ibid., pp. 128, 144, 153.

themselves with thoroughly traditional and antiquarian learning and arts: ancient Chinese and Japanese classics and poetry, court ceremonials, instrumental music, flower-arranging, the tea ceremony, and so on.[13] Buddhism had a minor role in stamping out Catholic influences, in that all citizens other than samurai were required to register at some local temple.[14] Propagation of new doctrines and sects was proscribed by the shogunate, and there was, in any case, little stimulus from China, where Buddhism on the whole was dormant.[15] Buddhism also failed to play any significant role in the intellectual outburst or in subsequent movements, except to react occasionally against novel ideas from the West (see p. 354).

TECHNIQUES

The intellectual outburst was not accompanied by a technical outburst. The seventeenth century yields no examples of technical processes, or the uses of mathematics in them, that affected the course of science or altered the intellectual atmosphere for or against science; nor were there any instances where technical needs prompted the formation of new training institutions. Chinese technical development had long since leveled off, though in Japan it was still necessary to refer to Chinese books in certain fields where the Japanese had yet to catch up.[16] But there was no longer any interest in what were described in the last chapter as "techniques strategic to modernizing processes."

---

[13] On the court nobility's life and customs, see "Kinchū narabi ni Kuge Shohatto" (Ordinances for the Imperial Household and Court Nobles) in Tsuji Tatsuya, pp. 365–371.
[14] Ibid., p. 408.
[15] The Ōbaku (Zen) sect, brought to Japan in 1659 by the Ming priest Yin-yüan (Ingen in Japanese), is a single exception. It exerted no profound influence on Japanese Buddhism as a whole, nor on intellectual currents of the time.
[16] An example is Sung Ying-hsing's *T'ien-kung k'ai-wu* (Exploitation of the works of nature; printed at the end of the Ming period in 1637), which was imported to Japan and widely used. It was reprinted in Japan in 1771. This work is now available in English translation; see Sun and Sun.

Gunnery and shipbuilding suffered most from isolation. Cannons were dismantled under order of the shogunate. Arquebus production was scaled down to meet only the limited needs of small, specialized platoons; the arquebus no longer appeared among Japan's export items. Combat skills yielded first place to administrative abilities, and hence competence in gunnery rapidly declined. Ironically, Japan first acquired mortars in 1639—the very year isolation became final—when three were manufactured at Hirado under the supervision of Hans Wolfgang Brawn (1609–1660), a German gunner on the Dutch ship *Buredan*. Although one of the three mortars exploded in a test firing at Azabu in Edo the same year, the others were kept in the shogunate's armory; and seven more were produced in Hirado in 1640. Interest in mortars continued for a while, and in response to a shogunate request, the head of the Dutch factory, François Caron (1600–1675), arranged for a group of four mortar technicians to spend eight months in Nagasaki in 1649. In addition to presenting two forty-pound cannons to the Japanese government, one of the technicians, Juliaen Schaedel, imparted knowledge of siege warfare to a shogunate military instructor, Hōjō Ujinaga (1609–1670), who recorded the information in his work, *Yurian kōjōden* (Report on Juliaen's siege warfare). Further knowledge of Western gunnery was preserved in *Kōmō kajutsuroku* (Notes on Dutch gunnery), complied by Furukawa Jiroemon from what he learned from a gunner on a Dutch ship. It listed contemporary European weapons and techniques such as rangefinding, sighting calculations, the crane, breechloading cannons, mortars, shells, explosives, and fireboats.[17] A few studies were made of the problems of the trajectory of a projectile, which could have revived interests related to dynamics. But this brief "gunnery boom" was soon silenced by the rapidly growing force of the isolation edicts, and possibilities in this area were stillborn.

As early as 1609 the shogunate confiscated ships of over 500 *koku* capacity (about 50 gross tons) belonging to lords in western

[17] Arima (3).

Fig. 26. Japanese coastal vessels of the late nineteenth century, which retain the main characteristics of Tokugawa era ships.

Japan. In 1635 construction of all ships over 500 *koku* was banned, though in 1638 this ruling was clarified as applying only to military vessels. Ships of purely Western or Chinese design, or even of blends of these with Japanese design, were all banned. Only purely Japanese-style ships were permitted, that is, keelless, flat-bottomed, single-masted ships with simple traditional rigging. Once voyages on the high seas were banned, use of Western navigational methods acquired during Western Wave I rapidly disappeared from practice. Associations between Japanese seamen and their Dutch or Chinese counterparts were strictly forbidden. Of course, some smuggling and other contacts inevitably occurred, but not on a scale capable of challenging the isolation decrees.

Gold, silver, and copper were in ever-increasing demand, for both domestic use and the limited overseas trade. Production of these metals rose steadily in the seventeenth century, but by its end serious technical limitations were faced, either in processing the increasingly low-grade ores available when only surface veins were mined, or in disposing of water seepage encountered as miners sought to go deeper with horizontal, inclined, or vertical mine shafts. Drainage problems were handled chiefly by labor-intensive means using various combinations of auxiliary shafts, pulleys and troughs, and scooping devices to raise water at necessary points in the drainage systems.[18]

Learning in the Seventeenth Century

The course of learning in Chinese Wave II was completely different from that in Chinese Wave I. The earlier wave proceeded from the top down, at the initiative and under the control of the government, which founded academic institutions, decided the content of courses, and cultivated a corps of scholars to teach them. Chinese Wave II arose, in the fifteenth and sixteenth centuries, as a ground swell of individually motivated scholars working on a variety of problems in various fields for reasons of their own. This latter wave reached its peak of creativity in the seventeenth-century intellectual outburst, when the academic disciplines gained enough recognition so that scholars could influence or even initiate public policy yet retained a certain autonomy to pursue interests unrelated to public affairs. Both the shogunate and domain governments were later to realize the importance of providing an institutional basis for such an enterprise.[19]

The Tokugawa household was not itself heir to a strong intellectual tradition. However, scores of learned men, nurtured in the years of Chinese Wave II prior to its establishment, were available to serve the Tokugawa shogunate as advisors and teachers. Foremost among them was Hayashi Razan (1583–

---

[18] On Sado island a Dutch pump with an Archimedean screw was imported in the next century (1782), but it did not then impress the Japanese as more effective in practice than their own man-powered pulleys, scoops, and drawtubes, partly because the pump frequently broke down. Asahi, pp. 401–404.
[19] Around 1600, Tokugawa Ieyasu had plans to establish near Kyoto a school similar to Ashikaga Gakkō (p. 118), but the plans were never realized. Wajima (1), p. 20.

1657) who, as an adviser to Ieyasu, was most influential in elevating Confucianism from a limited role in eradicating Catholic influences to a major function of ethical indoctrination in values which supported the sociopolitical system. More precisely, the philosophy of Chu Hsi (*Shushigaku* in Japanese), in its form as the Ming state orthodoxy, was made an official ideology. Due credit must perhaps be given Hayashi Razan's powers of persuasion. But early in the Edo era there were in Japan no strong advocates of other Confucian schools. *Shushigaku* had been the dominant school of thought in China since the end of the Yüan era, and the focus of Buddhist priests in Japan who were interested in Confucianism throughout Early Chinese Wave II. It was the starting point for Tokugawa Confucianists.

While promotion of this school rapidly expanded the numbers of learned men among the military, their interests could not be indefinitely confined to its norms. Some reacted to the rationalism of *Shushigaku* by embracing the more idealistic philosophy of Wang Yang-ming (*Yōmeigaku* in Japanese). What was perceived as the excessive innovation of both these schools drove others back to the more philologically oriented "Ancient Learning" (*Kogaku*) of the Confucian classics—as had happened already in China. The fifth Tokugawa shogun, Tsunayoshi (ruled 1680–1709), an ardent supporter of *Shushigaku*, met the challenge by giving fuller support to the Hayashi household and its academy for teaching *Shushigaku*. Other schools of thought—and, indeed, most of the important developments in various fields of learning and science in the seventeenth century—relied on schooling done in private academies.

Not a few scholars active in the intellectual outburst were samurai who remained in the service of the shogunate or some domain while pursuing academic interests. But there were other sources of budding intellectuals. Economic growth and the resultant prosperity of merchant households made it possible for many a son of merchant parents to seek a career in learning. The bright young townsman hoping for a career other than commerce often found his counterpart in the *rōnin* (lordless samurai) motivated by his samurai pedigree to participate in the shaping of his society yet frustrated by lack of an official relationship. In past generations many persons in like circumstances had sought learning as novitiates in Buddhist temples and then, as priests, had exercised considerable influence in cultural, intellectual, and even political life. But the private Confucian academies were now the favored avenue to a career other than an inherited vocation. It is not surprising, therefore, to find many of townsmen origin alongside those of samurai background among the major figures of the seventeenth-century intellectual outburst.

The only inviolable political condition was that a learned career contribute, if not to the enrichment, at least to the stabilization of the feudal order. Only a very careless scholar would contravene this norm—or else one deeply convinced that his apparent betrayal of it was actually in the best interests of his country and its people. Short of this extreme, a considerable latitude of intellectual judgment was quite permissible, though career possibilities were more limited. One could teach in a private academy, with or without shogunate or domain support; there were many private schools for the various arts and sciences that attracted enough paying students to support a teacher. And it was not extraordinary for a teacher to cultivate an avocation in some art or science which was his real interest, other than the one for which he was paid.

As already noted, Chinese books (excepting any related to Christianity) were imported in large quantities. Warmly welcomed and highly respected, they were read freely by learned Japanese according to Japanese syntax and pronunciation (few knew the Chinese pronunciations). Contacts with Chinese persons, however, were more limited. Only a few Japanese could observe the manners and customs of the Chinese merchants in Nagasaki. More important scholarly links developed when a number of exiled Chinese scholars fled to Japan as the Ming dynasty was replaced by Manchu rule in the Ch'ing dynasty

(1644–1912).[20] Such contacts helped sustain the enormous fascination of some Japanese with China and Confucianism but could not obscure the fundamental differences between dynastic China and feudal Japan. Chinese institutions of learning were to a large extent channels for the open selection of the ablest minds for public service, while learning in Japan was officially promoted to train only those whose right to rule was hereditary.

Isolation not only effectively restricted the flow of Western ideas, but also intensified anti-Western, and especially anti-Catholic, prejudices and fears: "Catholics are secret agents of a planned European invasion," "padres use sorcery and other black arts," "missionaries eat the flesh of little children," "Westerners' feet have no heels," and so on. Such distorted views were naively believed by many people throughout the Edo era. Japanese understanding of Western learning during the ninety-six years of Western Wave I was on the whole quite shallow and was quickly forgotten after isolation began. The Dutch in Dejima, in addition to the social restrictions already mentioned, were prohibited from bringing to Japan any books other than those related to navigation and medicine; they were not permitted to study the Japanese language; and the use of Dutch by the Japanese interpreters was to be confined to official business. The head of the Dutch factory on Dejima was held responsible for adherence to these conditions; one

---

[20] Chu Shun-shui (1600–1682; in Japanese, Shu Shunsui), for example, who fled China, came to Japan seven times while working for the restoration of the Ming dynasty. Last arriving in Nagasaki in 1659, he was invited in 1664 to reside in Edo as the guest of the Mito daimyo, Tokugawa Mitsukuni (1628–1700) and reached there the following year; he associated with Confucian scholars there and had a particular influence on those of the Mito domain. Chu Shun-shui, among his many other activities to promote the Confucian cult and its ideas, personally drew the plans for and supervised the construction of a Confucian temple and school building erected in Mito in 1672; the same plans were used in 1799 in the rebuilding of the Confucian temple in Edo (p. 243). The Taiseiden of Yushima Seidō thus stands as a monument to his efforts. Cf. Ching, pp. 188–189.

means of enforcement was a required annual visit to Edo (similar to that required of all daimyo).[21] Among those usually accompanying the factory head to Edo was the physician or surgeon stationed on Dejima, and he in particular was frequently questioned about conditions in the West (though the reverse flow of information about Japan to European head-

Table 19. Main Events of the Seventeenth-Century Intellectual Outburst

1605	Hayashi Razan entered shogunate service as an adviser, and promoted Chu Hsi's Confucianism (*Shushigaku*) for samurai training.
1620–1640	Appearance of various mathematical texts, printed in great quantities and sold commercially.
1630	Hayashi's Confucian academy (Kōbun'in) founded.
1640s	Calendrical studies initiated, centered mainly on the *Senmyōreki* and *Jujireki*.
1640s–1650s	Rise of Wang Yang-ming's Confucianism, beginning with Nakae Tōju and Kumazawa Banzan.
1657	Compilation of "History of Great Japan" (*Dai Nihonshi*) started under Tokugawa Mitsukuni, lord of Mito (not completed until 1906).
1660s	Rise of Confucian school of "Ancient Learning" (*Kogaku*), beginning with Yamaga Sokō, Itō Jinsai, and Ogyū Sorai.
1660–1670	Establishment of the Liu-Chang school of medicine, led by Aeba Tōan.
1670–1690	Breakthroughs in *wasan*, led by Seki Takakazu.
1680s	"Japanese Learning" (*Wagaku*) spread beyond court nobility, leading to "National Learning" (*Kokugaku*).
1684	*Jōkyōreki*, first calendrical reform done by the Japanese; establishment of the shogunate's Bureau of Astronomy (Tenmongata).
1690	Shogunate adopted the Hayashi academy as its official Confucian school; moved to Yushima in Edo and renamed Yushima Seidō.
1696	"Complete work on agriculture" (*Nōgyō zensho*) published by Miyazaki Yasusada).
1709	"Materia medica of Japan" (*Yamato honzō*) published by Kaibara Ekken.
1712–1715	Arai Hakuseki active in criticizing Neo-Confucianism (*Shushigaku*) from within and in revising Japanese views of the West.

Note: Dates in boldface type are for events of importance to science; others are for events of importance to general learning.

[21] From 1633 the visits were annual. After 1764 they were made every two years, and after 1790, once every five years. The last Edo visit was made in 1850. The delegation from Dejima usually stayed in Edo for two or three weeks. Iwao (4), pp. 396–399.

quarters was far greater; see, e.g., pp. 287, 340). The Edo visits and contacts with the Dejima community were not effectively utilized to explore Western learning and science until the eighteenth century, when Japanese attitudes toward the West were somewhat revised. Dutch intellectual influence was almost nonexistent, consisting only of offhand help to some of the Nagasaki interpreters interested in Western cosmology and surgery.

The intellectual outburst may be considered to have begun in the 1620s with the publication of some of the first mathematical texts which later became the springboard for the more sophisticated yet less practical Japanese mathematics called *wasan*. The listing of some of the main events of the intellectual outburst on the preceding page may provide a condensed image of the scope and pace of developments. A simple listing, however, does not adequately reveal that seventeenth-century Japanese scholars were, for the first time and on an unprecedented scale, dealing with complete systems of learning and science on their own terms; that they made many remarkable and a few original achievements; and that through these seventeenth-century achievements the foundations of all Tokugawa learning were laid.

TRADITIONAL LEARNING IN THE SEVENTEENTH CENTURY

The Confucian tradition that came first to Japan in Chinese Wave I was that formulated in the Han era and taught in the T'ang schools. Its scholarly style was one of carefully checking the meaning of sentences word by word with minute attention to details. This tradition was preserved in Japan after Chinese Wave I in the House Learning of the court nobility, who paid little attention to the Sung developments of Neo-Confucianism. It was Zen priests of the *gozan* temples who introduced Neo-Confucian thought during Early Chinese Wave II and bequeathed it to the seventeenth century.

By the Sung period Chinese knowledge of nature, man, and society had expanded so much that the unsystematized Han-

T'ang tradition was no longer satisfactory. The Sung philosopher Chu Hsi was most eminent among those who incorporated the many levels of Sung knowledge into a rational system. His goal was to achieve a unity of all knowledge, for which he posited a metaphysical dualism that presupposed a formal principle, *li* (*ri* in Japanese), the pattern which underlies all phenomena, whether natural, human, or social, and a pneumatic and energetic substratum called *ch'i* (*ki* in Japanese) which embodies and maintains this pattern in heaven and earth, man and society. Epistemologically, the *li* is to be grasped intuitively by exploring one's human nature and personal relations and by investigating the functional character of things. From the tradition of the Book of Changes he borrowed a further principle, the Grand Polarity (*T'ai-chi*), which gives unity to all things. The Grand Polarity created the *yin* and *yang*, and out of this complementary duality of function were formed all things. Chu Hsi presupposed not only *yin* and *yang* but also the functional complementarity of the Five Phases and the correspondence of the cosmos to the social order. He elaborately and consistently developed these major themes of natural philosophy and political cosmology, with repercussions for astrology and medicine.

Early Tokugawa Confucianists received Chu Hsi's philosophy as a total synthesis of knowledge organized on three main levels: *nature*, embracing the whole universe; *society*, mainly political, social, and economic theory; and *man*, especially his moral cultivation. In trying to relate Confucian categories to practical problems, Japanese intellectuals eventually came to realize that *Shushigaku* failed to provide adequate solutions over its entire range. Increasingly interest in the philosophic foundations declined, and active stress was put on either political economy or moral cultivation.

The most enduring and effective attack in China on the rationalistic Neo-Confucianism of Chu Hsi came from the idealistic philosophy of Wang Yang-ming. A contemporary of Chu Hsi named Lu Chiu-yüan (also known as Lu Hsiang-shan, 1139–1192) had already advocated "learning of the mind" (*Hsin-hsüeh*), an introspective and meditative approach to illumination in preference to Chu's more rationalistic investigation of phenomena along with investigation of the self in order to penetrate their *li*. It was Wang Yang-ming who most effectively rejected Chu Hsi's metaphysical dualism, and his activist idealism based on intuition and illumination gained a dominant position in the sixteenth century. This school of thought entered Japan in the middle of the seventeenth century and became a serious competitor to *Shushigaku*. It was easily reconcilable with Buddhism and appealed particularly to those concerned primarily with personal and social morality.

In Japan the most forceful attack on the theories of both Chu Hsi and Wang Yang-ming came from such thinkers as Yamaga Sokō (1622–1685), Itō Jinsai (1627–1705), and Ogyū Sorai (1666–1728), beginning in the 1660s. They argued that the excessively theoretical orientation of all Neo-Confucianism prevented effective application of basic Confucian insights to the practical problems of man and society; and they therefore advocated a return to the "Ancient Learning" (*Kogaku*) of the early sages. There were precedents for this line of argument in contemporary thinkers in Ch'ing China, such as Ku Yen-wu (1613–1682), who felt that the excessive rationalism and rigidity of the Chu Hsi orthodoxy and the passive and self-contemplative tendency of the Lu-Wang school were among the reasons that Ming China in 1644 fell under Manchu domination. The reversion of Ch'ing Confucianists to the pure and direct learning of the Han period and to direct reconsideration of classical texts was to develop into a school of textual criticism that would also be taken up in Japan in the eighteenth century.

OFFICIAL PROMOTION OF SHUSHIGAKU

Zen priests who had studied Neo-Confucianism as part of their general cultural quest believed in the unity of Confucianism and Buddhism. The break away from this position was made first by Fujiwara Seika (1561–1619) and his pupil Hayashi Razan (1583–1657). By their efforts Neo-Confucian learning was separated from Buddhist learning and institutions. Confucianism came to be taught publicly, breaking the monopoly of the nobility's House Learning.²² A new group of secular Confucian scholars appeared, some of whom eventually replaced learned Buddhist priests as advisers to the shogunate and daimyo. It was a most propitious time for the rise of such a secular and autonomous Confucian school to be favored by the ruling class; there were many functions to perform. There was, of course, in the first formative years of the Tokugawa period, the fight to extirpate Catholicism; and on a more continuing basis, the need for shifting from naked force to civil administration and the necessary indoctrination of the samurai class in the thought and values proper to their new roles.

Fujiwara Seika, son of a court nobleman, first studied *Shushigaku* at the Shōkokuji temple in Kyoto, one of the *gozan* group. Seika became a Buddhist priest devoted to the study of Chu Hsi's philosophy in the "Japanese tradition" of assuming the essential unity of Confucianism and Buddhism. Coming into contact with a Korean diplomatic mission which visited Japan in 1590, shortly before Hideyoshi's first expedition to Korea (1592), he was greatly shocked by a Korean scholar's insistence on the incompatibility of Buddhism with Confucianism.²³ At that time, Chu Hsi's system was at its peak in Korea, though in Ming China the idealism of Wang Yang-ming was already overshadowing it (except as state doctrine). Seika set out for China to study Confucianism further, but not far south of Kyushu his ship turned back due to foul weather.²⁴ Returning to Kyoto, he met a Korean named Kang Hang who had been taken captive in Hideyoshi's invasion; from him he received further instruction in Confucianism. Seika not only dissociated his Confucian standpoint from Buddhism but also left the priesthood and returned to secular life. Though once invited by Ieyasu to serve as an adviser, he declined and lived a rather secluded life, especially in his later years.

Hayashi Razan, a townsman's son born in Kyoto, also studied in one of the *gozan* temples, the Kenninji, but never

²²Sagara, pp. 23.
²³Ibid., p. 20.
²⁴Wajima (1), pp. 3–4.

became a priest. After studying under Fujiwara Seika for a time, in 1605 he began service as an adviser to Ieyasu,[25] a role he filled for a total of fifty-two years under four shoguns—Ieyasu, Hidetada, Iemitsu, and Ietsuna. He not only advised the shoguns on general affairs but also engaged in such concrete matters as drafting laws and diplomatic documents and instituting official ceremonies. Razan's knowledge, and particularly his familiarity with the Confucian classics, was prodigious; and while in retrospect we can see clearly that Chu Hsi's tenets well suited the Tokugawa feudal order, some allowance must be made for Razan's influence in shaping the order to suit the tenets.

Though Razan disclaimed Buddhism, in 1607 he took on the Buddhist clerical habit—shaving his head and donning the priestly robe—by Ieyasu's order; this was against his will, but he retained the habit for the rest of his life. In 1629 he received the highest of the Buddhist ranks. This constituted quite a concession to Confucianism, as there was at the time no system of ranking to acknowledge excellence in Confucianism commensurate with the court titles awarded to Buddhist priests.

Hayashi Razan in 1630 established an academy which, though not made an official shogunate institute at that time, was given financial and moral support. Named Kōbun'in (in 1663), it enjoyed a quasi-official status until 1690, when Shogun Tsunayoshi provided land for a new facility at Yushima on the Shōheizaka slope just outside the outer moat of Edo Castle. The new school erected there became the official shogunate school for *Shushigaku*. The whole complex, including the main hall (Taiseiden), lecture halls, dormitories, and so on, was called Yushima Seidō, though the school was generally referred to more simply as Shōheikō (Shōhei School).[26] Razan's descen-

---

[25] Razan was not, of course, the sole learned adviser; nor were they all, in the formative stage of the Tokugawa shogunate, Confucianists. Two Buddhist priests served in this capacity in Razan's time.

[26] Wajima (1), pp. 27–44, 107–108. This school was later officially renamed Shōheizaka Gakumonjo (Institute of Learning at Shōheizaka). This renaming is variously dated: by one source as 1790 (year of the Prohibition Against

dants succeeded by hereditary right to the directorship of the academy. There were also many other private academies in various places that adhered to *Shushigaku* and had direct or indirect master-disciple relationships with the Hayashi academy in Edo.

Some of those who studied at the Hayashi school in Edo went on to found their own academies for *Shushigaku* training, as for example, Kinoshita Jun'an (1621–1698), who had two distinguished pupils in Arai Hakuseki (1657–1725) and Muro Kyūsō (1658–1734). Born the year Razan died, Hakuseki served seven years (1709–1716) as the chief policy adviser to two shoguns, Ienobu and Ietsugu (ruled 1709–1712 and 1713–1716 respectively). Hakuseki severely criticized the theories of Hayashi Razan, making it easier for other scholars to engage in freer, open discussion. Shogun Ietsugu's successor Yoshimune (ruled 1716–1745) seems to have been less impressed with Hakuseki, whom he dismissed in favor of his "classmate" Kyūsō, a more loyal Hayashi supporter.

Early in the seventeenth century Neo-Confucianists were active in combating Catholicism; Razan utilized *Shushigaku* in writing an anti-Christian tract entitled *Hai yaso* (Refutation of Jesus), and Mukai Genshō (1609–1677) did likewise in compiling *Kenkon bensetsu* (Critical commentary on cosmography; ca. 1650). Razan also wrote against nonorthodox Confucianists, such as Kumazawa Banzan (1619–1691), a follower of Wang Yang-ming, and perhaps helped set the style of open criticism later directed toward his own school.

Attacks on those deviating from orthodoxy also came from scholars who received their training in *Shushigaku* elsewhere, for example, from Yamazaki Ansai (1618–1682), son of a Kyoto townsman, who studied *Shushigaku* in Tosa province and in

---

Heterodox Teaching; see *Nihonshi jiten*); Wajima gives 1797 (year the management of the institution was taken over directly by the shogunate from the Hayashi family); and another source gives 1843 (which, if correct, was perhaps associated with reforms of the Tenpō era, 1830–1844; see Rekishigaku Kenkyūkai, ed. *Nihonshi nenpyō*).

1655 founded his own academy in Kyoto. Academically and politically rather narrow-minded, Ansai in his later years became increasingly nationalistic and more interested in Shinto than in Confucian thought. While still an orthodox Confucianist he is thought to have been responsible for the shogunate's action against Kumazawa Banzan and Yamaga Sokō (see below). Before Razan's death, criticism tended to be directed toward ideas considered threatening to Tokugawa policy and power, whether Catholic or Western, on the one hand, or schools other than *Shushigaku*, on the other. Attacks on *Shushigaku* became more frequent after the master's passing, but even then rarely without backing by some powerful political figure.

EMERGENCE OF OTHER CONFUCIAN SCHOOLS

The Chu Hsi synthesis of natural knowledge, political economy, and moral cultivation seemed necessary to leaders like Razan who were responsible for the formation of a new social order but became less convincing to some after the major guidelines of society were more or less settled. Inevitably some thinkers became more concerned about cultivation of the inner qualities of those who made up that society. Nakae Tōju (1608–1648) and Kumazawa Banzan (1619–1691) were among the first to turn away from the tenets of Chu Hsi to those of Wang Yang-ming, which stressed cultivation of the mind and insisted that knowledge and action be unified, as against Chu Hsi's dictum that knowledge must precede action. As we have seen, Wang Yang-ming assumed an essential unity in all reality and thus espoused a more illuminationist approach as against the more rationally worked out objective synthesis of Chu Hsi.

Both Tōju and Banzan were of samurai background; but in 1634 Tōju (at age twenty-six) resigned from samurai service to the Ōsu fief on Shikoku island, and settled in Ōmi province near Lake Biwa as a *rōnin* (without stipend) to care for his aged mother who had refused to go to Shikoku; there he later opened a private academy. Banzan was the most prominent of Tōju's many disciples. Originally a lordless samurai (*rōnin*), Banzan served from 1650 as a political adviser to Ikeda Mitsumasa (1609–1682), lord of Okayama domain, whose chief administrator he eventually became. Banzan incurred the shogunate's displeasure and was put under house arrest in 1687, possibly because of unacceptable policy proposals made to the shogunate. His criticism of *Shushigaku* had drawn forth Razan's retaliatory attack on his essay *Sōzokuki* as a "variation of Catholicism" (*Yaso no henpō*), which aroused suspicions as to Banzan's reliability as a teacher of other samurai.[27]

In contrast to other Confucian schools, the Wang Yang-ming tradition in Japan lacked the continuity of master-disciple connections. Its emphasis on the internalization of Confucian values and on the integrity of thought and action, however, proved attractive to many men of different rank and station throughout the Tokugawa period.[28]

By the second half of the seventeenth century there appeared leading thinkers who rejected Neo-Confucian theories as arbitrary deviations from the original spirit of Confucianism and, in the growing complexity of administrative duties, as really not very useful. In advocating a return to the teachings of the Confucian sages, they expected scholars to read the early classics and to think for themselves to see reality. Their efforts, known together as the "Ancient Learning School" (*Kogakuha*), served as a correction to the narrowly moralistic mentality of some thinkers and as a forceful recognition that thought must account for historical changes taking place in society. In the hands of lesser disciples, however, this school itself later tended to fall into a rather dogmatic rationalism or pragmatism. It was not, in any case, a tightly-knit school of thought, as dif-

---

[27] Sagara, pp. 78–79.
[28] Unusual sensitivity to the plight of ordinary citizens burdened by rapidly escalating rice prices in the 1830s was shown by the shogunate samurai Ōshio Heihachirō (1793–1837), a former Osaka police officer (*yoriki*) and a recognized student of Wang Yang-ming's philosophy, who in 1837 led an unsuccessful one-day revolt by 300 persons, mostly poor townsmen and farmers aided by a few samurai, against the shogunate's troops in Osaka. Sakuma Shōzan (1811–1864), an important figure in the development of "Western learning" in the late Tokugawa era, was also attracted to this school of thought after first studying Chu Hsi.

ferences in emphasis among its leading founders evolved along three main lines of interpretation.

Yamaga Sokō (1627–1685) was the earliest of the Ancient Learning advocates. Having first learned Chu Hsi's philosophy under Hayashi Razan, Sokō in 1665 published his *Seikyō yōroku* (Essentials of revered learning), which criticized the abstract Neo-Confucian theories and called for a return to the Confucian sages. His scorn of Neo-Confucianism earned him banishment from Edo in 1666; he was allowed to return in 1675. There is some circumstantial evidence that it was one of Yamazaki Ansai's prize pupils, Hoshina Masayuki (1611–1672; third son of the second shogun, Tokugawa Hidetada, and from 1615 to 1669 guardian of the fourth shogun Ietsuna), who was most offended by Sokō's critique. On the other hand, Sokō was originally a *rōnin*, and there may have been some apprehension in government circles that someone who had deviated from orthodox ideology might, with other lordless samurai, attempt revolt against political authority. In any event, Sokō remained loyal to the martial tradition throughout his exile and after returning to the capital devoted himself to teaching military strategy and tactics, a special interest of his.

Another convert from *Shushigaku* was Itō Jinsai (1627–1705), who also advocated a return to Confucius and Mencius. Of the three levels of the Chu Hsi synthesis, Jinsai most clearly rejected natural knowledge and political economy, stressing rather moral cultivation. Son of a Kyoto townsman, he founded an academy called Kogidō (Hall of Ancient Principles), which lasted more than two centuries, into the early Meiji era, headed throughout by his descendants. Jinsai is said to have had three thousand disciples among samurai and townsmen alike, and among his students were some of the doctors of the "Ancient Practice School" (*Kohōha*) of medicine—which had affinities with the Ancient Learning School of Confucian thought as well as precedents in China.

The most influential school of Ancient Learning was that of Ogyū Sorai (1666–1728), who was inspired by, but was not a direct disciple of, Itō Jinsai. The son of a physician serving the shogun, Sorai in 1696 became a teacher and an adviser to the shogun's highest administrative council. Sorai's criticism of *Shushigaku* was so forceful as to uproot its authority among most Confucian scholars. He insisted that the proper focus of Confucian study was political economy, not cosmological and natural studies or moral cultivation. The authority of the Confucian sages for the discussion of politics and economics he took for granted. The method he advocated for ascertaining the original truths of the sages was critical linguistic and textual study (*kobunjigaku*), a method later adapted by Motoori Norinaga (1730–1801) to achieve a breakthrough in National Learning (*Kokugaku*).

Though Sorai proposed that the essential work of Confucian scholarship be confined to the single area of political economy, his own studies covered a vast range of academic interests, and his intellectual perspectives had similarly wide breadth. An intensely devoted Sinophile, Sorai greatly enhanced the Japanese understanding of things Chinese. He attracted many brilliant disciples, some of whom followed Sorai's admonition that the social system should change with the times and thus proposed various reforms not bound to the *Shushigaku* format. Other disciples of Sorai fell below his standards, inviting criticism from more conservative scholars for their excessive indulgence in literal exegesis while neglecting moral cultivation, or for advancing extreme and unbalanced opinions merely to curry favor or advance personal reputations.[29]

FROM "JAPANESE LEARNING" (*WAGAKU*) TO "NATIONAL LEARNING" (*KOKUGAKU*)

In the twelfth and thirteenth centuries there developed a tradition of studying the traditional Japanese verse form called *waka*, as part of the court nobility's culture of the Semiseclusion Era. This tradition, called "poetics" (*kagaku*, lit., "verse learning"), began with noblemen such as Fujiwara Shunzei (1114–

---

[29] Hattori Nankaku (1684–1759) was most typical of this trend. Sagara, pp. 206–207.

1204) and Fujiwara Teika (1162–1241). From the fourteenth century the essential traditions of this study were preserved by the court nobility through secret transmission from master to disciple (cf. *iemoto seido* in Chapter 3). In Early Chinese Wave II (fifteenth and sixteenth centuries), certain courtiers who became well versed in the Chinese classics also undertook study of the Japanese classics, Shinto, and court practices, widening the range of studies far beyond the original interest in *waka*. Consequently, investigations in this general direction came to be known as "Japanese Learning" (*Wagaku*).

In the period of the intellectual outburst the tradition of *Wagaku* was further enlarged to encompass Japan's history, literature, and social system. The methods of study became increasingly critical and systematic. An early student of this tradition was a *rōnin* named Toda Mosui (1629–1706) who challenged the nobility's traditional authority in interpreting *waka*. Other early Tokugawa representatives of this trend advocated freedom of interpretation not bound by "tradition."

"Japanese Learning" at this stage still had no clear ideological principles or methodology, and no academies to promote its cause. Kada no Azumamaro (1669–1736), a Shinto priest and student of the *Kojiki* and *Nihon shoki*, appealed for recognition and promotion of "National Learning" (*Kokugaku*) as a system of learning devoted to Japan's classical literature, history, and social system. A distinct movement with this name and purpose did not get under way, however, until later in the eighteenth century.

The Sciences

In the exponential expansion of intellectual activity that took place in seventeenth-century Japan there was no one single source of scholarly initiative and inspiration. Least of all should there be any image of a dominant Neo-Confucian mood moving all parties to new effort. Competition among various Confucian schools certainly enhanced the growing ethos of critical scholarship; and Ogyū Sorai's skeptical attitude toward Neo-Confucian speculation and his stress on critical examination of

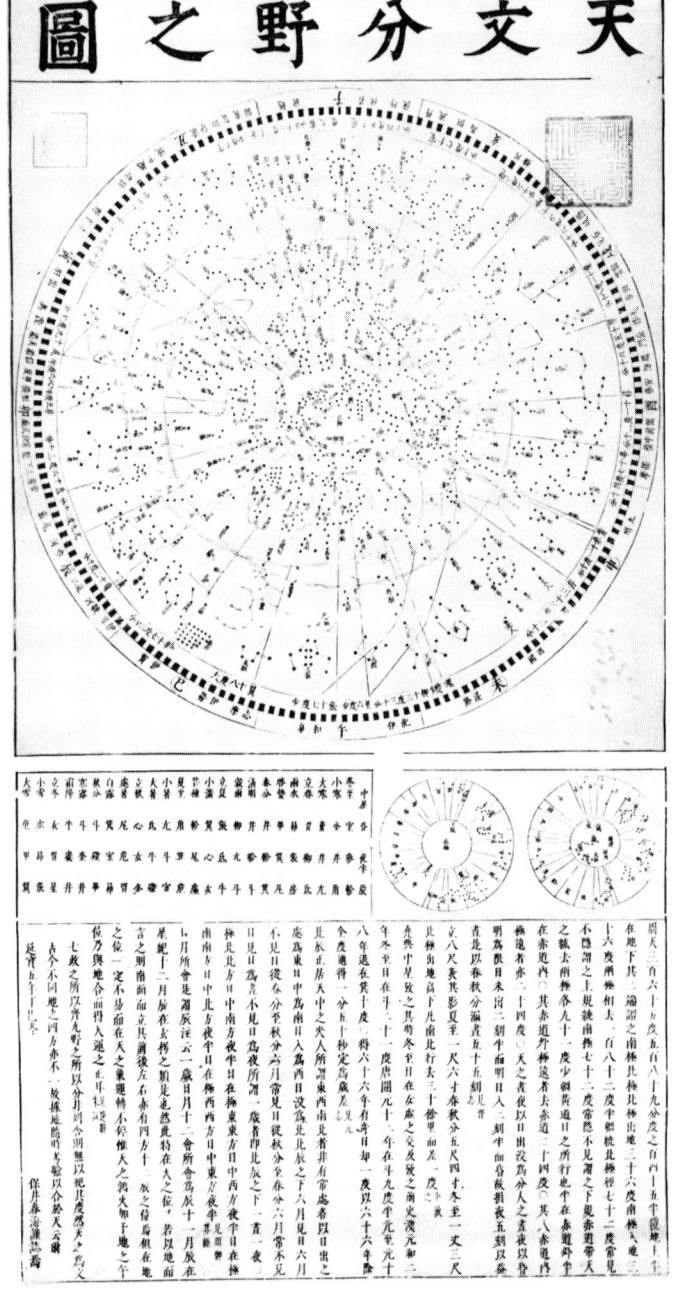

Fig. 27. A star chart, entitled (at top) "Tenmon bunya no zu," made by Shibukawa Harumi in 1677. Rectangular space at left center gives the twenty-four fortnightly periods; text below provides explanations of, for example, gnomon-shadow lengths at times of solstices and equinoxes. (The National Archives, Japan)

authoritative texts were an encouragement to medical scholars working along similar lines. However, seventeenth-century scientists did not merely coast forward on a Confucian wave. Medical scholars certainly gave as much, if not more, than they received from Confucian studies. They achieved a comprehensive grasp of the Neo-Confucian framework in the Li and Chu medical traditions long before *Shushigaku* as such gained prominence. Physicians, and particularly sons of physicians such as Kaibara Ekken, Ogyū Sorai, and Muro Kyūsō, were among the leading Confucianists.

Interdisciplinary interests were not a major motif of the seventeenth-century intellectual revival, but the contribution of trained mathematicians to the renewal of calendrical research was far from small (though separation of mathematics from astrology and calendrical astronomy later became the norm). The ability to select problems of crucial significance in one's field, to persevere to some breakthrough, and to search out and utilize new sources in doing so—usually without support from official quarters—characterized the best scientific labors. Work in most fields leveled off, it is true, after the century ended; but the levels that had been attained were much higher than when that century began, and the scholarly community had acquired far greater depth in its critical research.

ASTROLOGY AND CALENDRICAL ASTRONOMY

Interest and activity in the related fields of astrology and calendrical astronomy revived almost simultaneously with the beginning of isolation and conformed fully to the Chinese tradition of focusing primarily on calendrical astronomy. Residual knowledge of Western Wave I preserved by a small group of interpreters in Nagasaki centered more on cosmological theories; but it was only a minor by-stream, only partially understood, and it never entered the mainstream of traditional science in the seventeenth century. Private research in traditional science led to an epoch-making event: the first calendrical revision made by the Japanese, on their own, since adoption of the last Chinese calendrical system eight centuries earlier. It was an epochal event in another sense: the Tokugawa shogunate was moved by this accomplishment to incorporate—again, for the first time since Chinese Wave I—part of the scientific enterprise into its administration.

It needs to be reiterated that all Sino-Japanese astronomy was calendrical in import; there was no observational astronomy independent of the astrological and calendrical arts. "Calendar reform" means here the working out of a complete and integral set of step-by-step computational techniques by which all possible celestial phenomena (including those of planets as well as eclipses) could henceforth be predicted a year at a time by someone with no advanced knowledge of astronomy. In East Asian astronomy, the product of a calendar reform is, properly speaking, a new astronomical system, and the annual product of an astronomical system is the civil calendar (ephemerides is the more technical term). With the rise of competence in calendrical astronomy from the seventeenth century on, its specialists were less and less interested in astrology, and those concerned with astrology (and other forms of divination) were increasingly removed from serious work in astronomy.

In the 1640s civilians in various professions, especially mathematicians, began studying the traditional computational system, the *Senmyōreki*, used in Japan since 862. It was printed with all its integral astronomical tables and made available outside official circles for the first time in 1644. This was followed in 1672 by the reprinting of the *Shou-shih li* (in Japanese, *Jujireki*; discussed in Chapter 2), the best system of Chinese computational techniques produced by purely traditional methods. Ordinary intellectuals could understand Chinese cosmology and, if interested, the introductory explanation of Western cosmology that lingered from Western Wave I. Given some basic skill in calculation, they could even perform the simple task of calculating an annual calendar by use of a computational system. But constructing such an astronomical system—or, revising one, as was necessary in Japan at the time—was a task proper only to those well trained in mathematics. In the

Table 20. Proliferation of Books on Traditional Calendrics in the Seventeenth Century

Date	Author	Title	Remarks
1612	Abe no Seimei	Hōki naiden	Reprint of text attributed to Abe no Seimei; first printed book in Japan on calendrics (mostly superstitions)
1642	Imamura Tomoaki	Nichigetsu kaigō sanpō	First printed book in Japan on calendrics proper; authored by mathematician who wrote Jugairoku and Inki sanka
1644	Hsü Ang (Jokō)	Hsüan-miag li-ching	Printed Chinese text of Senmyōreki with all astronomical tables; prompted serious study of the Senmyōreki system
1645	Yoshida Mitsuyoshi	Wakan gōun	Printed; authored by mathematician who wrote Jinkōki
1648	Yoshida Mitsuyoshi	Koreki tenran	Printed
1658	Enami Kazusumi	Rekigaku seimō	Printed; explanation of computational methods of Senmyōreki
1663	Andō Yueki	Chōkei Senmyōreki sanpō	Printed; explanation of computational methods of Senmyōreki in kanamajiri style
1672	Kuo Shou-ching (Kaku Shukei)	Shou-shih li-ching	Printed Chinese text of Jujireki with all astronomical tables; prompted serious study of the Jujireki system
1673	Ogawa Masaoki	Shinkan Jujirekikyō	Printed; with excerpts from the Jujireki's computational tables
1680	Seki Takakazu	Juji hatsumei	Manuscript; author was leading mathematician of the seventeenth century.
1680	Amoghavajra, transl. (Pu-k'ung)	Hsiu-yao ching	Printed Chinese text of the Sukuyōkyō canon on Buddhist prognostication, with marks for Japanese reading appended by the Buddhist priest Engu
1682	?	Hōki genkai taizen	Printed; text in the tradition of the 1612 Hōki naiden; market for this kind of book persisted.
ca. 1690	Takebe Katahiro	Jujireki keigi	Manuscript; most thorough study of Jujireki, authored by mathematician of the Seki school of wasan

Sources: Nihon Gakushiin (3); Nakayama (6); Araki, pp. 103, 117.

Table 21. Calendrical Systems Used in Japan: From the *Senmyōreki* to the *Tenpōreki*

Calendrical system	Period used in Japan	Number of years used	Producer	Calendrical book used as a base	Tropical year (days)	Synodic month (days)	Remarks
Senmyōreki	862–1684	823	Hsu Ang	*Hsüan-ming li-ching*	365.2446	29.53060	Used in China: 822–992 (71 years)
Jōkyōreki	1685–1754	70	Shibukawa Harumi	*Jōkyōrekisho*	365.2417	29.53059	Adapted the *Jujireki* (see below) with his own observations and adjustment of longitude and latitude to Japan (Kyoto)
Hōryakureki	1755–1797	43	Tsuchimikado Yasukuni et al.	*Rekihō shinsho*	365.2416	29.53059	Minor revision of *Jōkyōreki*, but inferior
Kanseireki	1798–1842	45	Takahashi Yoshitoki	*Kanseirekisho*	365.24235	29.530584	Based mainly on *Li-hsiang kao-ch'eng hou-pien*
Tenpōreki	1843–1872	30	Shibukawa Kagesuke and Yamaji Tomotaka	*Shinpō rekisho*	365.24222	29.530588	Based mainly on Lalande; best of traditional calendrical systems
				Present-day values:	365.24220	29.530588	
Jujireki (*Shou-shih li*)	Not used in Japan		Kuo Shou-ching	*Shou-shih li-ching*	365.24250	29.530593	Used in China: 1281–1368 (88 years)

Sources: Nihon Gakushiin (3), pp. 250, 285–286; Nakayama (6), Appendix 2.

half-century from the 1640s to the 1680s a number of commentaries on the *Senmyōreki* and later the *Jujireki* appeared, and this display of interest was accompanied by definite advances in mathematical ability. Some of the participating mathematicians were those already noted for their production of standard mathematical texts in the period 1620–1640; for example, Yoshida Mitsuyoshi, author of the *Jinkōki* (1627), and Imamura Tomoaki, author of the *Jugairoku* (1639) and the *Inki sanka* (1640), both of whom were disciples of another important mathematician, Mōri Shigeyoshi, author of the *Warizansho* (1622).

In the latter half of the seventeenth century knowledge of traditional calendrical astronomy became a kind of widespread fashion. It also became widely known that there was almost two days' discord between the official calendar and the tropical year. Between 1673 and 1684 solar eclipses predicted in the calendar failed six times to materialize; predicted lunar eclipses failed four times to occur in the same period. Around 1672 a mathematician named Ikeda Shōi, a disciple of Imamura Tomoaki and a student of calendars, urged that the *Jujireki* be adopted to replace the unreliable *Senmyōreki*. Though his advice was not heeded, the Tsuchimikado household was itself unable to revise the faulty calendar. Revision was not accomplished until the advent of a rather unheralded amateur, Shibukawa Harumi.

THE FIRST JAPANESE CALENDRICAL REFORM

Shibukawa Harumi (1639–1715), known also as Shunkai (alternate reading of the characters for Harumi), a master of the game of *go* in the shogun's service (four households had the hereditary right to supply the shogunate's *go* masters), attended to his duties in Edo in fall and winter and in Kyoto in spring and summer. He studied mathematics and astronomy in Edo under Ikeda Shōi, and astronomy in Kyoto under Okanoi Gentei, a physician who had received instruction in the *Jujireki* methods by a Korean named Yong Na-san who came to Japan in 1643. Amateur astronomer Harumi became increasingly interested in official calendrical reform. Through his services as a *go* master he cultivated a number of influential patrons in high positions in both Kyoto and Edo, who later supported his reform proposal. Among these patrons were Hoshina Masayuki and the Mito daimyo Tokugawa Mitsukuni, both of them grandsons of Tokugawa Ieyasu. Even with such patrons, Harumi's progress toward revision was not entirely unimpeded, for resistance by the Tsuchimikado officials in Kyoto, proud hereditary holders of the court office for annual calendar production, forced Harumi to make repeated and persistent appeals to the court.

Harumi made the first of his appeals that the *Jujireki* be adopted without alteration in 1673 after a lunar eclipse predicted by the *Senmyōreki* for December, 1672 failed to occur; this appeal was rejected. In May, 1675 a solar eclipse unpredicted by the *Jujireki* was correctly predicted by the *Senmyōreki*, undercutting Harumi's continued campaign in behalf of the *Jujireki*, which he then began trying to correct by conducting his own observations and devising his own methods. A solar eclipse not predicted by the *Senmyōreki* occurred in November 1677, but Harumi did not renew his appeal until 1683, at which time he urged use of his own revised calendrical system, called *Yamatoreki*. In March 1684 the court decided, however, to adopt the Ming calendrical system, the *Ta-t'ung li* (used from 1368 to 1628 in China), though this decision was never implemented. Harumi appealed once more to the court to adopt his revised system. This time the court agreed, though first he had to present his proposed revision to the Tsuchimikado officials in Kyoto for inspection. It passed the test and was put into official use from January 1685 with the name *Jōkyōreki*. Unrevised in Japan for 823 years, the *Senmyōreki* was retired to the historical archives.

In December 1684 Harumi resigned his post as *go* master (for which the Buddhist clerical habit was mandatory), discarded his Buddhist robe, and let his hair grow, but did not leave the shogun's service. Quite the opposite; he became head of a new office in Edo established within the shogunate, called

Tenmongata (Bureau of Astronomy), and appointments to this office became hereditary for his descendants.[30] In 1702 Harumi changed his surname from Yasui to Shibukawa.

The Tenmongata was the astronomical bureau of the shogunate only; the court astronomer's office was not abolished. Rather, the two offices maintained their prerogatives by working in tandem for the annual production of the civil calendar. By the fall of each year the Tenmongata made the purely technical calculations (e.g., determination of long, short, and, if necessary, intercalated months; prediction of eclipses; etc.) and sent a provisional calendar to the Tsuchimikado officials in Kyoto, where hemerological notations were appended. Then the draft went back to Edo for final editing and production of a clearly written copy which was sent back to the Tsuchimikado household for formal presentation to the court. Court officials took it to the Grand Shrine at Ise for formal announcement that the calendar for the coming year was ready.

Distribution followed a similarly complicated course. Tsuchimikado agents had proofs of the final copy run off by the official court printer; they were proofread by the Tenmongata, after which the shogunate's Office of Temples and Shrines (*Jisha bugyō*) distributed approved copies to town commissioners (*machi bugyō*), who in turn made them available to local almanac dealers who printed and marketed them.

Later revisions of the calendrical system were made when the Tenmongata initiated proposals under shogunate order, but plans were always presented to the Tsuchimikado household for testing against observational data, and Kyoto's approval was required. The Tsuchimikado house had clearly lost the initiative in calendrical research; even so, it continued to employ personnel who engaged in observations and calcula-

---

[30] All appointments to the Tenmongata were hereditary. A total of eight households was assigned to the Tenmongata throughout the rest of the Tokugawa era. Only three of these (including the Shibukawa household) survived to the time of the *Tenpō* calendrical revision (last revision of the traditional system) in 1841 (p. 359). Nihon Gakushiin (5), p. 227.

tions. Despite its lack of competence, the Tsuchimikado house remained proud of its authority, and formally the Tenmongata was but its servant.

Theoretically, the *Jōkyōreki* was based largely on the *Jujireki*. Harumi had improved on its application to Japan by correcting for differences of longitude and latitude between Peking and Kyoto. He had established his own calendrical parameters. His observational instruments were inferior to the Chinese ones; Kuo Shou-ching, producer of the *Jujireki*, had used a giant gnomon forty feet tall with a pinhole device to produce a sharply defined shadow from which to determine solstices, while Harumi used a simpler instrument only eight feet tall. Consequently, his observations were less precise.[31] In time, defects in the *Jōkyōreki* were bound to surface.

RESIDUAL KNOWLEDGE OF WESTERN ASTRONOMY

In general, Western learning and science vanished with the persecution of Christians and isolation; but a small group of interpreters in Nagasaki kept alive a spark of knowledge of Western astronomy. This residual knowledge was later to play a small part, as a kind of subplot, in the rise of Western Wave II (1720–1854). Actually this "knowledge of Western astronomy" was merely introductory cosmology; it involved no observations of heavenly bodies by the Japanese nor any attempts to predict solar or lunar eclipses. Indeed, the Edo and Kyoto specialists viewed the Nagasaki group as mere bookish philosophers. Japanese astronomers active in the intellectual outburst, like their Chinese predecessors, used algebraic techniques which did not depend on any model of the cosmos. The "barbarian" residue based on geometric models of a Ptolemaic universe could make no direct contribution to their work.

A treatise entitled *Kenkon bensetsu* (Critical commentary on cosmography) was one form this Western-style knowledge took. According to the preface, it was based on a Latin text in the

---

[31] There is some doubt whether in fact Harumi used his own observational values or not. Cf. Nakayama (6), pp. 126–127.

possession of a missionary who tried to smuggle himself into Japan on board a vessel that shipwrecked in 1643.[32] Sawano Chūan (1580–1650), an apostate Jesuit (formerly Christavão Ferreira) who, aged and ailing, had recanted after several hours of torture and was now in the shogunate's service,[33] wrote a translation in roman letters, which Nishi Kichibei (? –1666), a Nagasaki interpreter, read aloud to Mukai Genshō, a physician and Confucian scholar. (A former vice-provincial of the Japan mission, Chūan could speak Japanese fairly well but could not write proper Japanese.) Genshō rewrote the treatise in Japanese, interspersing his own commentaries. The original text has been identified with Christopher Clavius' *In sphaeram Ioannis de Sacro Bosco, commentarius* (1607), which Matteo Ricci had used in his Chinese writing. Noting that the 1607 edition of Clavius included an attack on Copernicus' "absurd" hypothesis which is absent from the Japanese work and that the latter contains untranslated Portuguese words and the names of Japanese provinces and the Philippines, it has been recently suggested as "highly probable" that Chūan consulted works by Clavius or based on him but that the preface is "either incorrect or, as has been suspected, a forgery, and that the *Kenkon bensetsu* was written by Ferreira himself."[34]

In any case, the content is a general introduction of Western explanations of the cosmos, earth, eclipses, and meteorological occurrences, to which Mukai Genshō's commentaries are not infrequently refutations based on Neo-Confucian *yin-yang* and Five Phases metaphysics. Many of the "barbarian" explanations were regarded as tedious and overly materialistic; and indeed they would have been obvious to anyone familiar with the *li* and *ch'i* principles. Other purely technical explanations, such as those of seasonal differences in day-length, solstices,

---

[32] Ibid., p. 89.
[33] Boxer, p. 90. Chūan also made some contribution to the Japanese practice of Western surgery (p. 284).
[34] Nakayama (6), p. 89.

eclipses, meridians, and so on, were acknowledged with frank appreciation. The earth's sphericity, which had caused some serious contention a generation earlier, went uncontested as only a roundabout explanation of a fact grasped intuitively in the East in the egg yolk model (see p. 54). There was no mention of the heliocentric system.

The real clash between the main body of the *Kenkon bensetsu* and the interspersed commentaries was between the Aristotelian four elements assumed in Western physical theories and the *yin-yang*, Five Phases, and *li-ch'i* assumptions of Genshō.[35] Neither Chūan nor Genshō was a professional astronomer, and the numerical values given in the text were often confusing and unsatisfactory for use in calendrical calculations.[36]

A similar work that appeared sometime in mid-seventeenth century was the *Nigi ryakusetsu* (Outline theory of terrestrial and celestial globes) by Kobayashi Yoshinobu (1601–1684). Yoshinobu was imprisoned in 1644 when his teacher Hayashi Kichizaemon (? –1646) was executed on suspicion of being a Christian; released in 1667, he later lectured in Nagasaki. In 1683 his work was confirmed when he pointed out an erroneous solar eclipse prediction of the official calendar. The content of the *Nigi ryakusetsu* was much the same as the *Kenkon bensetsu*, though its source has been identified recently as a text called *De sphaera* (ca. 1593) prepared by the Spanish missionary Pedro Gomez (1535–1600) for use in the Jesuit college—one of the few traces of Western Wave II after isolation. Compared to the *Kenkon bensetsu*, its style is entirely different and more readable, while its astronomical content is more elaborate. But its astronomical calculations were only approximations and of little use in computations.[37]

Concurrently, the Jesuits in China were quite active in translating and publishing many scientific texts in the Chinese

---

[35] On Genshō's commentary, see Ibid., pp. 90–98.
[36] Ibid., p. 98.
[37] Ibid., pp. 99–100.

language, and there were many commentaries on them. Isolation restrictions prevented these texts and commentaries from being imported into Japan, though a few seem to have been smuggled in and read secretly. The Chinese text most influential in Japan was the *T'ien-ching huo-wen* (Queries on the classics of heaven; ca. 1675), authored by Yu I, who studied under the prominent scholar Hsiung Ming-yü, a close friend of the Portuguese Jesuit Manoel Dias (1574–1648), himself author of *T'ien-wen lueh* (An outline of celestial phenomena; 1615).[38] The *T'ien-ching huo-wen* was brought to Japan after its first printing in China and was already widely known before the first Japanese reprint appeared in 1730. Not meant for practical astronomy, it was a synthesis of Chinese and obsolescent Western theories. Its scope of cosmological coverage was broad but not deep, and it was not highly regarded in China. Very popular among Japanese intellectuals, it was reprinted many times with commentaries and was viewed as a classic in astronomy—partly because of the scarcity of such works.

The popularity of the *T'ien-ching huo-wen* contrasts sharply with the fate of the *Kenkon bensetsu* and the *Nigi ryakusetsu*, neither of which was printed or widely distributed. Only a few had access to handwritten copies, and open discussion of them was hardly encouraged. However, the influence of Kobayashi Yoshinobu, author of the *Nigi ryakusetsu*, did extend in the direction of Western Wave II through one of his students, Nishikawa Joken (1648–1724) and his son Nishikawa Seikyū (1693–1756).

In his *Tenmon giron* (Discussions of the principles of astronomy; 1712) Joken developed a moral-physical dualism of the two heavens; the heaven of *meiri* was understood as the realm proper to speculative philosophy, comparable to the Neo-Confucian *ri* (*li* in Chinese), while the heaven of *keiki* was for empirical research, corresponding to the Neo-Confucian *ki* (in Chinese *ch'i*).[39] Though held ultimately inseparable and com-

---

[38] Ibid., p. 80. Manoel is sometimes referred to by the Latin form (Emmanuel) of his Christian name, usually with the Spanish form (Diaz) of his surname.
[39] Ibid., pp. 107–108.

plementary, this dichotomy helped free empirical investigation of the *keiki* realm from the metaphysical norms of the *meiri* realm. In a subsequent work, *Ryōgi shūsetsu* (An explanation of collected materials on celestial and terrestrial globes; 1714), the study of *keiki* was further defined as the exclusive work of astronomers, which included geographical measurements for use in determining such necessary referents as the meridian, the equator, the ecliptic, the seasons, and the lengths of day and night. Joken knew the available materials on Western astronomy and learned something of navigational techniques from the Dutch in Dejima. Having freed himself conceptually for objective studies, Joken was able to make a cautious appraisal of the points where Western astronomy was superior to traditional East Asian science without having to accept the medieval European concepts which were not.[40]

In 1720 Shogun Yoshimune, with a specific interest in calendrical reform, relaxed the ban on the import of Jesuit translations of Western books into Chinese—from which time we date Western Wave II. Nishikawa Joken had appeared before Yoshimune by invitation the preceding year (1719) to explain Western astronomy, though he did not play so large a role in Yoshimune's plans as did his son Seikyū, who was asked to assist in the calendrical reform project of Yoshimune and in 1747 was made a supernumerary Tenmongata official. All in all, the situation at the end of the intellectual outburst held some promise: some Japanese specialists had come to master the traditional methods of astronomy, and calendrical reform was under way, while for the first time in nearly a century there were new possibilities for exploring Western sources as well.

MATHEMATICS

Traditional mathematics experienced great progress in the last two decades before isolation, when persecutions and isolationist activity were at their peak, and continued to develop steadily

[40] Ibid., pp. 109–111.

in the peaceful decades following closure of the country. Unaffected by the coming of Western mathematics or by its passing, it would appear that traditional mathematics was totally without relation to the vital social currents of its time. It had, rather, nonpolitical but important roots in society. The mathematical texts produced in the initial period of 1620–1640 resulted from technical, commercial, and social concerns that did not fade away simply because the nation was now secluded from most of the world. Practical mathematics as a utilitarian tool continued in the service of society.

The more sophisticated mathematics called *wasan* (lit., "Japanese mathematics"), for which the developed stage of practical mathematics was the springboard, has a slightly different history. Participation in calendrical studies by some of the best mathematicians of the seventeenth century helped preserve to some extent the practical orientation of *wasan*, at least in that century. Most of the samurai intellectuals who came to dominate certain levels of *wasan* felt that the utilitarian image of ordinary mathematics was beneath their dignity, partly because of the low regard their Confucian colleagues had for mathematics. *Wasan* developments subsequently veered away from concrete interests and eventually degenerated largely into trivialities—a high price to pay for sophistication. We shall return to the intellectual milieu and social aspects of *wasan* after first tracing its internal evolution.

PROBLEMS AND BREAKTHROUGHS

Private academies served to transmit the best mathematical knowledge, and calendrical studies stimulated the best mathematical scholars, like Yoshida Mitsuyoshi and Imamura Tomoaki. The most important vehicle for highly developed mathematical concepts and skills in the seventeenth century, however, was a process initiated by Yoshida Mitsuyoshi called "problem succession" (*idai keishō*). It was a process by which problems left unsolved (*idai*) were appended to a mathematical treatise or text, as a challenge to other mathematicians. When someone managed to solve a set of problems, he published his solutions and left his own *idai* for someone else. The succession ran from Yoshida to the recognized chief of all *wasan* scholars, Seki Takakazu.[41] (See Table 22 for the major sequence.)

To the *Shinpen Jinkōki*, a 1641 revision of his earlier *Jinkoki* (1627), Yoshida Mitsuyoshi appended twelve unsolved problems, which were not actually very difficult. Isomura Yoshinori was only one of many who responded to Yoshida's challenge, but the 100 problems which Isomura posed in his own *Sanpō ketsugishō* (Selected solutions of mathematical methods; 1661) were more complicated. Isomura was known as a "master of the abacus" and persisted in its use even after the "heaven-origin method" (*tengenjutsu*) became popular among other mathematicians.

To digress for a moment, it will be recalled that the heaven-origin algebra had been forgotten in China once the abacus came into widespread use—which was inevitable because the two-dimensional matrices could not be set up as on the counting board. The point is that Chu Shih-chieh's *Suan-hsüeh ch'i-meng* (the simpler of his works, with no detailed explanation of heaven-origin algebra[42]) was among books brought back to Japan from the late sixteenth-century Korean expeditions at about the same time that Ch'eng Ta-wei's *Suan-fa t'ung-tsung* (1593) on abacus calculation entered Japan. Yoshida had written the *Jinkōki* after mastering the latter text, so it may be said that competence in abacus calculations got a head start in Japanese mathematics. The decade between Isomura's 1661 publication and Satō Masaoki's *Sanpō kongenki* (Recorded bases of mathematical method; 1669), however, was one of intense study of heaven-origin algebra as a new weapon to use on Isomura's 100 problems. The heaven-origin method is a notational technique which uses contiguous files on the counting board to hold and operate on the numerical coefficients of the constant term, x, x^2, and higher powers of x in an equation on the counting board.[43] Quite different from the one-dimensional

---

[41] The longest sequence of "problem-succession" lasted for 172 years, i.e., from Yoshida Mitsuyoshi's *Shinpen Jinkōki* (1641) to *Sangaku kōchi* (Inquiry into mathematics; 1813) by Ishiguro Nobuyoshi. Hirayama Akira (2), p. 22.
[42] Needham, 3:46.
[43] Smith and Mikami, p. 86.

Table 22. Succession of Problem-Solving in Seventeenth-Century Japanese Mathematics

Key: *Title*
(date / number of problems)
Author

- *Shinpen Jinkōki* (1641/12) Yoshida Mitsuyoshi
  - *Kakuchi sanpō* (1657/0) Shibamura Moriyuki
  - *Sanryōroku* (1653/8) Enami Kazusumi
    - *Sanpō shigenki* (1673/150) Maeda Noriyuki
  - *Kaizanki* (1659/11) Yamada Masashige
    - *Sanso* (1663/0) Muramatsu Shigekiyo
  - *Enpō shikanki* (1657/5) Hatsusaka Shigeharu
    - *Hōen hikenshū* (1667/0) Tagaya Tsunesada
      - *Sanpō hatsumōshū* (1670/0) Sugiyama Sadaharu → *Wakan sanpō taizen* (1695/0) Miyagi Kiyoyuki
    - *Ketsugishō ippyakumon tōjutsu* (1672/0) Seki Takakazu
      - *Sanpō chokkai* (1670/0) Higuchi Kanetsugu → *Sanpō meikai* (1679/0) Tanaka Yoshizane
    - *Dōkaishō* (1664/100) Nozawa Sadanaga
      - *Kokon sanpōki* (1671/15) Sawaguchi Kazuyuki → *Hatsubi sanpō* (1674/0) Seki Takakazu
  - *Sanpō ketsugishō* (1661/100) Isomura Yoshinori
    - *Sanpō kongenki* (1669/150) Satō Masaoki

Sources: Shimodaira (1), p. 56; Hirayama Akira (2), p. 20; and Asahi (1), p. 74.

numerical procedures of the abacus, it was not easily grasped by early Edo mathematicians. Satō Masaoki studied the *Suanhsüeh ch'i-meng* and solved the 100 problems of Isomura; though it appears that he failed to understand *tengenjutsu* fully due to the lack of detailed explanation in the text. Even so, Satō left 150 problems of increased complexity.

In his seven-part *Kokon sanpōki* (Old and new methods of mathematics; 1670) Sawaguchi Kazuyuki devoted parts four to six to answering Satō's *idai*, for which he made use of heaven-origin algebra—though his solutions were not given in full.[44] What is clear is that Sawaguchi fully grasped the *tengen* method, and heaven-origin algebra came to be regarded as necessary to mathematical training as the abacus. A number of texts on *tengenjutsu* came out in the 1690s. In the meantime, certain defects of both heaven-origin algebra and of its basic tool, the counting board, became evident.

According to *tengen* methods, simple equations (of only one degree) are solved by division. Equations of the second or higher degrees are solved by taking trial unknowns one at a time to get successively closer approximate values. The degree (or power) is not the major problem; rather, it is that in Chu's introductory book only one unknown could be accommodated in an equation at one time, due to the nature of the positional matrix set up on the counting board. Two simple equations such as $x + y = 7$ and $xy = 12$ can be handled by converting the first and incorporating it into the second as $x(7 - x) = 12$, or $x^2 - 7x + 12 = 0$.[45] This works so long as the problems remain relatively simple. But all consecutive conversions and calculations must be kept in the mind as the operations are carried out manually on the counting board, with no record of what has gone by unless written records are kept.

Seki Takakazu (? –1708) provided the breakthrough, but

---

[44] Ibid., pp. 86–90; examples of Sawaguchi's solutions to Satō's problems are given herein.
[45] Asahi, p. 79.

Fig. 28. Portrait of Seki Takakazu. (Courtesy of the Japan Foundation)

not by simple score-keeping methods; he invented a method using algebraic notations for unknowns (*tenzanjutsu*) that permitted written calculations to replace the counting board. The manipulation of counting rods on the board tended to keep one's attention riveted to the operation at hand; Seki's invention allowed the mind to leave an operation once written down, ponder its meaning abstractly, and conceive further possibilities. Moreover, there was no limit, in principle, on the number of unknowns to be treated in a given equation. This was the first and the most creative achievement of Japanese mathematics, one made, of course, on the basis of full mastery of the

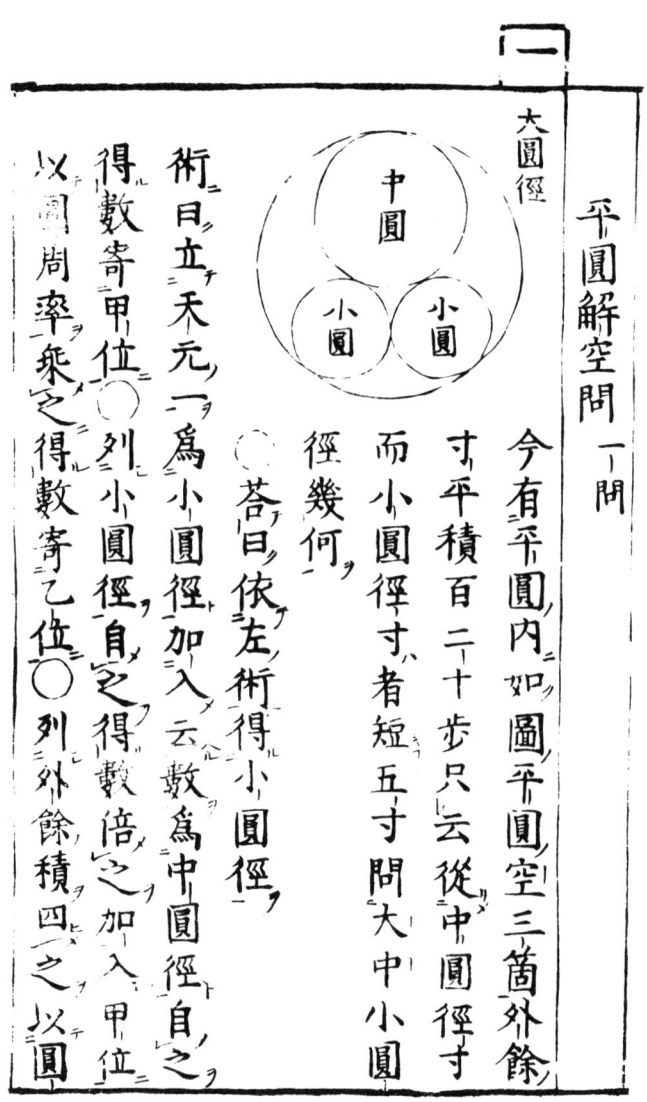

Fig. 29a. A problem, with procedural explanation (from right to left), in Seki Takakazu's *Hatsubi sanpō* (1674), here reprinted as volume 1 of Takebe Katahiro's *Hatsubi sanpō endan genkai* (1685). (University of Tokyo General Library)

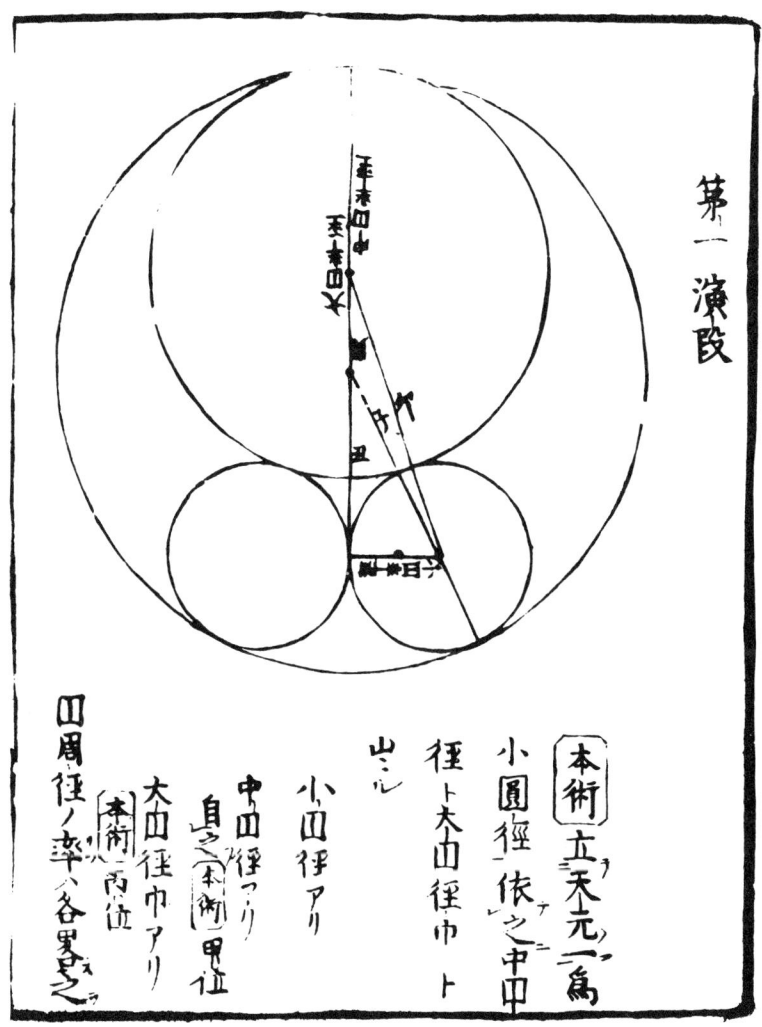

Fig. 29b. Two consecutive pages (pp. 269–270) from volume 2 of the *Hatsubi sanpō endan genkai*, giving detailed procedures for solving the problem in Fig. 29a. Vertical bars in the text on p. 270 are part of Seki's notation system (shorter diagonals through those bars note negative values). (University of Tokyo General Library)

立元二爲大徑○─── 内減小徑余爲二ケ子
小┼───自之得小┼小┼───内減小徑冪余爲
四段丑冪
段寅冪
中┼┼中
○　　　　寄角位
列大往内減中徑余爲二ケ寅中┼───自之爲四
中┼┼中
列中徑加入小徑爲二ケ卯ト中
　　　　寄充位
　　　　　　　自之得
　　　　　　　　木中中木
　　　　　　　　　内減
小徑冪余爲四段丑寅和冪

Chinese tradition. It even transcended the capacity for four unknowns in Chu Shih-chieh's more advanced counting board techniques, which were unknown in Japan.

When Seki in his *Hatsubi sanpō* (Detailed explanation of mathematical method; 1674) recorded his solutions to the problems that Sawaguchi had posed—a mere fifteen altogether —he did not give the intermediate calculations, only the initial problems and final answers. This was the usual practice. Other mathematicians, therefore, did not at first know how he had arrived at this solutions, much less that he had done so by using his own system of algebraic notations. There were many complaints, so his disciple Takebe Katahiro (1664–1739) in his *Hatsubi sanpō endan genkai* (Detailed explanation of calculations in the *Hatsubi sanpō*; 1685) explained the procedures used, from each initial problem step by step to final solution. From this secondary source it was learned that written calculations had been used.[46]

Seki Takakazu made many other creative contributions to *wasan*, principally his use of determinants to solve equations with two or more unknowns. The growing complexity of *wasan* problems had produced a variety of methods used in different equations to resolve unknowns. Seki's method of determinants could be used in all cases. He called this procedure *fukudai* ("covered problem"), which is explained in a revised version of *Kai fukudai no hō* (Method for solving *fukudai*), dated 1683. Thus his invention is considered to have preceded that of Leibniz by at least a decade.[47]

Problems related to the circle had interested mathematicians from the beginning of the seventeenth century, as evidenced by such earlier works as *Warizansho* and *Jinkōki*. The quest for greater accuracy in measurements of circumference, arc and chord lengths, areas, and so on, was called the study of "circular principles" (*enri*).[48] Precedents for this line of inquiry had long

---

[46] Ibid., p. 80.
[47] Smith and Mikami, p. 126; Fujiwara, p. 125; and Pledge, pp. 76, 179.
[48] Smith and Mikami, Ch. 8.

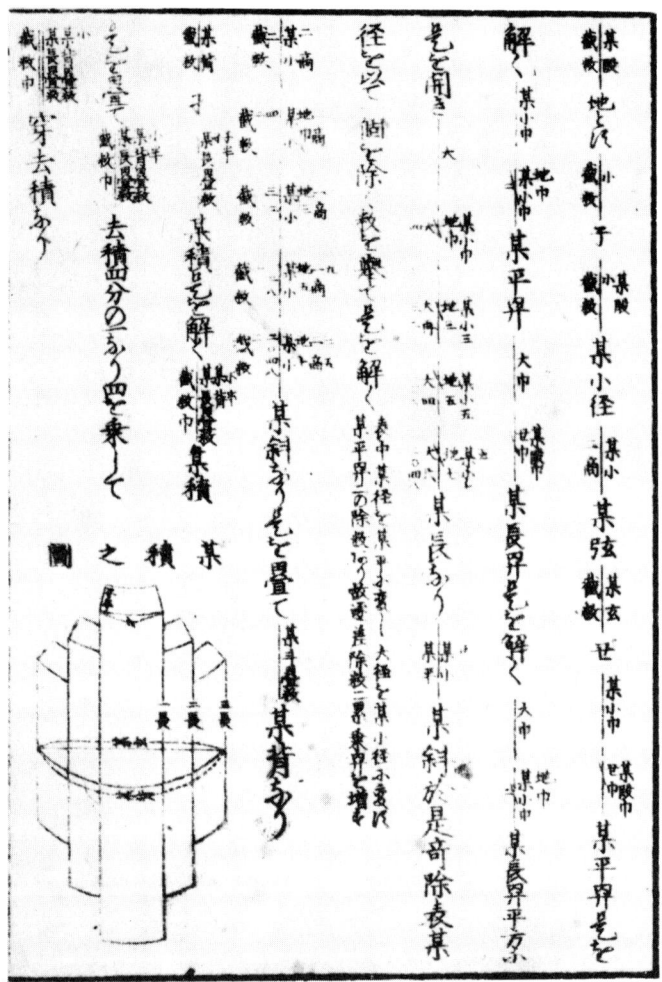

Fig. 30. A problem posed by intersecting cylinders, from the *Sanpō shinsho* (New text on mathematical method; 1830), by Chiba Yūshichi, compiled under the supervision of Hasegawa Hiroshi, a leading figure in the Seki school of *wasan* (Table 25). The first page (p. 273) gives the problem and answer (from right to left), followed by diagrams for the procedures, which are then explained on the next page (p. 272). (The National Archives, Japan)

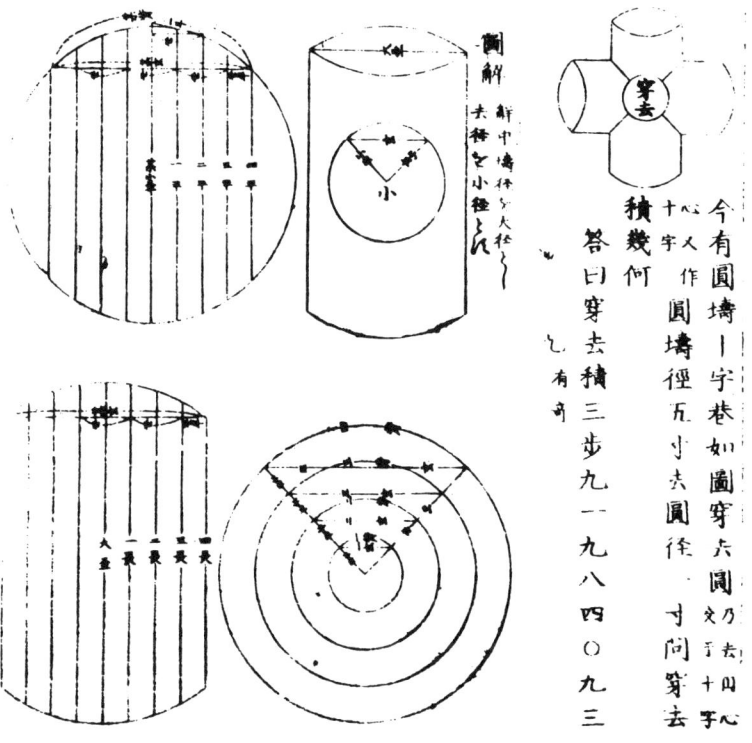

今有圓墻一字巷如圖穿去圓劐于十字心人作圓墻徑五寸去圓徑一寸問穿去積幾何

答曰穿去積三步九一九八四〇九三乙有奇

existed in China. Beginning with a square inscribed within a circle, Muramatsu Shigekiyo (fl. late seventeenth century), for example, doubled the number of sides of the inscribed figures, and all arcs into half, to the 2^{15} degree, yielding a 32,768-sided polygon that approximated the length of the circumference.[49] Seki went a step further, using approximations of 2^{15}, 2^{16}, and 2^{17} to determine a closer approximation of the circumference, which more refined approximations may approach indefinitely but cannot reach.[50]

THE INTELLECTUAL MILIEU OF *WASAN*

Early in the Edo period samurai did not necessarily feel ashamed of engaging in mathematical work.[51] By the middle of the seventeenth century, however, the status hierarchy was fairly well established and a new generation of samurai had been indoctrinated in the conscious distinctions of status groups. Confucian thought not only supported this system, its traditional alliance with agrarian values made it scornful of the calculating mentality of merchants. The disdain for practical mathematics became extreme as the gap steadily widened between the rising wealth of merchants and the declining fortunes of samurai. One view held that "the *soroban* [abacus] should never be touched by a samurai,"[52] and Ogyū Sorai remarked caustically, "Mathematicians boast of their exacting achievements, but in reality they are absorbed in mental acrobatics and contribute nothing to society."[53]

By the end of the seventeenth century, then, there were essentially two kinds of mathematics—one highly practical, the other proudly impractical—and each was content to evolve on its own terms, for its own purposes. Needham makes the point that Chinese mathematics was quite closely tied to the practical needs of the bureaucracy: land mensuration and surveying, granary dimensions, construction of dykes and ca-

[49] Ibid., pp. 77–79.
[50] Ibid., pp. 109–112.
[51] Sugimoto (2), p. 142.
[52] Hirayama Akira (2), p. 86.
[53] Ogyū Sorai, p. 145.

nals, taxation, rates of exchange, etc.; and that "of mathematics 'for the sake of mathematics' there was extremely little."[54] Japanese practical mathematics was in the service of bureaucracy and economy no less than its ancestor. But *wasan* was precisely "mathematics for mathematics' sake"—or at least for the sake of mathematicians, as a hobby. Needham makes his point while pondering why the Chinese search had not led along paths similar to those that yielded abstract systematized mathematical truths to the Greeks. Among other reasons, one he raises is whether the counting board and abacus might have been inhibiting factors, as they both allow calculations to vanish without a trace. What is perhaps more fundamental is that the Chinese never evolved a demand for rigorous proofs which would have focused attention on the sequence of steps in the solutions of problems.

Seki Takakazu overcame at least the problem of lost calculations by inventing step-by-step notations. But in actual fact neither the counting board nor the abacus completely disappeared. This was partly because Seki's inventions were not made fully accessible to all mathematicians until much later (see p. 365), and even then in violation of the spirit of protective secrecy surrounding sectarian schools of *wasan*. Beyond that, one may suppose Seki's inventions might have formed a more exacting system if *wasan* had been in closer touch with the technical, economic, and social conditions of the day.

In speculating, one might recall that mathematicians were quite active in calendrical reforms during the seventeenth century. After the calendrical reform of Shibukawa Harumi, however, both *wasan* and calendrical astronomy tended to become specialized and separated. Calendrical work was ignored by *wasan* devotees as part of practical mathematics.

SOCIAL ASPECTS OF *WASAN*

Education for commoners was available in the *terakoya* (lit., "temple schools," but mostly no longer connected with tem-

---

[54] Needham, 3:153.

ples), where the main curriculum was reading, writing, and arithmetic.[55] Use of the abacus was included in arithmetical training, as many of the pupils were from townsmen's households and would need abacus skills. Those who for business or personal reasons wanted further instruction in practical mathematics could easily obtain it in the many private academies for that purpose.

Private academies for *wasan* were often separate from those for practical mathematics but were equally available in large towns; smaller towns were visited by itinerant *wasan* teachers. But *wasan* on sophisticated levels was strictly an intellectual pastime. *Wasan* hobbyists looked down socially upon the prudential calculations of merchants, though others of the intelligentsia did not necessarily regard *wasan* as a high-class hobby. Most devotees were otherwise employed; Seki Takakazu, for example, was an official in the shogunate administration,[56] and not a few of his disciples were samurai.[57] On the other hand, many of the masters of the *wasan* academies were *rōnin* or townsmen, for whom *wasan* was a profession. Neither the shogunate nor any of the domains founded or supported schools or courses of study for *wasan* until much later in the Tokugawa period.

The point at which *wasan* became most visible to the general public was around 1670 when the custom began of offering votive plaques (*sangaku*) to some temple or shrine.[58] On each plaque was inscribed some *wasan* problem and its solution, without intermediary calculations. Hung in full public view, the plaques probably served to mystify, but hardly to edify, ordinary citizens. The original purpose of offering votive plaques was to express gratitude to some Shinto deity or to the

[55] Ogata Hiroyasu, pp. 150–153.
[56] From 1704 he assumed a relatively low-ranking post. Hirayama Akira (1), p. 18.
[57] The most prominent of his disciples, the Takebe brothers, Kataaki (1661–1716) and Katahiro (1664–1739), were samurai of the shogunate. Araki Murahide (1640–1718), first Sōtō (Master) of the Seki school, was a samurai, though he became a *rōnin* in his later years. Ibid., p. 47.
[58] Hirayama Akira (2), pp. 22–24.

Buddha. It became fashionable also as a means of publicizing one's ability and of advertising the achievements of sectarian schools. As the intermediate calculations and the attendant reasoning were omitted, this custom was of negligible significance in the development of mathematics, except as a challenge to other devotees; and though this practice served to attract new students to *wasan*, in time, like *wasan* itself, it degenerated into trivialities.

MEDICINE

In the two centuries from Manase Dōsan's time in the mid-sixteenth century to the middle of the eighteenth century Japanese scholars managed to explore fully the whole corpus of Confucianism and Chinese medicine that had developed over three millennia. Dōsan and his teacher Tashiro Sanki had already introduced the latest of the major developments in that long history, the Li and Chu medical schools of the Yüan period (1279–1368). The seventeenth century was a time of introducing and reintroducing, comprehending and establishing other major medical traditions, working back, as it were, in time. That is, the Liu and Chang medical schools of the Chin period (1126–1234) established themselves in Japan about mid-seventeenth century, from which time also appears a reaction against what was seen as the excessive Neo-Confucian dogmatism of the Liu-Chang and Li-Chu traditions. The reaction looked back to the ancient medical classics (whose conceptual foundations were conveniently overlooked) and emphasized practical approaches to cures. Known as the "Ancient Practice School" (*Kohōha*), it had close affinities with the similar reaction to Neo-Confucianism proper, the Ancient Learning School (*Kogakuha*).

Only in the field of medicine were there such close parallels with developments in Neo-Confucian philosophy. This was not because Confucianists were particularly interested in medicine; some, like Itō Jinsai, while not forbidding students a medical career, nevertheless felt it less than a Confucian scholar's proper vocation.[59] The major lines of influence ran the other way: medical scholars were first to comprehend the Neo-Confucian system as the background for the Chin and Yüan medical traditions. In the middle of the eighteenth century the level of debate, attack, and refutation among major medical traditions and within them reached new intensity. An Eclectic School (*Setchūha*) emerged as one attempt to mediate between the highly theoretical positions of the Chin and Yüan traditions and the more practically inclined Ancient Practice School. The eclectic approach concentrated largely on textual criticism and failed to introduce any new generative ideas into Japanese medicine. It did, however, eventually become the mainstream of Japanese medicine and, thus, the greatest obstacle to the introduction of Western medicine.

In the great debate of the late eighteenth century was woven a slim strand of Western medical influence, almost imperceptible at first but soon triggering the famous "Dutch learning" (*Rangaku*). Dutch studies widened afterward into "Western learning" (*Yōgaku*), especially when national defense needs became urgent, and the foundation was laid for the Meiji transformation when the whole of Japanese medicine, at least officially and on most scholarly levels, would become Western. Just as medicine constituted the primary agency for the exploration and eventual assimilation of Confucianism two centuries earlier, so also was medicine the primary stimulus and route for the reintroduction and assimilation of Western learning and science in the late eighteenth and early nineteenth centuries.

Chapter 5 will begin by picking up the great debate and then trace the course of renewed Western influx; the present chapter concentrates on the extension of Li and Chu medicine, the introduction of Liu and Chang medicine, the reaction against these traditions by the Ancient Practice School, and the residual knowledge of Western surgery kept alive by Nagasaki interpreters. The time span differs slightly from

[59] Fujikawa Yū, pp. 349–350.

other fields, running roughly to the middle of the eighteenth century.

THE EXTENSION OF LI AND CHU MEDICINE

After Manase Dōsan and his early disciples succeeded in making the Li and Chu traditions dominant in Japan, his school continued to attract many able followers, and not a few of them had among their clientele members of the new political elite—shoguns, daimyo, and high-ranking retainers —as well as dignitaries of the traditional aristocracy, the imperial household and court nobility. The most prominent physicians received high recompense for cures and not infrequently were awarded the higher Buddhist titles. Among the most famous disciples of Dōsan's son Manase Gensaku were, for example, Okamoto Genya (1587–1645), who treated the emperor and the first three Tokugawa shoguns and is said to have had about a thousand disciples; and Inoue Gentetsu (1602–1686), whose disciples were also counted by the thousands.[60] Furubayashi Kengi (1597–1657), a second-generation Dōsan disciple, established a private medical academy in Saga (Kyoto) that reportedly had three thousand students.[61] The scholars in this tradition were devoted teachers and prolific writers; one of them is said to have written more than a dozen works. Little wonder that the Li-Chu tradition became the mainstream of Japanese medicine and contributed greatly to its growth.

(The Li and Chu traditions in Japan came to be called together the *Goseihōha*, meaning rather literally, "the school of a later age" [i.e., the Yüan period], in contrast to pre-Sung practice. If in fact it was coined at that time, the term *Goseihō* probably had the connotation "modern school of medicine" and hence was attacked precisely for its "modernism" by the *Kohōha* which led the movement to reaffirm the primacy of the ancient medical classics. The first challenge to the *Goseihō* school in Japan came from the Liu and Chang traditions

[60] Ibid., pp. 100–101.
[61] Ibid., p. 102.

which, though historically older [i.e., Chin period], were introduced at a later date. The Liu-Chang line came to be known as the *Goseihō beppa*, that is, "variant modernist school." It is possible, however, that these terms were all coined much later and were never used by the Edo era medical scholars themselves.)

THE INTRODUCTION OF LIU AND CHANG MEDICINE

The voice of criticism against the Li-Chu position in favor of the Liu-Chang line was first raised by Aeba Tōan (1615–1673) in the 1650s. He and his followers, the most prominent of whom was Okamoto Ippō (1686–1754), did not attack the theories of the *Goseihō* school so much as its laxity in following those theories. The *Goseihō beppa* emphases on phase energetics were much stronger than in the *Goseihōha*, whose proponents were accused, for example, of simply adding more ingredients to medicinal prescriptions to be on the safe side rather than focusing on specific and usually stronger drugs. Use of strong drugs was especially characteristic of Chang's approach to therapy. Though disciples were attracted to the Liu-Chang position and the *Goseihō beppa* became a major school, it never gained sufficient strength in Japan to compete with the Li-Chu medicine of the *Goseihōha*. The Chin and Yüan traditions together became instead the target of growing frustration with, and rebellion against, their complex theories, which were increasingly seen as being of little help in actual therapy. Those who mounted this attack called for a return to the spirit and methods of the ancient medical classics by claiming their authority as a basis for the Ancient Practice School.

THE REACTION OF THE ANCIENT PRACTICE SCHOOL

In attacking Chin and Yüan medicine, the Ancient Practice advocates were not totally rejecting the broad basic assumptions of correspondence between the human organism and the cosmos, or even the use of *yin-yang* and Five Phases concepts to describe bodily functions—these ideas were deeply imbedded in the ancient medical classics and were taken for granted in China and, by this time, in Japan as well. What they objected to in the "modernist" traditions were schemas constructed with more concern for metaphysical symmetry and function than for observable (but seldom observed) physical structures and action, as well as therapeutic emphases derived in one-sided ways from such elaborately abstract theories (though the therapies were not as innovative as sometimes represented). The mastery of medical practice was not, in the growing mood of those rejecting modernist medicine, to be equated with mastery of Chin and Yüan theories. The Japanese "ancients" were at least partly right in charging the "modernists" with having corrupted much of the simplicity of early Chinese medicine. The ancient medical classics they claimed to be returning to were not by any means void of theoretical bases. The Ancient Practice School mythologized the past to create a new and more empirical approach to therapy which was unprecedented in China.

The major shift away from Chin and Yüan medicine in China was made by the "Conservative School" (*Hsin-ku p'ai*) led by Yü Ch'ang, whose 1648 publication "'On Cold Damage Disorders' Reaffirmed" (*Shang lun p'ien*) vigorously advocated a return to Chang Chung-ching's third-century classic (p. 85). Yü opened up the promise of a restored classic purity by showing how greatly the *Shang-han lun* had been altered by its fourth-century editor Wang Shu-ho. Yü did not, however, take the radical positions that his school encouraged in Japan. He still treated the Yellow Emperor's Inner Classic as an important authority (suggesting some important reinterpretations of it) and of course used the *yin-yang* and Five Phases concepts in his discussions of etiology and pathology. It is only in his delineation of the role of various sorts of *ch'i* in the functional systems of the body that we see the beginning of a more materialistic approach. Although this concept is mainly energetic in traditional medical discourse, Yü treats it much more pneumatically than usual. His major influences from the *Shang-han lun* were in the reclassification of certain disorders and in adopting its practice of making diagnoses in terms of entities named for the prescriptions that counteract them—a somewhat more direct linkage than in the much more abstract set of eight polar entities used in the Inner Classic's diagnosis.

Nagoya Gen'i (1629–1696) in Japan based his attack on this book but reinterpreted its message to castigate in his contemporaries what he felt was undue indulgence in theoretical considerations. His advocacy of the Ancient Practice position in medicine came, incidentally, a decade before Yamaga Sokō and Itō Jinsai broke with Neo-Confucianism and called for a return to Ancient Learning[62] (just as Manase Dōsan's turning to Li and Chu medicine came a half-century before Hayashi Razan established Neo-Confucianism in Japan). While Nagoya Gen'i's attack on the *Goseihōha* and *Goseihō beppa* schools was not decisive, it came at a time when their creativity was waning. Gotō Gonzan (1659–1733) as a youth failed to win acceptance as a disciple of Nagoya Gen'i but followed his precepts and succeeded in establishing the Ancient Practice School.[63] Gonzan practiced and taught in Kyoto, attracting about two hundred students. Like Gen'i he rejected Chin and Yüan theories and had a particular aversion to phase energetics theory. But neither Gonzan nor Gen'i renounced the simpler metaphysics of the Yellow Emperor's Inner Classic. Gonzan still sought some basic unity for medical theory, which he expressed as the "stagnation of the unitary *ch'i*" pathological principle, according to which in sickness the basic vital pneuma (*yüan-ch'i*; in Japanese, *genki*), that should circulate freely in the body, stagnates (i.e., loses its dynamic functions).[64] His therapy, therefore, was "to induce free circulation of the *ch'i*," and his prescriptions favored stimulants, such as moxibustion, bear's gall, cayenne pepper, and hot spring baths, "to release the *ch'i*."

The independent Gonzan often relied on methods not highly regarded by others. He despised and openly criticized the generally accepted custom requiring medical practitioners to shave their heads and don Buddhist robes. He especially reproached his colleagues for accepting honorary Buddhist

[62] Ibid., p. 296.
[63] Ibid., p. 343.
[64] Ibid., pp. 344–347.

titles from the court and challenged them to return to the "ancient custom" of practicing medicine as secular men. His disciples and some sympathizers followed his advice, and secular dress gradually replaced the clerical habit as the influence of the Ancient Practice School spread. There were, however, institutional limits on individual choice in the matter; medical men serving the shogunate or daimyo with assessed domain revenues of 100,000 *koku* of grain (1 *koku* equals approximately 5 bushels) or more were required to wear the clerical robe and shave their heads.[65]

Gonzan's most noted disciple was Kagawa Shūan (1683–1755), who also studied Ancient Learning under Itō Jinsai. Shūan went beyond his predecessors; he not only rejected Chin and Yüan medicine but even the Yellow Emperor's Inner Classic and On Cold Damage Disorders as well.[66] He disclaimed the use—in either pathology or prescriptions—of all theories having to do with *yin* and *yang*, the Five Phases, and the ways they were related to each other and the rest of creation. He insisted that the way to confirm the effects of drugs was one's own experience and observations. The elaborately schematized theories he regarded as baseless, evil, and disgusting. Shūan was not without all sense of need for unity in interpretation; he allowed that Confucianism and medicine were one in essence, a sentiment reflected in his literary name Ippondō (Hall of the Single Basis) and thus in the title he gave to a collection of prescriptions, *Ippondō yakusen* (Selected drugs of Ippondō; printed in 1729).

Shūan had about four hundred disciples, and their abilities in both practice and education were to make this school the dominant one from the middle of the eighteenth century. To become what we frequently call the "mainstream" is, however, to attract the keenest minds and hence eventually to invite criticism from within. This is precisely what happened in the 1750s, and the great debate that arose around Shūan's suc-

[65] Yamazaki Tasuku (2), p. 451. See also p. 289, this chapter.
[66] Fujikawa Yū, pp. 350–352.

cessors, men like Yoshimasu Tōdō (1702–1773) and Yamawaki Tōyō (1705–1762), was symbolic of a deeper ferment that was to push Japanese medical scholars once again into the forefront as pioneers in the introduction and assimilation of a new system of learning, this time based on renewed influx from the West in the form of *Rangaku* (Dutch learning).

RESIDUAL WESTERN MEDICINE

The practice of the "Southern barbarians' surgery" (*Nanban geka*) was, along with the survival of interest in Western cosmology, the only other slim thread of Western science not cut off completely by isolation. It was, however, quickly renamed "Surgery of the Red-hairs" (*Kōmō geka*), that is, of the Dutch, who were legitimately in Japan and had no Catholic connections. In the beginning there was, of course, essential continuity between *Nanban* and *Kōmō* surgery, with no particular progress beyond what was known in Western Wave I. The only carry-over in preisolation foreign personnel was the apostate Sawano Chūan, who wrote, or had written for him, some materials on surgery, but they represented Western surgery prior to Paré.

Kōmō surgery was promoted mainly by Dutch interpreters whose principal duty was to serve the Dutch-Japanese trade. At first the interpreters were not highly proficient in the Dutch language—indeed, most never became so—though most of them knew some Portuguese, with which many Nagasaki residents had become familiar during Western Wave I. Most of the interpreters interested in the Red-hairs' surgery learned what techniques they could by observing operations and, if they had access to them, from Chūan's monographs. One of the interpreters, Narabayashi Chinzan (1643–1711), not only acquired uncommon skill in the Dutch language but also was able to study under a Dutch doctor, Willem Hoffman, who stayed at Dejima from 1671 to 1675. Chinzan translated portions of Paré's book on surgery into Japanese under the title "Orthodox tradition of red [-hair] surgery" (*Kōi geka sōden*; 1706). It was widely copied by many schools of *Kōmō* surgery and passed on as the "secret manual" of their trade. The same content was often put into different forms, with various titles.

This work was one of the first two translations by a Japanese of any Western medical book—indeed, of any Western scientific book.[67]

Compared with the place of traditional medicine, and especially with that of the dominant *Goseihō* school, the existence of small numbers of *Kōmō* surgeons here and there was meager indeed. Theirs was but the unsophisticated practice of unlearned men who had picked up a few techniques but knew little or nothing of the complex systems of internal medicine. But, because traditional medicine regarded surgery as the lowest of all its skills and largely neglected it, the *Kōmō* surgical practice was generally superior to that of learned traditional practitioners, and some of the *Kōmō* surgeons were called into the service of the shoguns and daimyo.

As a whole, though, the *Kōmō* practice was rather loosely organized. Many an interpreter of Dutch got some knowledge from a European doctor stationed at Dejima and claimed to be a "master" of *Kōmō* surgery. Some practiced it as an avocation, while remaining interpreters; others resigned from interpreting to practice it full-time. Not a few of the full-time surgeons made *Kōmō* surgery a kind of private "house learning" taught in private academies. They rarely published what they knew; their secrets were transmitted privately only to their disciples—sometimes in the form of handwritten manuals for restricted circulation. As the reputation of the *Kōmō* surgeons grew, many aspiring students came to Nagasaki from other parts of the country and, after rather brief instruction under one of the *Kōmō* masters, returned home and hung out their shingles as *Kōmō* surgeons. Regardless of their low standards and random competence, the neglect of surgery by traditional medicine left the field wide open to the *Kōmō* surgeons.

A few of the schools and some of their activities bear mentioning. One of the schools went by the name "Caspar school" (*Kasuparuryū*), as it claimed to have gotten its knowledge of

---

[67] Koga (1), pp. 139–145. The other was Motoki Ryōi's (1628–1697) translation of an anatomical text (see p. 382).

surgery directly from Caspar Schamberger, who stayed in Japan from 1649 to 1651. As this was very early in the isolation period, the "Caspar school," with many minor branches, dominated the initial stage of *Kōmō* surgery.[68] The Arashiyama school was founded by Arashiyama Hoan (1633–1693) of the Hirado domain; he had received instruction from three of the Dejima doctors and in 1683 wrote two volumes of "Classified records of the barbarians' therapy" (*Bankoku chihō ruijūteki den*; unpublished). The Arashiyama family won a hereditary post as *Kōmō* surgeon serving the Hirado lord. A disciple of the Arashiyama school, Katsuragawa Hochiku (1661–1747), acquired a hereditary position as *Kōmō* surgeon in the Edo shogunate. Katsuragawa Hoshū (or, Kuniakira; 1751–1809), in the fourth generation of the Katsuragawa household, was to participate in the epoch-making translation of the *Kaitai shinsho* (see Chapter 5); he and his son Katsuragawa Hoken (1786–1844) became prominent scholars of "Dutch learning." This linkage is reminiscent of Nishikawa Joken and his son Seikyū, who came from the rather insignificant by-stream of Western cosmology but managed later to enter briefly the mainstream of Japanese calendrical astronomy by dint of their own talents. Nishi Genpo (?–1684), son of Nishi Kichibe (p. 259), resigned his hereditary interpreter's position in 1669 to devote himself full-time to *Kōmō* surgery; from 1673 he was employed by the shogunate, and his surgical services were later officially recognized with an award of a high Buddhist rank. Genpo wrote a work on surgery in which the illustrations were taken from the Dutch version of Paré's work.

The total number of European doctors who were stationed at Dejima during the isolation period, or, more precisely, from 1641 to 1858 when the last one left Dejima, exceeds a hundred;[69] most stayed only two or three years. While it is likely that not a few of them had received authentic medical education in Europe (some were only barber-surgeons), it may be assumed

[68] Ibid., p. 58f.
[69] Ishihara (2), pp. 130–131.

that most of them never gained any real proficiency in the Japanese language. Whatever medical knowledge was transmitted went through the agency of those Japanese serving as official interpreters. So it was that *Kōmō* surgery emerged from the interpreters' community. The Dejima doctors sometimes presented their Japanese "students" with "diplomas" that naturally enhanced their prestige as surgeons.[70]

Some of the Dejima doctors were very eager to study Japan. Though confined to the small enclave on Dejima island, except for the required annual trips to Edo,[71] they succeeded from time to time in collecting a considerable amount of information on things Japanese and in publishing it after returning to Europe. Notable among such men during the period here reviewed were Willem ten Rhijne (1647–1700) and Engelbert Kaempfer (1651–1716). Ten Rhijne was at Dejima from 1674 to 1676 only. He studied acupuncture and moxa therapy and after his return home introduced the Japanese forms of these techniques to Europe for the first time. Kaempfer was in Japan two years, 1690–1692, and traveled to Edo twice. He accumulated a great deal of research material with the help of such men as Narabayashi Chinzan, Motoki Ryōi, and others, who studied surgery under him. Though Kaempfer died in 1716, a completed manuscript of his was published in English translation as *The History of Japan* (2 vols.; 1727), and his collection of Japanese books, maps, and specimens was later acquired by the British Museum (opened in 1759).[72]

In their practice the *Kōmō* surgeons also made use of traditional Chinese-style surgical methods and the "combat surgery" developed during the Sengoke era, and they relied upon simple instruments and medications. Beyond observing Western surgical operations and receiving simple instructions in techniques, they did not engage in studies of contemporary European physiology, pathology, pharmacology, and so on.

[70] Ibid., pp. 134–135.
[71] See n. 21 above.
[72] Though his collection of books included no medical texts, he was active in introducing acupuncture to Europe. See Bowers, pp. 38–58, esp. 55f.

In the political climate of the seventeenth century such initiatives were too risky, and the Dejima doctors were quite cautious about exposing themselves to penalties or expulsion. (This was no idle speculation, as we shall see in the case of Siebold, who was deported from Japan for his indiscretions; see pp. 338–344.) The *Kōmō* surgeons, in fact, lacked authentic training in either traditional or Western medicine. They were in no position to serve as a channel for the systematic reintroduction of Western medicine. That channel was to be opened by traditional medical scholars, and then only after they had first fully mastered and explored the limits of Chinese medicine. Then it was that some scholars working within the traditional framework became critical of it and began to look to Western medicine for more than surgical techniques.

SOCIAL ASPECTS OF MEDICINE

By the end of the seventeenth century the shogunate had endorsed the Yushima Seidō as its official institute for Confucian studies, and in its last two decades fifty-five official domain schools (*hankō*) had been founded.[73] A parallel growth in medical schools was to come much later, when the private medical academy (est. 1765) of the Taki household was made a shogunate institute of medicine late in the eighteenth century (p. 375). In the first half of the Tokugawa era there were no institutes for medical education founded with support from the public sector—shogunate, domains, or religious bodies. The Ten'yakuryō was formally maintained by the court to serve court circles; headed hereditarily in Tokugawa times by members of the Nakarai family (descendants of the Wake household), they were occasionally unable to fill their hereditary post, and prominent medical doctors from outside these households received the appointment and a high Buddhist rank with it. Otherwise, medical education was handled entirely by private academies, which represented the different varieties of traditional medicine. Most of the academies ceased to exist after the founding master died; maintaining a parti-

[73] Dore, p. 25.

cular tradition of medical thought and practice depended upon one or more of his disciples starting their own academies.

Specialization of medical practice characterized previous periods, but in the seventeenth century this came to be a matter of course. A more interesting facet of the seventeenth century, perhaps, was the legal ranking applied to specialists serving the shogunate. The ranking retained the traditional emphasis of Chinese-style medicine on internal medicine as the core of all theory and practice, i.e., the "main way" (*hondō*) to health.[74] All other specialists ranked below *hondō* physicians, whether they were trained in pediatrics, eye treatment, oral hygiene, and so on, or were practitioners of acupuncture or moxa therapy, or simply surgeons (lowest in the scale).

Many physicians served the shogunate and domains in official and usually hereditary capacities; the surgeons mentioned earlier were only a small minority of those in official employ. Moreover, the reason for the requirement that doctors serving the shogunate and daimyo with revenue over 100,000 *koku* wear clerical dress was to make it clear that they stood outside the status hierarchy of samurai-farmer-artisan-merchant. Most important, they were not accepted as samurai; though Confucian scholars, who in Hayashi Razan's time were similarly treated, later could drop the clerical robe and role and carry out their professional duties in a samurai status. Medical practitioners in towns and villages usually wore the Buddhist robe, though some followed the lead of Ancient Practice advocates in discarding it.

It is remarkable that so much was accomplished in the field of medicine, with few medical academies lasting beyond the lifetime of a single master, with little public support for the academies, and with most documents at this stage not printed but only copied and handed down privately from master to disciple. There were no equivalents of a medical association or

---

[74] *Hondō* is the direct successor of *Tairyō*, central course in the Ten'yakuryō curriculum (Table 5).

a medical journal. There appears to have been no equivalent even of the "problem succession" (*idai keishō*) process found in *wasan*. Yet scholarly interest was sustained at such a high level that medicine in the late eighteenth and early nineteenth centuries would again pioneer, as it had repeatedly in Japanese history, in opening up new vistas and discovering new knowledge and techniques relevant to the Japanese situation.

5

The Shift from Traditional to Modern Science

Challenge to Isolation: 1720–1854

The eighteenth-century impulses that eventually led to a shift away from traditional science toward Western science initially arose as largely academic concerns within traditional learning and science. Problems unsolved by traditional disciplines in time led some scholars to Western sources in search of answers. From early in the nineteenth century foreign threats to Japan's isolationist stance increased the urgency of the quest for Western science and technology, and the new knowledge was rapidly coopted by the shogunate and domains for defense purposes. Through these processes the foundations were being laid for the overall adoption of modern science in the Meiji transformation.

Following the seventeenth-century intellectual renaissance many Japanese scholars grew dissatisfied with traditional learning. Official promotion of one branch or other of learning or science served only to accentuate its limitations in solving scholarly problems or resolving social tensions. As Chinese Wave II appeared to be exhausting its capabilities, the introduction of Western science was naturally enhanced. In 1720 Shogun Yoshimune was persuaded to relax restrictions on imports of foreign books as a means to expedite his interests in calendrical reform; and he also sanctioned Dutch-language study by selected scholars. The new opening to Western knowledge thus created was then effectively exploited in the latter half of the eighteenth century by certain medical scholars to learn more about Western medicine. From this there developed the studies and translations known as "Dutch learning" (*Rangaku*).

Western science did not long compete with its traditional Chinese-style counterpart on purely intellectual terms; political priorities soon entered to dominate the process. Once isolation excluded the Portuguese and Spanish presence and confined Japan's European contact to the Dutch company on Dejima, a gradual though radical change took place in the Western approach to Asia. European states with modern economies and modernized armies and navies engaged in aggressive competition for colonies as sources of raw materials and as potential markets. When this competition later extended to Japan, it was the British, not the Dutch, who led the field.

The challenge to isolation came first, however, from the Russians, who as early as 1778, but repeatedly around the turn of the century, sought entry to Japanese ports for trade. While Japanese officials debated defense of the northern territories, the challenge assumed added seriousness when the British frigate *Phaeton*, in connection with the Napoleonic wars, called at Nagasaki harbor in 1808 in search of Dutch ships (none were there at the time). The first real shock to Japanese officialdom came, however, from the British victory over China in the Opium War of 1840–1842. Although the King of Holland in 1844 sent a message to the shogunate to explain the changing world situation and to recommend that Japan be opened to international intercourse, isolation was adamantly maintained until Commodore Matthew C. Perry of the United States succeeded in forcing open Japan's ports in 1854.

The United States had a far more active interest than had England in the challenge to isolation because her Pacific whaling fleets needed shelter ports and provisions. The Dutch governor of Batavia (Indonesia) had in 1852 already informed the shogunate of the American plan and advised acceptance of treaty proposals for trade and amity relations with the United States and afterward with other Western powers. This advice received little consideration until Perry's small fleet of four ships (including two steamships) arrived off the coast of Uraga at the mouth of Edo Bay on July 8, 1853 (Gregorian calendar; June 3 by the Tenpō calendar), and threatened to use force if the shogunate refused to receive—at Uraga, instead of

Nagasaki as the shogunate proposed—a state letter from the president of the United States.

When isolation was instituted two centuries earlier, the shogunate and domains possessed sufficient firepower to enforce it. By mid-nineteenth century, scientific and technological progress in the West had wrought vast improvements in Western military capabilities. Not only the capital of Edo but also vital coastal shipping lanes were now threatened; stubborn adherence to the isolationist line only invited possible colonization. The state letter was received on February 13 (Gregorian; January 16, Tenpō) of the following year (1854) when Perry returned with eight ships; a treaty of amity with the United States was signed on March 31 (Gregorian; March 3, Tenpō) at Kanagawa (present-day Yokohama).

Captitulation to the American demands might not have occurred so readily had not Japan also faced mounting domestic crises. One of these was a financial crisis; annual tax receipts, gleaned largely from agricultural production, increasingly failed to keep pace with swollen shogunate and domain expenditures.[1] The overtaxed rural sector was itself undergoing gradual disintegration, one aspect of which was the polarization into a small class of wealthy landowners and an impoverished majority of peasants who, with increased frequency, reacted with uprisings to natural calamities, low producers' prices for rice, and local maladministration.[2] While

---

[1] Repeated programs of retrenchment and reform sought to curtail extravagance and to instill greater discipline at all levels, notably in the Kyōhō (1716–1736), Kansei (1798–1801), and Tenpō (1830–1844) eras. They were conservative reforms aimed at reinstating an idealized agrarian society. In fact, though, tax claims which gave the shogunate and domain governments about 40 percent of production early in the Tokugawa period were adjusted to a fifty-fifty basis by the beginning of the eighteenth century, and later in the Tokugawa era the earlier ratio was reversed, giving the governments 60 percent and leaving only 40 percent for the peasants. Cf. Reischauer and Fairbank, p. 603; Yazaki, p. 246.

[2] In the 268-year period from 1600 to the end of the Tokugawa era, there were about 3,200 peasant revolts (*hyakushō ikki*). Relatively infrequent before the eighteenth century, peasant revolts occurred with ascending frequency and on larger scales in the Kyōhō and Tenmei (1781–1798) eras, and especially in the final days of collapsing Tokugawa power. Cf. *Nihonshi jiten*, p. 805.

most poor peasants remained in the rural sector as tenant farmers[3] or hired hands, many chose to migrate to the cities. The cities, in turn, had only limited capacity to absorb the influx of poor peasants,[4] so that urban riots also marked the declining years of the Tokugawa order.[5] Erosion of the feudal system was further evident in the status conflicts brought about as more and more low-ranking samurai sought to supplement their meager incomes by taking secondary jobs as simple craftsmen, while wealthier merchants could buy their way into privileges previously enjoyed only by samurai.

Relaxing the ban on books in 1720 was comparatively easy; ending isolation in 1854 involved much greater risks. But the Japanese leaders faced the difficult juncture of the mid-nineteenth century with much broader perspectives on the outside world and, as this chapter shows, a fledgling academic foundation of modern scientific knowledge with which to risk competing with the West on its own terms. In the post-Perry adoption of modern Western science and technology, learning and science in academic institute, manufacturing shop, and administrative bureau were wedded to defense needs. Rapid progress, especially after the Meiji Restoration of 1868, was possible only because of the preparatory work done in Western Wave II. This in turn was dependent on the legacy of traditional science and learning cultivated in and since the seventeenth-century intellectual renaissance.

---

[3] The rate of tenancy is estimated to have approached one-third of the peasant population late in the Tokugawa period. Yazaki, pp. 246, 249; Reischauer and Fairbank, p. 631.
[4] One indication of the problem is the percentage of city-dwellers occupying rented quarters, which, from the end of the seventeenth century to the end of the Tokugawa period, climbed to between 50 percent and 70 percent in most cities, and as high as 80 percent in Nagasaki and Sakai. Yazaki, p. 257.
[5] Over two hundred urban riots are said to have occurred in the eighteenth and early nineteenth centuries, with only ten erupting in the Kyōhō era, but as many as sixty in the Tenmei era and forty-five in the Tenpō era. The last twenty years of the Edo era saw thirty-five urban riots, half of them in a desperate three-year period of excessive increases in taxes and the cost of rice. Ibid., p. 258.

Learning and Science

From early in the eighteenth century to the middle of the nineteenth century official and private activities in traditional learning and science expanded greatly. Already in 1690 the shogunate had adopted the Hayashi academy as its official institute for teaching *Shushigaku*; in 1797 it took over direct control of this academy from the Hayashi family. Meantime, in 1791 the private medical academy of the Taki household was recognized by the shogunate as its official institute for medical education. By the end of the Tokugawa era there were well over two hundred domain schools (*hankō*)[6] and approximately fifteen hundred private academies of all sorts.[7] Some of the latter were strictly for samurai, though many served to educate commoners, as did the thousands of *terakoya* (not usually related to Buddhism by this time). The social impact of the lower-level schools is suggested by the estimate that at the end of the Tokugawa era nearly one-half of school-age boys and about one-tenth of the girls received three to four years of schooling. That the literacy rate ran high seems evident from another estimate that already in 1710 there were over six hundred publishers and booksellers in Japan.[8]

While traditional learning spread widely and became better understood, by late in the eighteenth century it began to lose its creativity and authority. The shogunate sought in 1790 to salvage the authority of *Shushigaku* by imposing a prohibition on the teaching of competing schools of Confucian thought at the Shōheikō, but in fact eclecticism was the dominant trend. Split by criticisms within, traditional systems were also increasingly challenged by the renewal of Western studies—prompted first by Yoshimune to aid his program for calendrical reform and carried farther by later calendrical reforms and especially by activities among medical scholars. Initially confined to "Dutch learning" (*Rangaku*), by the time Commodore

[6] Dore, p. 31.
[7] Ogata Hiroyasu, p. 149.
[8] Dore, p. 20.

Perry forced an end to isolation in 1854 the scope of Western studies had already begun to include other European languages and certain disciplines completely new to the Japanese.

In the period 1720–1854 there were more interrelationships between the enterprises of learning and science than in any previous historical period. It is therefore appropriate to discuss the more general social and intellectual aspects of learning and science together, leaving treatment of the internal developments of the specific sciences to the concluding section of this chapter.

LATE CHINESE WAVE II DEVELOPMENTS IN TRADITIONAL LEARNING

The central issue of learning in the eighteenth century was the challenge to the Chu Hsi synthesis. Once its position began to weaken under attack, the authority of all other schools, and even of the venerated sages, came under question. Called upon to exercise their critical judgment, many scholars opted for the more easily defended compromises of eclecticism. Scientists, for their part, could no longer assume an unchallenged framework for their labors. One option was to reinterpret traditional concepts to support their continuing work, while another was to function as practically as possible on the basis of one's own experience and observations. Both approaches, by tacitly conceding inadequacies in traditional disciplines, actually favored the reintroduction of Western science, particularly when, in the early half of the nineteenth century, national defense considerations lent a growing sense of urgency to the quest for Western knowledge.

CHALLENGE TO THE CHU HSI SYNTHESIS

The fate of the Chu Hsi synthesis in Japan is a fair example of what can happen when governmental promotion rather than free intellectual competition determines the survival of a school of thought. Chu Hsi's philosophy certainly came to be fully and widely understood among scholars; but it also came to be severely criticized—from within by the Confucianists themselves and from the outside by those increasingly inclined to Western ideas.

The synthesis of Chu Hsi, it will be recalled, sought to integrate all cosmological, political, and ethical knowledge (of the Sung period) into a comprehensive system. In Japan this synthesis was attacked first by Nakae Tōju, an advocate of Wang Yang-ming's idealism, but later and most vigorously by Ancient Learning advocates Ogyū Sorai and Itō Jinsai. The latter two agreed in rejecting speculative metaphysics as a system, though Sorai wished to confine Confucian studies to political economy, while Jinsai favored emphasis on moral cultivation. Once *Shushigaku* was attacked on the basis of the classics, which were also criticized, the resultant long-range breakdown of any authority by the final years of the Tokugawa era left a situation wherein only a weak resistance to the growing Western challenge was possible.

The challenge from outside Confucianism came in more general terms. Through various Western sources it became clear to many learned men that the Chinese world was not the only world, that its learning and science were not the only learning and science, and that, at least in specific cases, Western achievements were more advanced. These ideas emerged first among professional scholars working to translate and introduce Western learning but were more publicly advocated by "enlightened" thinkers such as Miura Baien (1723–1789), Yamagata Bantō (1748–1821), Shiba Kōkan (1738–1818), Hiraga Gennai (1728?–1779), and Honda Toshiaki (1744–1821). All of the "enlightened" thinkers acted privately on their own initiatives, not in official capacities (Baien was a medical doctor in Kyushu, Bantō and Kōkan were merchants, and Gennai and Toshiaki were *rōnin*). Though their grasp of Western learning was often imprecise, they succeeded in the elementary task of introducing the Western world and its learning and in convincing some of their compatriots of the advantages of Western learning over its Chinese-style counterpart in Japan.

An almost exact parallel to Confucianism can be seen in traditional medicine in the same period. After Li-Chu medicine was established as the mainstream in Japan, it was vigorously attacked by advocates of the Ancient Practice School. Thereafter, some of the medical scholars in the latter school began paying more attention to Dutch medicine. A less precise parallel occurred among calendrical specialists in the shogunate's Bureau of Astronomy (Tenmongata) who came to favor certain methods of Western astronomy. Though their concern was mainly with computational techniques, they were also interested in some of the cosmological concepts of Western astronomy, such as the earth's sphericity and a heliocentric universe. For many scientists, Chi Hsi's cosmology, and particularly the assumption of correspondence between celestial phenomena and human affairs, became untenable. (No such parallel comes from the field of mathematics, as the problems of orthodoxy and internal criticism were not of the sort to lead Japanese mathematicians to look outside their own tradition.)

Under these circumstances it became impossible to claim that the Chu Hsi synthesis integrated all knowledge—natural, public, and personal. To many scholars, the best way to salvage Confucian studies seemed to be eclecticism—and restraint in pressing its uses and claims. Not a few followers of Ogyū Sorai, sensing, as did Katayama Kenzan (1730–1782), that the school of Ancient Learning had gone to unacceptable extremes, still sought to preserve what they favored eclectically. Eclectic activities reached their peak in the 1780s. In reaction to this compromising trend and as part of the conservative reform mood of that time, the shogunate in 1790 issued a Prohibition Against Heterodox Teachings which required that teaching in the Shoheikō be confined to the Chu Hsi synthesis. This served, in effect, to inhibit fief schools from teaching other schools of Confucian thought. The decree was inspired by a small band of *Shushigaku* devotees who succeeded in persuading the shogunate that it was the right move, but many Chu Hsi followers, including Hayashi Nobutaka, current head of the Shoheikō, did not favor it.[9]

[9] Sagara, pp. 232–237.

The result, in the short run, was devastating. Many shogunate and domain samurai ceased patronizing teachers who were critical of *Shushigaku*, and many professional Confucian teachers abandoned other Confucian schools in favor of the Chu Hsi line. So, at least formally, Chu Hsi orthodoxy was established. In reality the tide of critical thinking was not reversed; scholars were merely forced to pay lip service to the Chu Hsi system. In the long run, scholarship remained in fact largely eclectic, as there were many who continued to study Ancient Learning and Wang Yang-ming's idealism.

In the first half of the nineteenth century, when the feudal system was increasingly shaken by internal and external crises, *Shushigaku* enjoyed greater official support than ever, but it had less and less to contribute toward solving problems in either philosophy or politics. Even sporadic attempts by Chu Hsi enthusiasts to suppress students of Dutch learning and their allies among enlightened thinkers had little success. In the Meiji Reforms, the Shōheikō was abolished without being incorporated into new Meiji institutions, and it was not the Chu Hsi synthesis as such, but rather Chinese-style learning in general, that was the main Tokugawa intellectual legacy to the Meiji transformation.

CONFUCIANISM AND TRADITIONAL SCIENCE

Early in the Edo period Hayashi Razan and many others could wholeheartedly accept the unified Neo-Confucian synthesis of man and nature and use it to refute Western notions such as a spherical earth. Such refutations helped domesticate the residual interest in Western cosmology. However, once Catholicism was successfully suppressed, there was no strong foe to subdue. Calendrical specialists in the seventeenth-century intellectual renaissance were not much drawn to cosmology anyway; some were interested in Western methods given in the translations by the Jesuits in China, but not all were, and certainly not Shibukawa Harumi. As attitudes toward Confucianism itself were shifting, it was eventually suggested— first by Arai Hakuseki (1657–1725), a *Shushigaku*-trained scholar—that attitudes toward the West might also be modified. Hakuseki confirmed this view after his interrogation of the Jesuit missionary Giovanni Battista Sidotti (1668–1715), the last European to smuggle himself into the closed country (in 1708). Later eighteenth-century specialists in calendrical astronomy acquired and confirmed the usefulness of some Dutch-language astronomical books, and from these experts the enlightened thinkers mentioned above obtained some of their knowledge of the West.

Within Confucianism, Ogyū Sorai's skepticism about fathoming heaven's secrets and his restriction of Confucian studies to political economy were carried even farther by Miura Seiin (fl. mid-eighteenth century), a second-generation disciple of his school. Seiin not only refuted the speculative cosmology of the Neo-Confucianists but also insisted that it was the opinion of specialists that should be heard in such matters.[10] Just because the sages did not say the earth is spherical, he argued, is no reason to ignore evidence of its sphericity from lunar eclipses and from differences in the altitude of the pole star according to the latitude. Sung Confucianists were mistaken, he felt, in explaining lunar eclipses by the irrelevant *yin-yang* and Five Phases theories. The effect of such views as Seiin's was to give specialists a somewhat freer hand in their fields, even though the importance of those fields was not necessarily endorsed.

The recognition accorded Western astronomy among learned Japanese increased from the middle of the eighteenth century. Chu Hsi enthusiasts refrained from making direct attacks against it, permitting quiet acceptance of Western ideas without serious repercussions. Institutionally, the Tenmongata early in the nineteenth century became the base for specialists interested in further exploration of Western astronomy. This official agency of the shogunate never sought help from *Shushigaku* scholars to refute Western concepts. Indeed, the occasional claim of an orthodox scholar that Western learning actually had its historical origins in Chinese

[10] Nakayama (4), pp. 161–162.

learning, though intended to bolster the latter's prestige, served to facilitate acceptance of Western ideas.[11]

In any case, orthodox Confucianists reacted far less emotionally than a few Buddhists, particularly monks like Monnō in the eighteenth century and Entsū (1754–1834) in the nineteenth, who feared Western astronomy would undermine the already weakened appeal of Buddhist cosmology (p. 354). Shintoists, on the other hand, had no firm scholarly tradition to defend and welcomed the opportunity to utilize Western theories to diminish Confucian and Buddhist influence. Unperturbed by debates among religionists, the shogunate in 1811 opened an Office for the Translation of Barbarian Texts (Bansho wage goyō) in the Tenmongata for the simultaneous cultivation and control of Dutch learning. This office developed into the main center for late Tokugawa studies of Western learning, science, and technology and went through several modifications to become in 1863 the shogunate's Center for Western Learning (Kaiseisho; discussed later in this chapter).

Practical mathematics was confined to financial, commercial, and technical affairs, and neither *Shushigaku* nor other Confucian schools recognized any intrinsic value in the higher mathematics known as *wasan*. After all, it claimed to be no more than refined amusement and its own best defense was simply that it is "useful for no use" (cf. p. 365). This could hardly impress scholars concerned for political economy or moral cultivation. Among enlightened thinkers there were some—notably, Honda Toshiaki, himself a *wasan* practitioner—who acquired a fresh perspective on mathematics as a valuable means, especially in navigation, for extending Japan's wealth and sphere of influence abroad. But on the whole, *wasan* was limited to a coterie of people indifferent to its practical applications.

Instruction in *wasan* was available only in private academies or from private tutors; there were no *wasan* institutes of the

[11] Ibid., p. 165.

shogunate or domains (only very late in the Edo era did *wasan* win a place in the curricula of some fief schools). In contrast to the early Edo tolerance of mathematical skill, its eventual identification with the affluence of merchants and, by implication, with the impoverishment of many samurai, increasingly provoked the disdain of traditional scholars. When the decision was made to initiate studies of Western mathematics, just before the fall of the shogunate, in the Center for Western Learning and in the Nagasaki Naval Training Center (p. 397),[12] the reasons were entirely utilitarian and had nothing to do with the philosophic merits of either Confucian or Western thought. In fact, Western mathematics was urgently needed for the adoption of Western science and technology, which were essential to national defense.

The parallels between medicine and Confucianism go beyond that mentioned earlier. The attacks on Neo-Confucianism by the school of Ancient Learning had led to the compromises of eclecticism, provoking shogunate action to eliminate heterodoxy from the Shōheikō. Similarly, attacks around 1800 on *Goseihō* medicine, which was oriented toward Neo-Confucianism, by the Ancient Practice medical school precipitated the emergence of an Eclectic School of medicine; shortly before, in 1791, the shogunate had promoted the Taki household's private medical academy, the Seijukan (est. 1765), to official status as its own Institute of Medicine (Igakkan). In the shogunate's efforts to control and cultivate medicine, the Igakkan played a very important role. Sons of the shogunate's medical personnel were expected to study there, and its hereditary director (of the Taki household) enjoyed the privilege of recommending doctors for shogunate employment.[13] The rigidity seen in Confucian and medical scholarship contrasts

---

[12] As a matter of convenience, we use the term "center" for agencies related to the introduction of Western learning, science, and technology (except for the conventional usage for army and navy "academies") and reserve the term "institute" for those agencies related to traditional Chinese-style learning, science, and technology.

[13] Fujikawa Yū, p. 434.

sharply with the relative tolerance found in the Tenmongata and its Office for the Translation of Barbarian Texts.

Like the Shōheikō, the Igakkan's actual scholarly content and methods were eclectic. It produced a vast amount of detailed data but was lacking in creativity. Both schools undertook the publication of fairly reliable editions of classical texts. With the rise of Western learning, both the Shōheikō and the Igakkan were threatened by the new ideas and private academies of "Dutch learning" and "Dutch medicine," with which eventually they could not compete. Like the Shōheikō, after the Meiji Reforms the Igakkan vanished without a trace.

CONFUCIANISM AND MODERN WESTERN SCIENCE

The various responses of Confucianists as they came to recognize the nature and values of Western science during the eighteenth and early nineteenth centuries can be traced by noting successive phases in a tendency to equate Western science with one of the basic concepts of Chu Hsi philosophy, namely, *kyūri* (in Chinese, *ch'iung-li*).[14] The *ri* of *kyūri* is the pattern (*li* in Chinese) which underlies and unites all things in nature. *Kyūri* is the cumulative process of perceiving and comprehending the *ri*. This was to be accomplished by "exhaustively studying the characteristics (of entities) to grasp the *ri*" (*jinsei kyūri*) and "investigating things to penetrate the *ri*" (*kakubutsu kyūri*). In the original Chu Hsi conception this meant empirical inquiry into individual functions, synthesized by intuitive perception of larger and larger functional systems until one encompasses the one *ri* permeating all nature, including human nature.

Even in the early part of the Edo era, when the Chu Hsi synthesis was most fully accepted as a comprehensive system, natural science and even metaphysics were for the Japanese Neo-Confucianists rather peripheral concerns. They were certainly outside the interests of Ancient Learning advocates. In the seventeenth century no one attempted to equate the *kyūri*

---

[14] On some aspects of the relationship between science and Confucianism, see Craig, pp. 133–160, and Nakayama (4), pp. 157–168.

concept with Western natural science. Indeed, for one of the early apologists of Western astronomy, it was quite the opposite; Nishikawa Joken related Western science to the energetic and qualitative *ki* (*ch'i*), not the *ri* (*li*) side of his Neo-Confucian model (p. 261). This usage can also be seen as late as Arai Hakuseki.

Early in the eighteenth century there were a few Confucianists who favored an empirical understanding of *kyūri* in relation to the traditional study of materia medica (*honzō*). Kaibara Ekken (1630–1714) saw the study of *honzō* as the study of the *ri* of things, and Matsuoka Joan (1669–1747) wrote in a similar vein of a *honzō* text: "This is not solely a basic text for doctors, it is truly a part of *kakubutsu kyūri*."[15] This usage was rare and limited to that field. At the peak of the seventeenth-century intellectual renaissance, the school of Ancient Learning openly rejected the Chu Hsi concept of *kyūri*; Ogyū Sorai and Itō Jinsai are said to have despised it.[16]

With the emerging importance of Western learning in the latter half of the eighteenth century, there appeared a functional usage of *kyūri*, particularly among Nagasaki interpreters interested in Western astronomy. One of them, Motoki Ryōei (1735–1794), grandson of Motoki Ryōei by adoption (see pp. 287, 382), used the term *taiyō kyūri* for "tracking the sun's orbit."[17] For him, Kepler and Galileo were *taiju* (great scholars) who had formed the bases for *kyūri* research. Shizuki Tadao (1760–1806), a disciple of Ryōei and the first to introduce some of Newton's theories, repeated this usage in his *Rekishō shinsho* (New treatise on calendrical phenomena; 1802). At this stage of functional usage, investigation of underlying patterns was beginning to be equated with Western science.

Toward the end of the eighteenth century traditional learning was increasingly on the defensive, and the shogunate was concerned to protect the Chu Hsi position. As resistance

---

[15] Craig, p. 140.
[16] Nakayama (4), p. 161.
[17] Nakayama (6), p. 177; also Nakayama (1) and (2), as well as Hirose (1), pp. 74–76.

against Western learning mounted, the functional usage was modulated into an apologetic usage by some scientists and thinkers who positively equated *kyūri* with Western science in order to emphasize the value of its content, method, and spirit. Yamagata Bantō, who had studied *Shushigaku* at the Kaitokudō academy in Osaka and astronomy later under the ablest men of his time (Asada Gōryū [1734–1799] and his disciple Takahashi Yoshitoki [1764–1804]; see pp. 354–359) was the most representative of those who attempted to make a claim for Western science as the fulfillment of traditional research ideals.

As Western science was viewed early in the nineteenth century as a growing menace, the apologist frequently turned propagandist. The term *kyūri* was used as an ideological umbrella to recommend Western science and as a talisman to ward off adverse criticisms from Chu Hsi orthodoxy. This usage often appeared in the names for branches of science or in the titles of monographs. Hoashi Banri's (1778–1852) encyclopedic work on natural science (covering mostly physics, but also aspects of geometry, chemistry, zoology, botany, physiology, hygiene, etc.), completed in 1836, was entitled *Kyūritsū*, that is, "General investigation [of science]." In another instance, Kawamoto Kōmin (1809–1871), in the second translated work on physics, *Kikai kanran kōgi* (Extended explanation of *Kikai kanran* [see Table 18]; 1851), rendered physics as *kyūrigaku*, "the study of *kyūri*." These usages were not compatible with the original understanding of *kyūri*; they were reinterpretations created to meet the exigencies of the times, though flat denials of their propriety rarely came from Chu Hsi scholars in their eclectic mood. Ōhashi Totsuan (1816–1862) was one of the few to object, branding these reinterpretations of *kyūri* as "a means of destroying the *ri*."[18] Related more to the politics of science than to its internal development, these usages became widespread among the allies and foes of Western science.

[18] Craig, p. 154.

In the accelerated adoption of Western science from the middle of the nineteenth century to the early days of the Meiji era, there were many mixed uses of these converted senses of *kyūri*. The last phase in the process was largely a terminological quest for equivalents of Western scientific concepts. Since Western science could now be studied freely without the protective cover of the term *kyūri*, its use passed generally from the scene. *Kyūrigaku* for physics became *butsurigaku*, *jinshin kyūrigaku* for physiology became *seirigaku*, and so on. The long process of evolving usages of *kyūri* had its impact, however, on the terminological quest; the single component *ri* can be found in many terms used today, such as *butsurigaku* and *seirigaku* given above, and *byōrigaku* for pathology, *shinrigaku* for psychology, and so on. But many scientific terms were coined without the *ri* component, such as *kagaku* for chemistry, *seibutsugaku* for biology, *dōbutsugaku* for zoology, and *shokubutsugaku* for botany. As the names for branches of science often include *-rigaku*, the term *rigaku* is used today to denote a university faculty of natural science.

A striking variation on this record appears in the field of philosophy. Though the *ri* as understood in Chu Hsi's system is supposed to permeate society as it does individual man and all nature, the study of Western society, which started early in the nineteenth century in Japan, was never labeled *kyūri*. The word was used to express "philosophy" in Nishi Amane's (1829–1897) works—a usage closer to the original sense of the term. Nishi Amane experimented first with the term *rigaku* and turned finally to the term *tetsugaku*—the term that has continued until today.

Perhaps Chu Hsi's metaphysics was, after all, rather foreign to early Tokugawa scholars. Yet it is most interesting that it was so easily accepted by Hayashi Razan, only to be so easily denounced by Ogyū Sorai, so easily reinterpreted in the late Tokugawa era, and so easily renounced after the Meiji Reformation.

"NATIONAL LEARING" (*KOKUGAKU*) AND "JAPANESE MEDICINE" (*WAHŌ*)

From the activity in "Japanese learning" (*Wagaku*) of the seventeenth-century intellectual renaissance there evolved in the eighteenth century a new tendency to search the Japanese classics for the "original ethical way of the Japanese" and the "original aesthetic sense of the Japanese." This quest led to a distinct tradition of National Learning (*Kokugaku*) as a system of learning with an emergent ideology—though a much humbler one than that of Confucianism. It was heavily influenced by Ogyū Sorai's critical methods of textual research.

Kada no Azumamaro's wish (p. 249) for a recognized tradition was partially fulfilled by one of his disciples, Kamo no Mabuchi (1697–1769), son of a Shinto priest, who in 1738 opened the first private academy for National Learning in Edo, eventually attracting a total of 123 disciples.[19] Mabuchi championed the revival of the ancient Japanese way and spirit to counter Confucian and Buddhist influences. Devoting himself mainly to study of the *Man'yōshū* (Collection of Myriad Leaves; eighth century), he strongly emphasized the "natural way" (*onozukara no michi*) of the Japanese.

The National Learning movement took on added strength through the work of Motoori Norinaga (1730–1801), a townsman and disciple of Kamo no Mabuchi. Norinaga first studied Confucianism and medicine in Kyoto, then practiced pediatrics in Matsusaka in Ise district while engaging in *Kokugaku* studies, specializing in critical research on the *Kojiki* and writing an extensive commentary on it (*Kojikiden*; printed during the period 1786–1822). Recalling Ogyū Sorai's observation that the Confucian way was not the "natural way" of heaven, earth, and man, but the creation of the ancient Chinese sages, Norinaga advocated the view that the ancient way of the Japanese is the creation of the gods. He was most critical of foreign influences, both Chinese and Western. The temper of the times is reflected in the large following he had among rural landowners and among Buddhist and Shinto priests. His registry of disciples lists 493 persons.[20]

---

[19] In the "Directory of disciples of Kamo no Mabuchi" entitled *Agatai monjinroku*.
[20] The 493rd entry is Hirata Atsutane, who never saw Norinaga but claimed that he had spiritual communication with the already deceased master. Kitajima (2), pp. 351–352.

National Learning became more nationalistic early in the nineteenth century under the leadership of Hirata Atsutane (1776–1843), who attracted 553 registered disciples from a wide cross section of society.[21] Atsutane was regarded as heterodox by most of Motoori Norinaga's disciples, who themselves concentrated largely on critical studies of the Japanese classics. Hypercritical of both Confucianism and Buddhism, Atsutane was willing to concede that there were good points in them as well as in Western science and even Christianity; but he claimed that all these good points had their origin in Japan. He was the major figure in wedding Shinto sentiments to National Learning, but in so doing he rejected the Shinto of his time in favor of a fictitious and idealized "ancient Shinto" embodying the true Japanese spirit uncovered by *Kokugaku*. This combination of Shinto and National Learning came forward as one of the main contenders to guide the nation in its shift from a decadent feudal system to restored imperial polity early in the Meiji era; but after a short period of heavy influence it rapidly disappeared backstage to make one more grand appearance in the 1930s and early 1940s.

Kokugaku was also taught and studied with official support. The shogunate gave its backing to a Lecture Hall for Japanese Learning (Wagaku kōdansho) in Edo in 1773, under the leadership of Hanawa Hokiichi (1746–1821). The number of domain schools including *Kokugaku* in their curricula increased from seventeen in the period 1818–1829 to 152 just after the Meiji Restoration.[22] *Kokugaku* attracted thousands of followers among wealthy peasants, townsmen, and Shinto and even Buddhist priests. While their studies may have lacked intellectual depth, they were marked by sincere devotion and were known as "grass-roots National Learning" (*sōmō no Kokugaku*).

This brief background helps make plausible the emergence

---

[21] 553 is the number of entries in the Disciples Registry confirmed by Atsutane; Hirata Kanetane, adopted son of Atsutane, regarded his own disciples as the disciples of Atsutane, and this category of disciples numbered 1,330. Ibid., p. 354.

[22] Wajima (1), p. 168.

of claims advanced early in the nineteenth century that there was also an original Japanese tradition of medicine. Just as *Kokugaku* sought to uncover the original Japanese spirit, aesthetic sense, and moral attitudes predating Chinese and Buddhist influence, so the *Wahō* (Japanese Practice) school of medicine sought to locate its genesis in ancient Japanese medical books and in popular medicine used among the masses. The *Kokugaku* movement could, of course, refer specifically to the *Kojiki* and *Man'yōshū* to make claims for uniquely Japanese data, despite all possibilities of heavy influences from China and Korea. The *Wahō* school of medicine likewise wished to claim certain ancient treatises as its original literary corpus. *Wahō* teaching was based on such works as the *Daidō ruijūhō* (p. 88), compiled under imperial order in 808 by Abe no Manao and Izumo no Hirosada, and the *Kinranpō*, produced in 868 by Sugawara no Minetsugu and Mononobe no Hiroizumi. These documents were intended as records of treatments and medicinal preparations employed in Japan from antiquity. The existing texts are generally believed to be forged manuscripts of a much later period.[23]

The first doctor to become well known for his practice of *Wahō* medicine was Ōta Kenryū (1725–1812), who wrote *Shintō kireiden* (On the mystic spirits of Shinto) in 1807; this work was not printed. Other key figures in the development of *Wahō* medicine were Morikawa Sōen and Matsukawa Tsurumaro (1791–1831). The most prominent *Wahō* advocate was Gonda Naosuke (1809–1887), who studied *Kokugaku* under Hirata Atsutane. Naosuke not only tried to refute both Chinese and Western medicine, he even went to the extreme of claiming that the *Shang-han lun* (On Cold Damage Disorders) was merely a Chinese offshoot of the ancient Japanese medical tradition.[24] After the Meiji Reformation, when it became eminently clear that his position could not be defended, Naosuke gave up his claims and became a Shinto priest. The

[23] Fujikawa Yū, pp. 50–54.
[24] Ishihara (2), pp. 175–176.

effort to distinguish a *Wahō* school of medicine simply disintegrated.

LEARNING IN WESTERN CULTURAL WAVE II: 1720–1854

Despite diverse internal emphases, traditional learning broadened its base in Japanese culture and remained the central arena for most scholars throughout the eighteenth century. The intellectual carry-over from Western Wave I, limited to cosmology and surgery, was effectively held in check by anti-Western prejudices and policies. If Western learning was to be revived in Japan, it would have to be on an entirely new basis.

The first moderately effective effort toward overcoming entrenched suspicions of the West occurred early in the eighteenth century in unusual circumstances. In 1708 a Sicilian Jesuit named Giovanni Battista Sidotti smuggled himself ashore at Yakushima island, south of Kyushu, poorly disguised as a samurai with two swords. Arrested within a few days, he was taken to Edo for confinement in a special detention house for captured Christians. There he was interrogated in 1709 by Arai Hakuseki, who had just become a Confucian consultant to Shogun Tokugawa Ienobu (ruled 1709–1712). Contrary to prevailing assumptions, Hakuseki failed to discover any Catholic intentions to invade Japan or to promote indirectly an invasion. He found Sidotti a man whom he could respect, one whose knowledge of history, global geography, and astronomy was superior to the contemporary Japanese levels in these fields. Just the same, Hakuseki felt that the Catholics took unfair advantage of the unlearned masses and was perplexed that a man of Sidotti's stature could believe doctrines that to Hakuseki seemed preposterous.

From talks with Sidotti, Hakuseki arrived at a fresh view of the West which, by 1715, he recorded in his *Seiyō kibun* (Memorandum concerning the West). The work was in three parts: on the Sidotti investigation, world geography, and the nature of Catholicism. It modified the public outlook on Catholicism and insisted that the West's strength lay in its practical knowledge and techniques, not its theories, which he thought inferior. On Catholicism, it gave information on the Church and its doctrines as well as Sidotti's own criticisms; but, because all mention of Catholicism was taboo at this time, this memorandum was not made available for wider distribution beyond a few high officials until nearly a century later.[25] Hakuseki's views on world geography (and customs) were further developed in his *Sairan igen* (Selected strange accounts; 1713), based on the second part of *Seiyō kibun* and on what he learned from the Dutch head of Dejima about Johan Blaeu's world atlas and other materials which had earlier been presented to the shogunate by the Dejima head.[26]

Hakuseki's summation of the Sidotti incident to the shogunate seems to have indicated that, although Sidotti could be sentenced to prison or even executed, he might also be released to return to Europe. His writings perhaps suggested that the book ban might also be relaxed. But Sidotti died in his prison cell, and something more practical than cultural perspectives was needed to create an opening for Western books.

PRELUDE TO *RANGAKU*: YOSHIMUNE'S CALENDRICAL REFORM PLAN

While Yoshimune (ruled 1716–1745) rejected Hakuseki's spirit of moderation in favor of a reactionary revival of an idealized form of Ieyasu's "founding spirit," he also promoted "practical learning" (*jitsugaku*), so called to distinguish it from Confucian learning, which decided larger issues. Practical learning embraced such traditional studies as materia medica but was broadened by Yoshimune to include some fields of Western science and techniques. Particularly pertinent to our account were his interests in gunnery and calendrical astronomy.

Yoshimune's desire to reform the *Jōkyō* calendrical system (in use since 1685) derived partly from his very Confucian understanding of a ruler's responsibilities and prerogatives, yet also from his special interest in astronomy. He had an

[25] *Nihon rekishi daijiten*, 2:128–129.
[26] Numata (2), pp. 25–27.

Fig. 31. Astronomical observatory of the Tokugawa shogunate's Bureau of Astronomy (Tenmongata), located in the Kanda district of Edo. On the left is a simple armillary sphere; the shed to the right housed a quadrant. Drawing is from the *Kansei rekisho*. (The National Archives, Japan)

astronomical observatory built on the Edo Castle grounds in 1744 (destroyed in 1757; rebuilt in Ushigome in 1765, and again in Asakusa in 1782) and often engaged personally in observations. Among those consulted about revising the calendar was the shogunate samurai Takebe Katahiro, an accomplished *wasan* scholar and author of a commentary on the *Jujireki* (see Table 20). Takebe in turn recommended one of his disciples, Nakane Genkei (1661?–1733), another shogunate samurai, who had some familiarity with a few forbidden works of the Jesuits in Chinese translation that were kept in the shogunate library. Genkei urged lifting the book ban enough to import Jesuit books in Chinese not directly concerned with Christianity. In 1720 Yoshimune issued such an order to the shogunate's magistrate in Nagasaki. Although the scientific treatises imported immediately following this order did not represent the latest trends in Europe, it is nonetheless true that for the first time Western knowledge on an advanced and professionally useful level was transmitted to Japan and had a decisive influence on the subsequent course of Japanese astronomy.[27]

These imported works were made the basis for technical revisions of the official astronomical system. Before the plan could be fully realized, however, most of the principal figures passed from the scene: Genkei died in 1733, Takebe in 1739, and Yoshimune himself retired in 1745 and died in 1751. The remaining members of the revision group, such as Nishikawa Seikyū, who had been brought to Edo in 1719 to confirm the value of Western astronomy (p. 262), were unsuccessful in incorporating Western methods into the preparation of the new calendrical system. Called the *Hōryakureki*, it was prepared by the Tsuchimikado officials in 1754 and completely followed traditional methods. But for the next computational system, the *Kanseireki* (used from 1798), Japanese specialists were able to incorporate Western measurements into an official calendrical reform for the first time.[28] Half a century later, in the last calendrical revision before modern times, the *Tenpōreki* was composed completely on the basis of Western astronomy.[29] Completed in 1842, it was formally adopted for use from 1843.

Use of Western methods in calendrical reform would have been impossible without persistent translating and studying of Western sources. The early role of Nishikawa Joken and his son Seikyū (p. 261) was carried forward by an official Nagasaki interpreter named Motoki Ryōei (1735–1794) whose translations were the first to include explanations of the Copernican heliocentric system, the most detailed exposition appearing in 1792–1793. In all references to the Copernican hypothesis Ryōei omitted theological arguments, presumably in deference to the anti-Christian atmosphere of his time; and his translations lacked detailed data of use to practical astronomers. One of his disciples, Shizuki Tadao (1760–1806), son of an inter-

[27] Nakayama (6), p. 167.
[28] Ibid., p. 194.
[29] Ibid., p. 201.

preter, quit his own job as an interpreter at age eighteen to concentrate on astronomy and the Dutch language; he became the chief person responsible for introducing Newtonian mechanics, spending more than twenty years on his principal translation.

Not only did the translations of Ryōei and Tadao condense or omit sections, they were never printed and were shared with only a limited number of their close associates, among whom were the enlightened thinkers who popularized Western astronomy. Thus, Shiba Kōkan drew mainly on Ryōei's translation and frankly acknowledged Western achievements without indulging in the apologies for traditional science that characterized Ryōei's and Tadao's translations; Yamagata Bantō advocated the new cosmologies while promoting Dutch learning in general; Hoashi Banri (1778–1852) was most active in propagating Newtonian concepts.

The ablest civilian astronomer of the eighteenth century was Asada Gōryū (1734–1799), though he made no direct contribution to the growing corpus of translated works. His most eminent disciples were Takahashi Yoshitoki (1764–1804) and Hazama Shigetomi (1756–1816); it was these two men who were responsible for the application of Western methods in the *Kanseireki* reform. Late in his life Yoshitoki turned to the task of translating part of J. J. L. de Lalande's *Astronomie* from a Dutch version, but he died the next year. His colleague Shigetomi continued the work of translation, but the final Japanese edition was completed by Shibukawa Kagesuke (1787–1856; Yoshitoki's son, adopted by the Shibukawa family) and published in 1836 as *Shinkō rekisho* (Astronomy by the new technique). This, along with a translation by a fellow Tenmongata astronomer, Yamaji Tomotaka, was made the basis of the *Tenpōreki*, the best of the Japanese calendrical systems before modern times. Yoshitoki and Shigetomi were active astronomers in the Tenmongata, which in the meantime had set up its Translation Office in 1811; its staff of translators no longer confined themselves solely to astronomy and geography as national defense concerns became more urgent.

While Motoki Ryōei and Shizuki Tadao were faithful translators more than advocates of new ideas, the later translators such as Takahashi Yoshitoki, Hazama Shigetomi, and Shibukawa Kagesuke were all professional calendrical specialists who more openly favored use of Western astronomy. Their understanding of its computational techniques (though apparently not of its physical concepts) was more sophisticated, despite little government encouragement; yet these specialists consistently avoided deviations from the traditional calendrical form. The developments from Yoshimune's initial sanction of Western astronomy as "practical learning" to the *Tenpō* calendrical reform constitute a perfect example of willingness to use Dutch learning tempered by reluctance to jettison the traditional framework. Mathematics presents a striking contrast; *wasan* specialists were totally uninterested in exploring Western sources. Only a few medical scholars showed a willingness to alter traditional positions as a result of engaging in Dutch learning.

Yoshimune's program indirectly contributed to the rise of Dutch learning among medical scholars. In 1740 he commissioned two men in his service, the Confucian scholar Aoki Kon'yō (1698–1769) and the physician Noro Genjō (1693–1761), to study Dutch. Making use of the Nagasaki interpreters to query the Dejima doctor annually visiting Edo about a Dutch book, Genjō by the following year produced a compilation of notes entitled *Oranda kinjūchūgyozu wage* (Japanese version of Dutch explanations of drawings of birds, beasts, insects, and fish), giving 81 animal names in Dutch and Latin, with Japanese equivalents. From 1742 he turned to a work (by the Flemish botanist Rembert Dodoens, 1517–1585) more directly related to medicine. Through the interpreters he interviewed the Dejima doctor every year but one up to 1750, when he completed a series of notes under the title *Oranda honzō wage* (Japanese version of Dutch materia medica), giving the Dutch, Latin, Chinese, and Japanese names of 118 plants with descriptions of their medicinal uses.[30] By using official interpreters to ask questions about these books presented to the shogunate—and because the Dejima doctor needed only to read the material under question—Genjō was able to produce concrete results; neither compilation was printed but only filed in the shogunate library, and thus the work had little direct effect on Dutch studies. Genjō himself does not appear to have gone further in his study of the Dutch language.

Aoki Kon'yō does appear to have studied Dutch, though he too relied heavily on official interpreters to compile notes. In 1742 he published *Oranda kaheikō* (On Dutch currency) and between 1744 and 1746 *Oranda monji ryakkō* (Primer of the Dutch language). The latter contained a collection of 721 Dutch words. This was not in itself innovative, as the Nagasaki interpreters had for years compiled such word lists, and Arai Hakuseki had made a list of about 340 Dutch words.[31] Kon'yō's departure from earlier practice was to write the Dutch words in roman letters, adding the *kana* script to indicate proper pronunciation; earlier word lists had given Dutch words only in *kana*. Kon'yō is not thought to have studied the Dutch language beyond this, and his grammatical understanding was probably minimal.

After Yoshimune died in 1751, government sponsorship of Dutch language study ceased. Medical studies of Dutch sources had to come from private initiatives, and yet it was the Confucianist Kon'yō, not the physician Genjō, who was able to aid the first efforts by teaching what he knew of the Dutch language to one of the principal figures, Maeno Ryōtaku (1723–1803).

MEDICINE AS THE VEHICLE OF *RANGAKU*

For substantial progress on a professional level, it was necessary for real competence in the Dutch language to be combined with mastery of traditional science and for the results to be publicly available. These conditions were met first and sustained longest in the field of medicine. The right conditions

---

[30] Kimura, pp. 110–111.
[31] Sugimoto (2), p. 248; Numata (2), p. 126.

first occurred in 1771 when a group of medical men in Edo collaborated in the translation into Japanese of an introductory Dutch-language work on Western anatomy.

The impetus from traditional medicine came from the work of a Kyoto physician named Yamawaki Tōyō (1705–1762), who in 1754 had occasion to observe a post-mortem dissection (treated later in this chapter). Anatomy enjoyed only a very low priority in traditional medicine, and dismemberment of human bodies was generally frowned upon. Hence this and most later dissections involved a condemned criminal. A disciple of the Ancient Practice school of medicine, Tōyō realized that the traditional anatomical charts did not accord with the empirical realities disclosed by the dissection he witnessed. He recorded his findings in his *Zōshi* (On the viscera; 1759), and one of his students, Asanuma Sukemitsu, made the drawings, *Zōzu* (Charts of the viscera), to accompany it. This pioneer work sparked a "dissection boom" from which several original achievements in traditional Japanese medicine issued (see later in this chapter).

Interest in dissections was one of the important lines that converged in the friendship formed between Maeno Ryōtaku (1723–1803), a student of Tōyō and a physician of the Nakatsu domain serving in Edo, and Sugita Genpaku (1738–1818), a practitioner of the Obama domain who had come to Edo to learn surgery from the Nishi school that served the shogunate. Ryōtaku had learned elementary Dutch from Aoki Kon'yō but, conscious of his inadequacy, had obtained his daimyo's permission in 1770 to go to Nagasaki to study under official interpreters, where he spent about three months. He brought back to Edo a copy of a Dutch translation, *Ontleedkundige tafelen*, of a German anatomical atlas with text and notes, *Anatomische Tabellen* (*Tabulae anatomicae*),[32] traditionally referred to in Japan as *Tafel anatomia*.

[32] The original was written by Johan Adam Kulmus (1689–1745), a German professor of a gymnasium in Danzig. Printed in 1722, this introductory textbook was clear and concise; it was highly regarded and was translated into Latin, French, and Dutch. The Dutch version was printed in 1734 in Amsterdam.

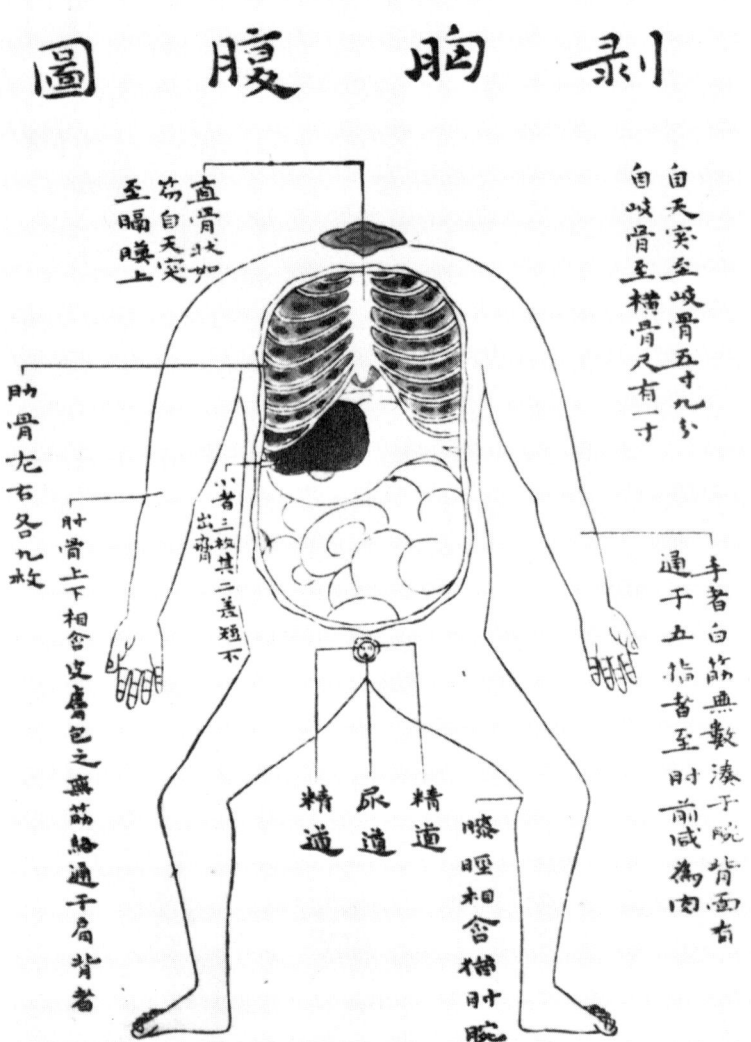

Fig. 32. Diagrams of the internal organs, from the *Zōzu* (Charts of the viscera) that accompanied Yamawaki Tōyō's *Zōshi* (On the viscera; 1759). (The National Archives, Japan)

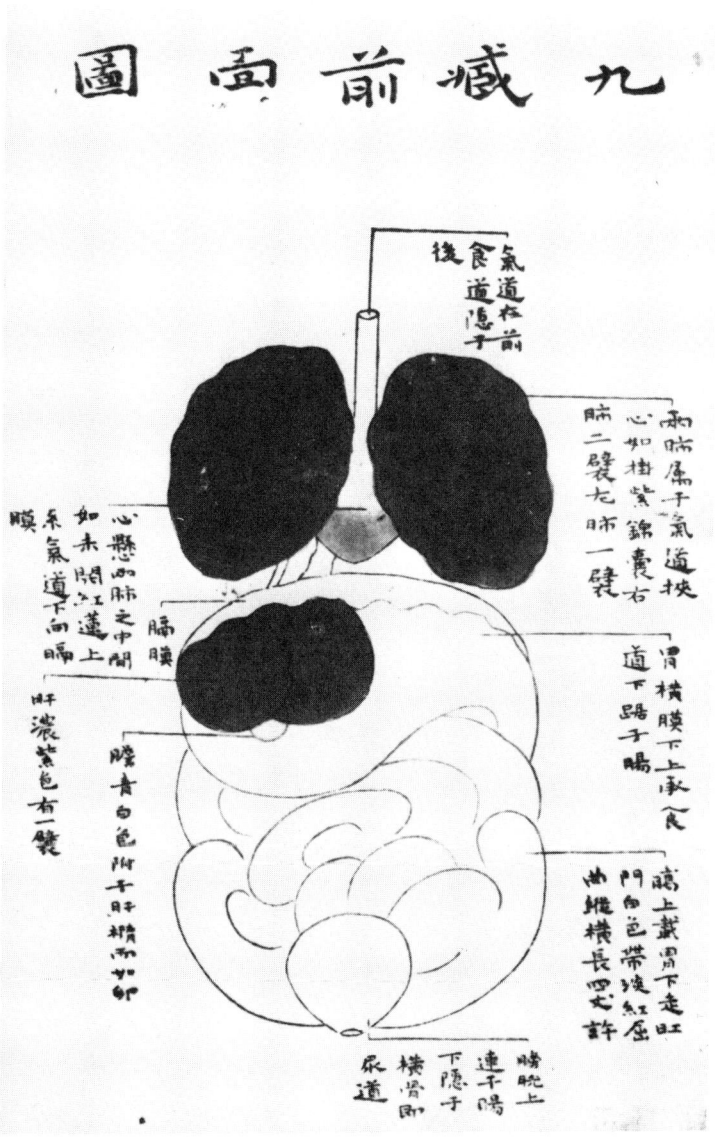

Fig. 33. Portrait of Sugita Genpaku. (Courtesy of the Japan Foundation)

Fig. 34. Portrait of Maeno Ryōtaku. (Courtesy of the Japan Foundation)

Genpaku meantime had become acquainted with Nakagawa Jun'an (1739–1780), a physician serving the Obama daimyo in Edo. An interpreter had loaned Jun'an two Dutch books on anatomy, hoping to sell them. Genpaku noticed that the illustrations showed the internal organs in shapes and relations completely different from those in traditional Chinese-style anatomical charts. He personally lacked funds to purchase one of them but prevailed upon an elder (*karō*) of his domain to make the purchase.

Ryōtaku and Genpaku attended a post-mortem dissection of a beheaded female prisoner at the Kozukahara execution grounds on the outskirts of Edo on the morning of March 4, 1771;[33] each brought his copy of the anatomical atlas—which happened to be the same—and both were amazed by the complete agreement between the diagrams in their copies of *Tafel anatomia* and the dissected body. They decided to immerse themselves in the laborious task of translating this atlas—with no workable dictionaries or grammars and only Ryōtaku's limited knowledge of Dutch. Besides Nakagawa Jun'an, they were joined by Ishikawa Genjō (1754–1816), Katsuragawa Hoshū (1751–1809), a shogunate *Kōmō* surgeon, and a few others. As the translation work proceeded, ever so slowly, the group's ability in Dutch improved. Even so, the draft was rewritten eleven times before they were satisfied. Including numerous woodblock illustrations (less precise in detail than the copperplate etchings of the Dutch original), the text was written in Chinese-style *kanbun*, as it was expected that the book would be read in China too. Though the original includes a short text and lengthy explanatory notes, the translation omitted the notes altogether and often failed to capture the meanings of the original.[34]

---

[33] Hōryaku calendar; April 18, 1771 in the Gregorian calendar.

[34] The story of this translation project was later recorded by Sugita Genpaku under the title *Rangaku kotohajime* (Beginning of Dutch studies; available in the Iwanami Bunko series, No. 5095, Tokyo, Iwanami Shoten, 1974, 18th printing); translated into English by Matsumoto Ryōzō and Kiyooka Eiichi under the title *Dawn of Western Science in Japan*.

Inclusion of Katsuragawa in the project had a tactical value because of his shogunate connections. There existed no sanction for publishing translations of Dutch books; so far only professional interpreters had produced any translations, and these had been restricted to very limited, private circulation in manuscript form. As recently as 1765 a harmless work, *Oranda-banashi* (Tales of Holland; 1765) by Gotō Nashiharu (1696–1771) had been confiscated and the printer's blocks destroyed merely for including the Dutch alphabet in the illustrations. The group's strategy was first to issue in 1773 a brief preview, entitled *Kaitai yakuzu* (Concise chart of anatomy), copies of which were presented to high-ranking shogunate officials in Edo and to leading court nobles in Kyoto. Only after it was clear that there would be no strong political repercussions did they, in 1774, publish all of their translated material, under the title *Kaitai shinsho* (New text on anatomy). Copies were presented to the shogunate by Katsuragawa to obtain official sanction before it was put on sale.

An anatomical atlas, though of no use in Chinese medicine, is one of the basic tools of modern medicine. The fact that the charts in the *Kaitai shinsho* were precise and accurate was readily recognized by Japanese medical scholars; this dealt a heavy blow to traditional medicine. An immediate chain reaction was set off in the Japanese medical community. The "dissection boom" was intensified, and zeal for studying Dutch medical books led to formation of a school of "Dutch medicine" (*Ranpō*). In 1791 the shogunate sought to buttress the position of traditional medicine by promoting the Taki household's Seijukan to official status as the government's Institute of Medicine. By the end of Western Wave II the position of Dutch medicine in Japan had not only grown stronger but also broadened its scope so that it was referred to more generally as "Western medicine" (*Yōhō*) and was to gain official shogunate support in the period of rapid assimilation of Western science after Perry.

At the heart of the subsequent expansion of Dutch medicine in Japan was the very work that got it started—the translation

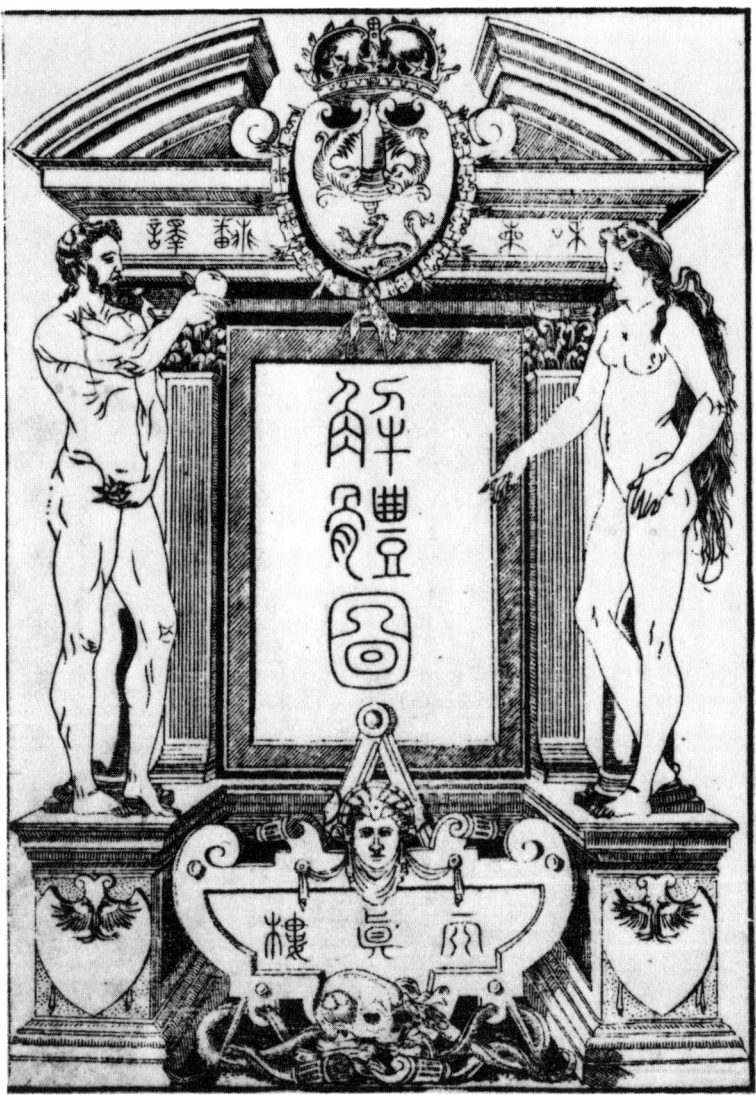

Fig. 35. Frontispiece and first page of the *Kaitai shinsho* (New text on anatomy; 5 vols., 1774). Names of four of the translators appear at the lower right of the first page (Maeno Ryōtaku's name is missing). (University of Tokyo General Library)

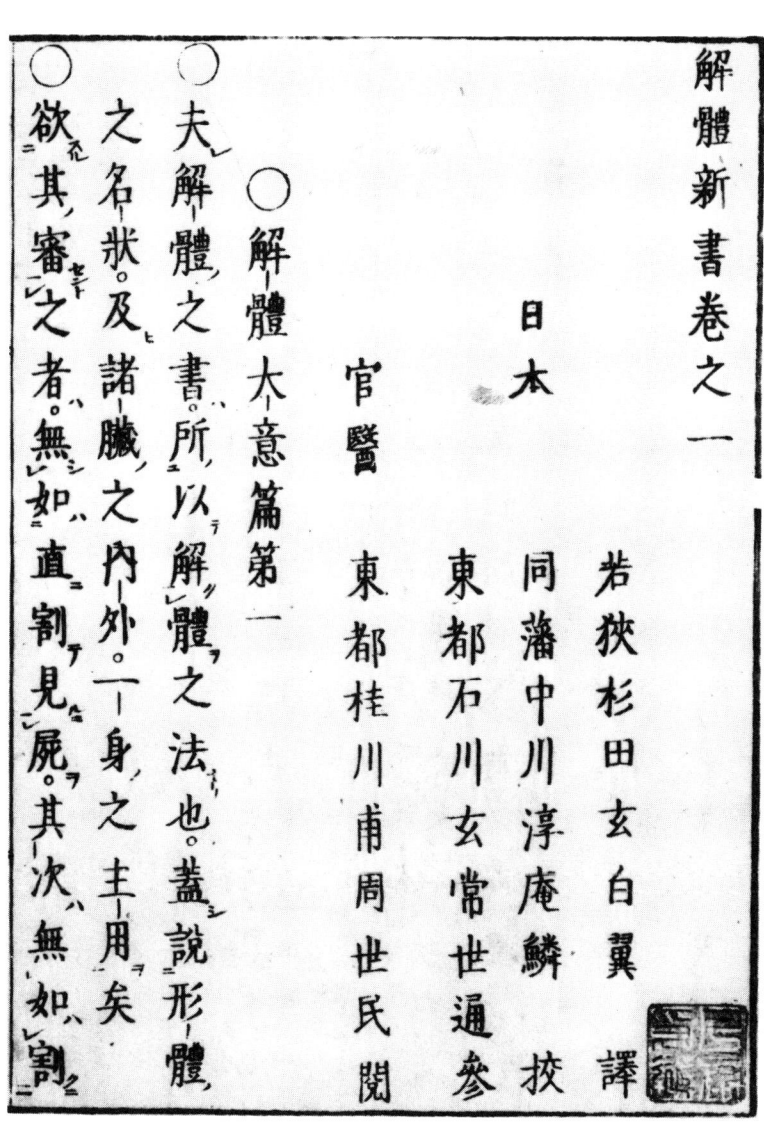

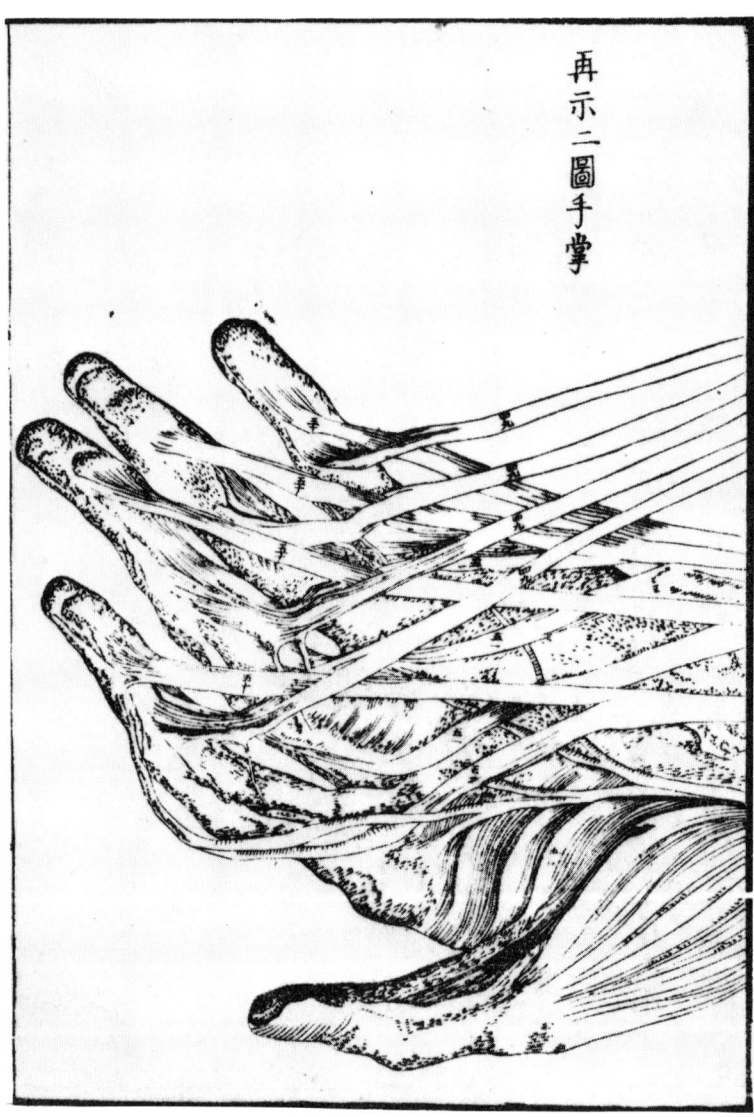

Fig. 36. Sketch of the left hand, palm up, in the *Kaitai shinsho*. (University of Tokyo General Library)

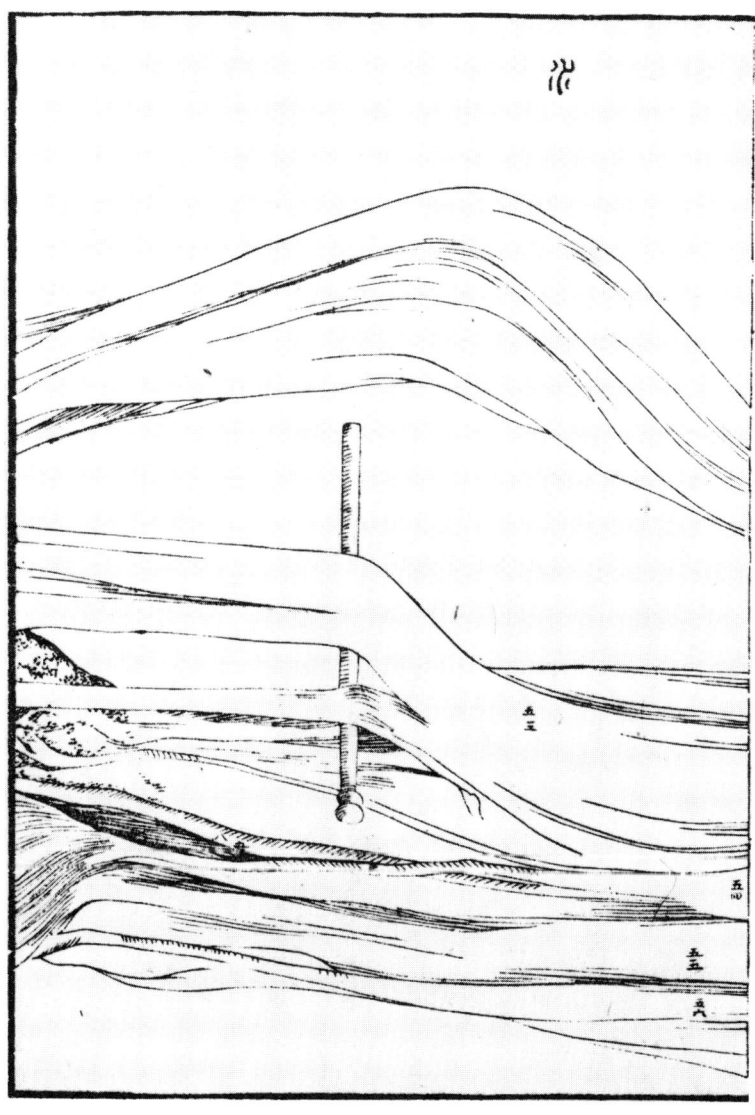

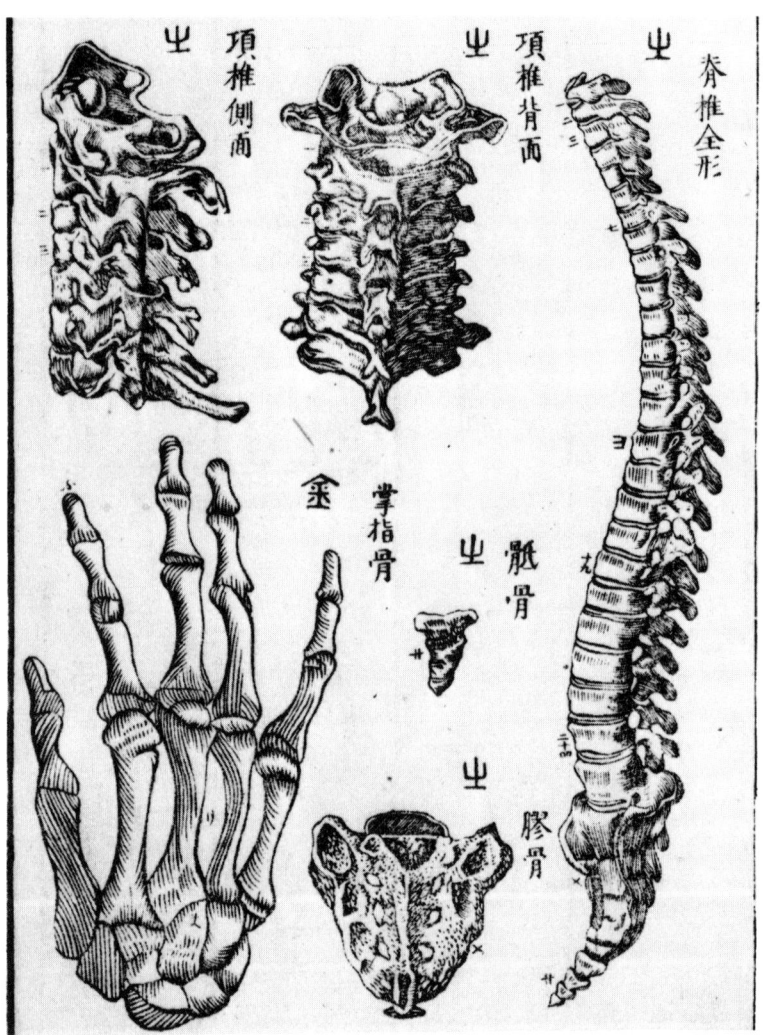

Fig. 37. Representative drawings of bone structures in the *Kaitai shinsho*. (University of Tokyo General Library)

of Dutch books. Counting only published books, translations of texts on basic medicine from 1772 to 1866 number seventeen, and those on clinical medicine run to thirty.[35] Moreover, in the process it became clear that texts in related scientific fields were needed. The accompanying brief list of some of the earliest translations (Table 17) will give some idea of the pace and scope of continued translation activity.

Udagawa Genzui's (1755–1797) 1793 translation, *Seisetsu naika sen'yō* (Essentials of Western internal medicine), was the result of ten years spent on the Dutch version of Johannes de Gorter's *Medicinae compendium* (Leiden, 1731). It was the first comprehensive text on Western internal medicine translated into Japanese and exercised a great influence on the spread of the *Ranpō* school. The 1825 publication of *Yōi shinsho* (New text on surgery) by Ōtsuki Gentaku (1757–1827) was a translation of the Dutch version, *Heelkundige onderwyzingen*, of Laurens Heister's original, *Chirurgie* (Nürnberg, 1718), on surgery. Heister was the outstanding German surgeon of the early eighteenth century; he was well known in Holland as he had studied in both Leiden and Amsterdam and from 1708 served as chief surgeon of the Dutch army.[36] In both instances these translations came long after the originals had appeared; the translators were not necessarily working on the most up-to-date texts, only with what was available by chance. Apart from isolation restrictions, Japanese scholars lacked a comprehensive knowledge of Western medicine as a whole and even the linguistic skills to examine widely and select the latest, most representative books.

When the language problem was, as in Chinese Wave I, to move from no written language at all, progress was of course exceedingly slow, even though there were literate Chinese and Koreans on hand to help; and only a very few Japanese, mostly noblemen and Buddhist priests, gained literary competence to deal with scientific materials. In Western Wave I some of the

---

[35] Ōtori, pp. 32–34.
[36] Bowers, p. 68.

Table 23. Early Translations by *Rangaku* Scholars: Pioneering and Influential Works in Various Fields

Date	Title	Author	Remarks
1773	Kaitai yakuzu	Maeno Ryōtaku, Sugita Genpaku, et al.	Preview of *Kaitai shinsho* (1774)
1774	Kaitai shinsho	Maeno Ryōtaku, Sugita Genpaku, et al.	First printed work in anatomy, with woodblock charts
1785[a]	Oranda yakusen	Katsuragawa Hoshū	First translation of "Dutch pharmacopeia"
1793—	Seisetsu naika sen'yō	Udagawa Genzui	First in internal medicine; published a few volumes at a time
1798[a]	Chōtei (or Jūtei) kaitai shinsho	Ōtsuki Gentaku	Revised edition of *Kaitai shinsho* (1774), with copperplate charts; printed 1826
1803[a]	Rarande rekisho kanken	Takahashi Yoshitoki	Extracts of calendrical section of Lalande's *Astronomie*
1805	Oranda naikei ihan teikō	Udagawa Genshin	Concise, simple text on anatomy, with reference to physiology & pathology
1808	Naishō dōbanzu	Udagawa Genshin	Excellent anatomical atlas, to accompany *Oranda naikei ihan teikō* (1805)
1811– 1830[a]	Kōsei shinpen	Ōtsuki Gentaku, Baba Sadayoshi, et al.	Encyclopedia for household use, by Translation Office of Tenmongata
1816	Ganka shinsho	Sugita Ryūkei	First in ophthalmology
1822	Botanikakyō	Udagawa Yōan	First in botany
1823[a]	Karin sankasho	Aochi Rinsō	First in obstetrics
1825	Yōi shinsho	Ōtsuki Gentaku	First specialized text in surgery; ms completed in 1792
1827	Kikai kanran	Aochi Rinsō	First in physics
1829	Taisei honzō meiso	Itō Keisuke	Introduction of Linneaus' classifications
1831[a]	Seikiron	Oka Kenkai	First in physiology
1832[a]	Jinshin kyūrigaku shōkai	Ogata Kōan	Second in physiology
1832	Igen sūyō	Takano Chōei	First printed work in physiology
1832	Yōka shinsen	Sugita Ryūkei	Most important text in surgery

Table 23 (continued)

Date	Title	Author	Remarks
1833	Shokugaku keigen	Udagawa Yōan	Reliable botany text, more advanced than *Botanikakyō* (1822)
1836	Shinkō rekisho	Shibukawa Kagesuke	Complete translation of calendrical section of Lalande's *Astronomie*
1836	Kyūritsū	Hoashi Banri	Extensive work on physics (major part), chemistry, geometry, zoology, botany, physiology, hygiene, etc.
1837	Seimi kaisō	Udagawa Yōan	First in chemistry
1842–1857	Fushi keiken ikun	Ogata Kōan	Most influential work in internal medicine before Meiji era
1843	Yōyō seigi	Horiuchi Sodō	Most influential in pediatrics
1849	Byōgaku tsūron	Ogata Kōan	First in pathology
1851	Kikai kanran kōgi	Kawamoto Kōmin	Second in physics, more advanced than *Kikai kanran* (1827)
Postisolation Works			
1856	Wātoru yakusei ron	Hayashi Dōkai	First in pharmacological theory; ms. completed 1850
1857	Yōzan yōhō	Yanagawa Shunzō	First in Western mathematics

N.B. This is only a *partial* listing of translations by *Rangaku* scholars; e.g., in the field of anatomy alone there were about a hundred manuscripts translated or compiled before the Meiji era (see Ishihara (2), p. 149).
[a] These dates indicate manuscripts not printed; hence "first" for other dates means "first printed work" in that field.

Jesuits attained a surprisingly high level of competence in Japanese and a few Japanese also acquired considerable facility in Portuguese, and some fairly reliable dictionaries, grammars, and translations were produced, all in less than a century. That Western Wave II depended initially on official interpreters is clear, but their contribution was greatly restricted by government regulations and by their own lack of academic training. Until the end of the eighteenth century there were only the unpublished and virtually inaccessible astronomical translations of Motoki Ryōei and Shizuki Tadao, and in medicine the early and equally inaccessible translations of Narabayashi Chinzan and Motoki Ryōi on surgery and anatomy respectively, the 1774 work by Maeno Ryōtaku, Sugita Genpaku, *et al.*, on the *Kaitai shinsho*, its revision by Ōtsuki Gentaku in 1798, and Udagawa Genzui's 1793 translation of the work on internal medicine (see Table 23). Medical scholars still lacked translated texts on diagnosis and treatment of diseases, not to mention works on related sciences. And there were as yet no Dutch-Japanese dictionaries or grammars of Dutch.

The first Dutch-Japanese dictionary worthy of the name was produced by the same kind of teamwork witnessed in production of the *Kaitai shinsho*. Nishi Zenzaburō (? –1768), a Nagasaki interpreter who tried to compile a Dutch-Japanese dictionary but died before finishing it, had used P. Marin's Dutch-French dictionary as a reference. Maeno Ryōtaku took up this task briefly but also was unable to complete it. It was not until Ōtsuki Gentaku recruited one of his leading students, Inamura Sanpaku (1759–1811), that a successful project got under way. Joined by a Nagasaki interpreter, Ishii Tsuneemon and, as work progressed, by Udagawa Genzui and Udagawa Genshin (1769–1834), the team continued using P. Marin's Dutch-French dictionary, *Woordenboek der Nederduitsche Fransche Taalen: Dictionnaire Flamand et Français* (Amsterdam and Utrecht, 1729) but added a second Dutch-French dictionary, *Nieuw Nederduitsch en Fransch Woordenboek* (Amsterdam, 1717), published by François Halma (1652–1722).[37] Only thirty copies were produced, each containing 80,000 Dutch words printed in a column, with the Japanese equivalents handwritten in characters and *kana* to the right. It bore the title *Haruma wage* (Japanese version of Halma), with the date February 18, 1796; Inamura Sanpaku was listed as the compiler. It came to be known also as the *Edo haruma* (The Edo "Halma") after another dictionary compiled independently from Halma's reference book was produced in Nagasaki.

[37] François Halma is not the name of the compiler, as sometimes represented, but of the publisher.

Teamwork was also the mark of the Nagasaki project, this time including Hendrik Doeff, who came to Japan in 1800. From 1803 to 1817 he served as head of the Dejima company and acquired some fluency in Japanese. Collaborating with several Nagasaki interpreters, he made Halma's dictionary the basis of a Dutch-Japanese dictionary known by various titles, including *Nagasaki haruma* (The Nagasaki "Halma"), to distinguish it from its Edo predecessor. Doeff left Japan in 1817 before its completion, but his collaborators kept at work until it was finally finished in 1833. This dictionary was not printed until Katsuragawa Hoshū (or, Kunioki, 1826–1881; seventh generation) obtained official permission for its publication, after expansion and revision by Kunioki himself, under the title *Orandajii* (Dutch glossary; completed in 1858).[38]

One extant copy of the *Edo haruma* consists of nine volumes with 64,035 words listed on 2,187 pages. The sheer bulk of this book and the *Nagasaki haruma* made them available to only limited circles. Thus, Fujibayashi Fuzan (1781–1836), a medical doctor and a student of Sanpaku, in 1810 published 100 copies of a smaller dictionary, *Yakken* (Translation key) that contained 27,000 of the most commonly used words selected from the *Edo haruma*. Other Dutch-Japanese dictionaries were produced, but the Edo and Nagasaki versions of Halma and the *Yakken* were the most widely used.[39]

Motoki Shōzaemon (1767–1822), eldest son of Motoki Ryōei, compiled a grammar of the English language in 1811 and three years later an English-Japanese dictionary, *Angeria gorin taisei*. Neither of these was printed.[40] Nor were they very influential, as Dutch learning continued to dominate the field until the middle of the nineteenth century. Takano Chōei (1804–1850), a prize pupil of the Dejima doctor Siebold (see

---

[38] The team had a third reference, a Dutch-Latin dictionary by S. Hannot, *Nieuw Woordenboek der Nederduitsche en Latynsche Taalen* (Amsterdam; n.d.). Cited in Bowers, p. 101.
[39] Numata (2), pp. 73–74.
[40] The first printed Japanese-English dictionary was J. C. Hepburn's (1815–1911) *Waei gorin shūsei* (Shanghai, 1867). Cf. *Kindai Nihon sōgō nenpyō* (Tokyo: Iwanami Shoten, 1968), p. 33.

below), is said to have acquired some proficiency in English, German, and French, in addition to his recognized fluency in Dutch. The systematic cultivation of competence in languages other than Dutch, however, came after the shogunate launched a more aggressive program for adopting Western science and technology after Perry.

The occasional help of a willing Dutchman was undoubtedly a great boost to the new *Rangaku* movement, though not all of the Dutch doctors were as academically qualified as Siebold. But Doeff was a rarity, with added leisure because the Napoleonic wars drastically reduced the Nagasaki trade. The main strength of the *Rangaku* movement came from its academic communities. The *Rangaku* scholars did not simply spin off from the schools of traditional learning and live as intellectual hermits or vagabonds. They cultivated groups of colleagues and disciples who shared their interests. To study *and* teach was the traditional way of scholarship, and it was faithfully followed in developing the many Dutch learning academies called *Rangakujuku*.

It is not clear how many *Rangakujuku* were founded, though they proliferated rapidly in the final years of the Edo period. They were certainly not unique as educational institutions in a society that had some fifteen hundred private academies of all sorts. But they were the main centers for all Dutch learning and trained many samurai in the Dutch language and, after national defense became a top priority, in Western technology. More pertinent to our study, the *Rangakujuku* constituted the basic communities in which the work of translating and introducing medical texts was sustained, and the number of medical personnel with a working knowledge of Western medical theory and practice rose sharply. From these academies were also recruited the ablest men for the Tenmongata's Office of Translation and later for the various shogunate agencies for science and technology founded in the late 1850s and 1860s.

Ōtsuki Gentaku (1757–1827) opened the first academy for Dutch medicine in 1786 in the Shiba township of Edo. It was called the Shirandō (*Shi* from Shiba, *ran* for Dutch, and *dō* for

Table 24. Leading Academies for Dutch Learning

Date established	Academy name	Location	Founder	Remarks
1786	Shirandō	Edo	Ōtsuki Gentaku	First of the *Rangakujuku*
1801	Kyūridō	Kyoto	Koishi Genshun	Founder studied at Shirandō
1801	Shikandō	Osaka	Hashimoto Sōkichi	,,
1805	Inamurajuku (?)	Kyoto	Inamura Sanpaku	,,
—	Udagawajuku (?)	Edo	Udagawa Genshin	,,
ca. 1817	Nakajuku	Osaka	Naka Ten'yū	Founder studied under Hashimoto Sōkichi; academy sometimes cited as Shishisaijuku from the literary name (Shishisai) of Ten'yū
1829	Nisshūdō	Edo	Tsuboi Shindō	Founder studied at Udagawajuku
1833	Shōsendō	Edo	Itō Genboku	Founder studied under Siebold
1836	Chōzendō	Osaka	Kō Ryōsai	,,
1838	Tekitekisaijuku	Osaka	Ogata Kōan	Founder studied at Nisshūdō
1839	Junsei shoin	Kyoto	Shingū Ryōtei	Founder studied under Dejima doctors Feilke and Bateij[a]
1842	Juntendō	Sakura	Satō Taizen	Founder studied under Dejima doctor J. E. Nieman[b]
1846	Taiseikan	Nagasaki	Narabayashi Sōken	Founder was an interpreter and *Kōmō* surgeon[c]

Sources: Nihon Gakushiin (1), V; Bowers, p.94.

[a] Ryōtei served as "acting physician" at Dejima briefly in 1817 between the death of Feilke and the arrival of Bateij, under "appointment" by Hendrik Doeff, head of Dejima at the time.

[b] Taizen in 1838 opened an academy in Edo called Wadajuku, but soon was called to Sakura (castle town of Chiba) where he founded the Juntendō, a combined *Rangakujuku* and medical clinic—predecessor of Juntendō Medical University and hospital in present-day Tokyo.

[c] Before Siebold secured his own house and land in Narutaki he gave clinical treatment and instruction in the Narabayashi and Yoshio households.

academy). Gentaku was himself an active translator as well as an energetic teacher. The Shirandō enrolled ninety-four students in the period 1789–1826 (not the entire period of its existence).[41] Among the students were Inamura Sanpaku, who worked on the *Edo haruma* and in 1805 founded his own academy. Perhaps Ōtsuki's ablest disciple was Udagawa Genshin (1769–1834). Genshin originally was surnamed Yasuoka but changed his surname to become the successor and heir of his first teacher, Udagawa Genzui, who inspired him to take up Dutch medicine before he studied under Gentaku. Both of the Udagawas were prolific translators, as was also Genshin's adopted heir, Udagawa Yōan (1798–1846). The academy started by Genshin in Edo attracted several hundred students, among whom was the influential *Rangaku* scholar Tsuboi Shindō (1795–1848). Genshin was one of the earliest Dutch learning scholars recruited for the Translation Office of the Tenmongata, serving there from 1813 to 1832.

There were many other *Rangakujuku* with prominent graduates that belong in any fuller account of the renewal of Western learning in Japan during Western Wave II. Perhaps it is fitting to conclude this brief survey with mention of Itō Genboku (1800–1871) and his Shōsendō academy started in 1833 in Edo. Genboku was one of the leaders in organizing a vaccination center called Shutōjo in 1858 in Edo, which two years later became the shogunate's Center for Western Medicine (see later in this chapter), a symbol of official recognition of the place won for Western medicine in Japan through the collective labors of so many devoted *Rangaku* scholars. That the struggle was vigorously contested by entrenched interests should be evident from the following review of government policy with respect to the rise of Western learning.

SHOGUNATE POLICY TOWARD WESTERN LEARNING

After Yoshimune died and official encouragement of Dutch study ceased, the shogunate assumed a hands-off policy toward *Rangaku* for the remainder of the eighteenth century. Tacit

---
[41] Numata (2), p. 69.

permission to publish the *Kaitai shinsho* and subsequent translations in the field of medicine, unlikely to threaten or otherwise damage the shogunate's position, did not constitute a positive policy. Once the usefulness of Western astronomy was demonstrated by Takahashi Yoshitoki and Hazama Shigetomi in the *Kansei* calendrical reform of 1798, official concern for the course of Dutch learning was soon rekindled. This concern was first publicly expressed in 1811 in the establishing of the Office of Translation in the Tenmongata.

The Translation Office had a clear mandate to cultivate the translation and study of Western sources in the Dutch language, and for this purpose staff personnel were recruited from among the ablest men active in Dutch studies. Two key figures in this project were Ōtsuki Gentaku, foremost among Western-oriented physicians, and Baba Sadayoshi (or Sajūrō, 1787–1822), a language student of Shizuki Tadao and one of the ablest of the Nagasaki interpreters.[42] Astronomers in the Tenmongata continued to promote the translation and use of materials such as went into the *Tenpō* calendrical reform based on Lalande's *Astronomie* and other sources. As first-rate medical scholars and interpreters were brought into the Translation Office, the interests and capabilities of the staff gradually expanded to encompass languages other than Dutch and disciplines other than astronomy and medicine. The work of the Translation Office usually proceeded at a rather slow pace, its primary role being to cultivate Dutch learning and only secondarily to control its course. There were occasional reactions from the side of traditional learning and even repressive measures when Dutch learning went beyond official expectations. But the Translation Office was never abolished; rather, with the crisis in national defense provoked by Perry in 1853, it was separated from the Tenmongata and made an independent agency for the rapid introduction of Western science and technology, the Center for Western Learning (Kaiseisho).

Between the more leisurely first few decades of the Trans-

[42] Numata (3), pp. 14–15.

lation Office's existence and the more aggressive role it had after Perry, two crucial incidents occurred to challenge the viability of the shogunate's moderate and utilitarian policy toward Dutch studies. One of the incidents involved the Dejima doctor Philipp Franz Balthasar von Siebold, who first came to Japan in 1823, and the second a group of Dutch learning enthusiasts who ran afoul of the shogunate's conservative foreign policy.

THE SIEBOLD INCIDENT: REACTION

For perspective on the Siebold incident it may help to sketch in the larger international background. During the Napoleonic wars Holland was conquered by the French, and the Dutch colonies overseas were occupied by the British. In 1811 the British occupied Batavia, formerly the base of the Dutch East India Company in Java, and for a while Dejima was the only place on earth where the Dutch flag was raised. With the end of the war in 1816, Holland recovered her independence and Java was returned to her by Britain. But the Dutch-Japanese trade could not easily be revived, due to Britain's rising power. The Dutch East India Company had been dissolved in 1800, and the Dutch government, having taken over trade operations in the East,[43] felt a strong need to study the situation in Japan in order to foster the once profitable Dutch-Japanese trade, which had fallen off sharply. Between 1809 and 1817 no cargoes had been shipped to Nagasaki.[44]

Siebold, having completed his professional education at Würzburg, Germany, nourished a strong desire to go to Japan and become a pioneer and authority in the study of Japan. The Dutch government assigned this talented and ambitious young medical doctor to Dejima with the expectation that he would gather new information on Japan. Soon after his arrival in 1823 Siebold quickly established his reputation as a doctor and gathered around himself a large group of able medical students from all over Japan. Allowed unprecedented liberties

[43] Kure (2), p. 54.
[44] Ibid.

Fig. 38. "Picture of Siebold performing a blood-letting operation" (*Shaketsu shujutsu zu*), by Kawahara Keiga (1786–?), who drew various illustrations for Siebold's books. (Original multicolored painting preserved in Nagasaki Prefectural Art Museum)

to engage in his activities by the Nagasaki magistrate, Siebold in 1824 was able to purchase a house and land in Narutaki, then on the outskirts of Nagasaki. This purchase was made in the name of one of the Nagasaki interpreters. Here he saw his Japanese patients and gave lectures to his students on clinical procedures and Western zoology and botany. There were two main buildings and three smaller structures on his property, and some of his students lived there, as did also his mistress Taki.

Siebold's instruction in medicine marked a real breakthrough in Dutch learning in Japan. During the fifty years since publication of the *Kaitai shinsho* Japanese medical scholars had come to understand something of the methods and content of Western medicine through Dutch books but had had only limited contacts with foreign doctors of Deijima in Nagasaki or on their Edo visits; they sorely lacked systematic instruction by a Western doctor. Direct clinical instruction by Siebold not only contributed to the advancement of Dutch learning but in the process made many of his students the leading figures in its development.

From the very beginning, the visits to Edo by the head of the Dutch company had been used by the shogunate and the Dutch alike as a kind of intelligence-gathering occasion. The shogunate's enthusiasm for this opportunity had apparently lagged, as the once annual visits (from 1609) had been reduced to once every two years in 1764, and soon after, in 1790, to once every five years (the last was in 1850). Already active in information-gathering in and around Nagasaki, Siebold used his chance to accompany the Dutch pilgrimage to Edo to engage in systematic and energetic foraging for all kinds of data. Specimens of flora and fauna were a particular interest of his, but no item seems to have escaped his inquisitive and acquisitive activity.

In his investigations Siebold relied on all available means— some open and legal, some cunning, and some downright illegal, given the shogunate's policy at that time. Many items were given to him by patients and friends, some were purchased, and some acquired through exchange. He also hired men to collect botanical and zoological specimens. Even his students were assigned essay themes designed to yield valuable information,[45] and as a reward for written essays, presented in Dutch, Siebold issued his students a kind of "diploma." At times, though, Siebold chose to circumvent regulations to get data; for example, on his way to Edo he used a compass and plumb line to survey the Shimonoseki straits, which were vital to Japanese coastal shipping and defense. He also used the Edo trip to survey the longitude and latitude of large cities along the way with a chronometer and sextant.

Siebold made only one trip to Edo, in 1826, but due to his reputation, many were eager to see him there. They represented many stations in life: feudal lords; shogunate officials in the Tenmongata; doctors of Dutch medicine serving the shogunate, various domains, and townsmen; as well as those simply obsessed with things Dutch. From the Tenmongata came Takahashi Kageyasu (p. 357), who was also head of the shogunate's library, and Ōtsuki Gentaku and Baba Sadayoshi, key men in the Translation Office. Among the medical doctors were Katsuragawa Hoken, Udagawa Yōan, Habu Genseki (1762–1848), and others. For these people Siebold lectured and demonstrated European therapy and surgery and received in exchange gifts and information concerning almost everything.

Among the many acquisitions Siebold made in Edo were some which were to cause him serious trouble. Foremost among these were his exchanges with Takahashi Kageyasu of

---

[45] Some of the themes were: "Essentials of obstetrics in the Kagawa school" (see p. 388) by Mima Junzō (1795–1825), which was printed in 1825 with Mima's name in a Batavian scholarly journal, and in German translation (by Siebold, with commentary) in 1826 in the obstetrics journal of Frankfurt; "Questions and answers on physiology" and "Description of diseases in Japan" by Kō Ryōsai (1799–1846); "Illustrated explanation of insects in Japan" and "Illustrated explanation of spiders in Japan" by Ishii Sōken; "On whales in Kii province" by Oka Kenkai; and "Cultivation of tea and its use" and "Southern islands of Japan" by Takano Chōei. Ogata Tomio, et al., pp. 61–274.

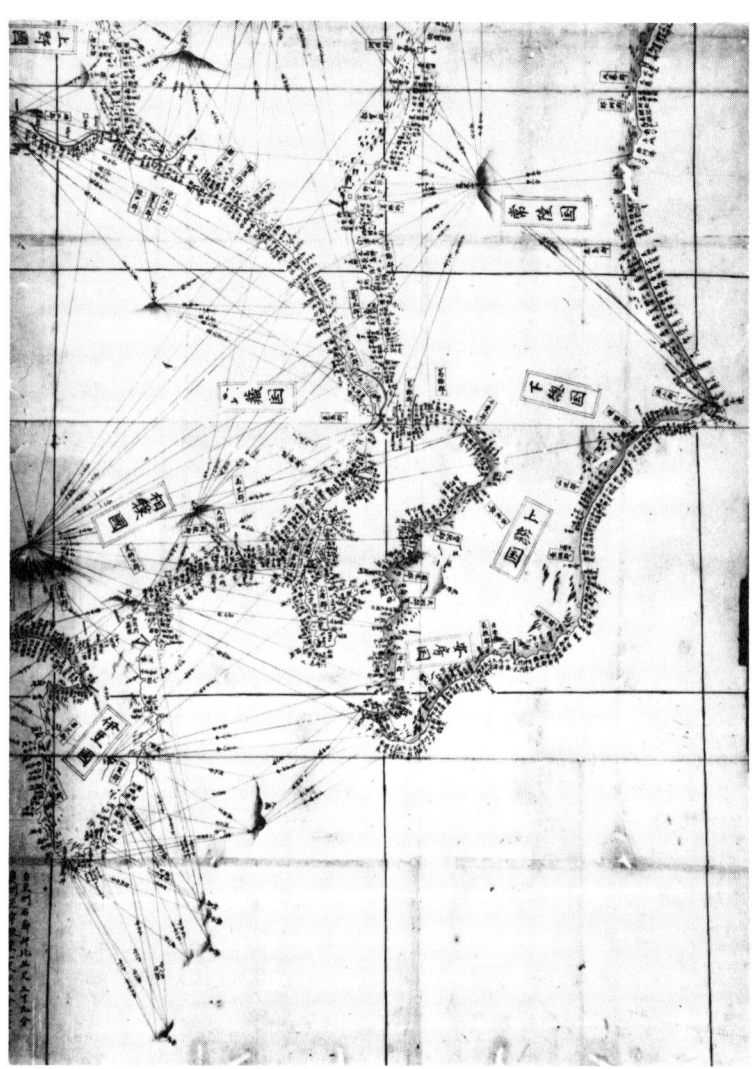

Fig. 39. Part of Inō Tadataka's coastline map of Japan, with Edo Bay at the center and Izu Peninsula at lower left. From *Nihon kairo sokuryōzu* (Surveyor's coastal map of Japan; scale, 1:432,000). (The National Archives, Japan)

the Tenmongata, to whom Siebold gave a book, *Voyage de 1803 à 1806* by a Russian admiral, A. J. von Krusenstern, and nine out of a set of eleven maps of the Dutch East Indies. From Kageyasu in exchange he received Inō Tadataka's (or Chūkei; 1745–1818) *Map of Japan* (scale: 1:864,000) and *Map of Hokkaido* (scale: 1:432,000). Kageyasu also loaned Siebold maps of Kyushu, the city of Kokura, and the Shimonoseki straits.[46] This was, of course, a clear violation of shogunate regulations. Moreover, for instruction on how to produce an ophthalmological drug for dilating the pupil of the eye given to Habu Genseki, an ophthalmologist in shogunate service, Siebold received a jacket (*kosode*) decorated with the hollyhock (*aoi*) design of the Tokugawa household, which Genseki had been granted by the eleventh shogun Ienari (ruled 1787–1837).[47]

Siebold's contravention of shogunate restrictions went undiscovered until 1828 when, preparatory to a planned return to Europe, he began loading his acquisitions on a ship that was grounded by a typhoon in Nagasaki before Siebold's departure date. Siebold's shipment was impounded and inspected, and certain restricted items were exposed. He was put under strict confinement on Dejima island, where he managed through friends to replace many of his lost specimens before being expelled from Japan the following year (1829). The various maps had not yet been stored aboard the ill-fated ship, so Siebold secretly copied them before he was ordered to surrender the originals. Japanese persons related to the incident were arrested and given various punishments. Takahashi Kageyasu—guilty of the most serious charge of providing the maps of Japan and Hokkaido—was imprisoned and died a natural death before sentencing; his two sons were exiled to a distant isle, and Habu Genseki was deprived of his position; three interpreters involved in the map exchange were given

[46] Kure (2), p. 218.
[47] Ibid., p. 232f.

life imprisonment; and many of Siebold's students suffered varying degrees of punishment.⁴⁸ Back in Europe, Siebold worked feverishly to establish himself as Europe's foremost expert on Japanese affairs, publishing a number of books introducing Japan. After the opening of Japan he also repeatedly sought and, in 1859, finally gained reentry to Japan, but his maneuvers were so embarrassing to both the shogunate and the Dutch consul-general that he was asked to leave for good, which he did in 1862.

The Siebold incident was not the only warning to the shogunate that the political implications of Dutch learning demanded a more decisive stand. The other incident that brought home the same message involved a group of Japanese citizens, one of whom was Siebold's former student, Takano Chōei (1804–1850).

THE "BARBARIAN CIRCLE" INCIDENT: REPRESSION

After his teacher's deportation, Takano Chōei continued to advance Dutch learning through scholarly translations which alone established him as one of the most capable *Rangaku* scholars. He transgressed conventional political wisdom by drawing from the new learning conclusions that the shogunate was unwilling to accept.

Chōei was not alone in this venture. He was only one member of an Edo group banded together under the leadership of Watanabe Kazan (1793–1841), an elder of the small Tahara fief and a highly respected painter. Chōei was a commoner, a physician, and a *Rangaku* devotee. Another key member of the group was Ozeki San'ei (1787–1839), a doctor serving the Kishiwada domain and formerly a staff member of the Tenmongata's Translation Office. Both Chōei and San'ei were disciples of Siebold and because of this the group's main source of information about the West.⁴⁹ Besides these and other scholars of Dutch learning, there were in the group a number of middle-level shogunate officials, including Egawa Hidetatsu

⁴⁸ Ibid., pp. 376–378.
⁴⁹ Satō Shōsuke, p. 150.

(or, Tarōzaemon; 1801–1855), all concerned about problems of Japan's domestic and international destiny. Other members included some domain samurai and commoners.

This group adopted the name of *Shōshikai* (Elders-respecting circle), and its aim was to study domestic and diplomatic affairs, particularly with the help of Dutch learning. They were not revolutionaries but an enlightened element within established society seeking its reform. They were intellectuals and samurai operating as a loyal opposition, but their open criticism of the shogunate's defense and foreign policy was quite unwelcome to some conservatives in government whose commitments were to traditional learning and established policies. Among such conservatives were Hayashi Jussai (1776–1841), hereditary head of the Shōheikō, and one of his sons, adopted by the shogunate samurai Torii family, Torii Yōzō (1804–1874). Both of them feared and despised Kazan's group.[50]

In 1825 the shogunate barred all foreign ships from Japan's shores, regardless of purpose. In 1837 the American merchant ship *Morrison* approached Japan's coast to return some shipwrecked Japanese fishermen rescued by other ships; but the American vessel was bombarded and thus withdrew. Both Watanabe Kazan and Takano Chōei criticized this treatment of the *Morrison*, insisting that such action might provoke an attack on Japan. Kazan's criticisms appeared in a book titled *Shinkiron* (On caution at a vital juncture; 1838) and Chōei's in his *Yume monogatari* (Tale of a dream; 1838). Though neither was printed, both were copied and read by many.

With growing concern over naval defense, the shogunate in 1839 ordered Torii Yōzō and Egawa Hidetatsu to survey the coast at Uraga, a vital defense point at the mouth of Edo Bay. Men working under Yōzō used traditional land-surveying methods, but Hidetatsu sought Kazan's advice. Kazan recommended Uchida Yatarō (or Itsumi, 1805–1886) and Okumura Kisaburō, minor shogunate officials who were also members

---

[50] Ibid., p. 220.

of the *Shōshikai* and had learned Western land-surveying methods from Takano Chōei. Yatarō was a *wasan* scholar—the fifth Master (see Table 25) of the Seki school—and though competent in mathematics, not as capable in land-surveying as Kisaburō.[51] Hidetatsu's results were found superior, causing a serious loss of face for Yōzō, who determined to seek revenge on the *Shōshikai*.

In 1839 Yōzō managed to have Kazan, Chōei, and other members of the *Shōshikai* arrested and indicted on such charges as criticizing the shogunate's foreign policy, planning to go to the Bonin Islands, and communicating with Ōshio Heihachirō (1792–1837), leader of an Osaka uprising in 1837.[52] During the subsequent interrogations, charges other than criticizing foreign policy were dropped as baseless. But in an age of mounting tension caused by increasing foreign pressures, the shogunate was incapable of tolerating policy criticism from private sources. Kazan was sentenced to house confinement in his domain and later committed suicide in 1841. San'ei had already committed suicide that year to avoid arrest. Chōei was sentenced to life imprisonment. In 1844, during a fire in the prison, he was released temporarily, according to custom, on the condition that he would voluntarily surrender himself within three days. Chōei failed to do so, roaming the country as far south as Kagoshima in Kyushu as a political fugitive, though he eventually returned to live in disguise in Edo. There he resumed medical practice and translating. It is said that his translations were so good one could easily guess they must have been done by Chōei. In 1850 when the police apprehended him, he killed himself while resisting arrest.

Chōei's own account of the persecution of the *Shōshikai* refers to the group as *Bangaku shachū*, or "Circle for barbarian learning," or briefly as *Bansha*, "Barbarian circle." The incident came to be known as *Bansha no goku*, "Prosecution of the barbarian circle."[53] The *Bansha* incident reflected the

[51] Ibid., p. 256.
[52] See Ch. 4, n. 28.
[53] Satō Shōsuke, p. 145.

mounting antagonism toward the growing capabilities of Dutch learning scholars. The most vigorous attacks after that incident issued from the citadel of traditional medicine, the shogunate's Igakkan, which tried increasingly to discredit Dutch medicine. The Siebold and *Bansha* incidents provided an opportunity for those frustrated in academic assaults to shift to political attacks. Through powerful connections in the shogunate a number of edicts were issued, beginning with an 1840 ban on all Western studies except in medicine. In 1842 it was made compulsory to get permission to publish translations from town commissioners (*machi bugyō*); this authority was centralized in the Tenmongata in 1845. An 1849 order gave the Igakkan control over the printing of all medical books, traditional and Western, and restricted the practice of Western medicine by shogunate medical doctors to surgery and eye treatment.

These were only stopgap measures, products of a conservative reform mood of the Tenpō era (1830–1844) or reflex reactions to failures of reform measures. They were neither systematically enforced nor notably effective. They reflected in part the agony of indecision in the face of much graver dangers: England's victory in the Opium War (1840–1842), and successive warnings of growing Western domination in Asia which threatened to force a reversal of the two-centuries-old isolation policy.

Developments in the Sciences

The late eighteenth and early nineteenth centuries were a time of ferment in the Japanese intellectual world in general. More particularly, there was intense competition between champions of Dutch Learning and the traditionalists who opposed its incursions—and nowhere so much so as in the sciences. Western science, after a false start in Western Wave I and then a stalemate by isolation, was reintroduced and partially understood in Western Wave II, a pattern similar to the Early Chinese Wave II rediscovery of Chinese science after a poor beginning in Chinese Wave I followed by inactivity during semiseclusion. The crucial difference was that the reintroduction of Chinese science in Chinese Wave II had no established tradition to resist it. Western science in Western Wave II had to wage an uphill battle against a tradition of Chinese-style science that was, or had been, a challenge to the best minds and a workable system for many problems. The issues of resistance versus accommodation were nowhere so evident as in mathematics and medicine—the former a case of "successful" resistance, the latter of "successful" accommodation. Between these two extremes lay calendrical astronomy, where cautious compromises merely postponed the day when clear-cut decisions would have to be made. And if medical scholars were ready to accommodate Western medicine at points where it was clearly preferable to their inherited tradition, Japanese mathematicians largely ignored Western mathematics, thus forfeiting their intellectual options ultimately to political and technical priorities.

CALENDRICAL ASTRONOMY

While Shibukawa Harumi's computational system, the *Jōkyōreki* (1684), was the outstanding achievement of the seventeenth century in calendrical astronomy, it lacked precision. Its defects led to Yoshimune's plan for revision, but the possibility that a new calendrical system would represent genuine improvement was severely inhibited by two factors, neither of which was readily apparent when Yoshimune proposed reform. One of these had to do with the amount of Western knowledge available to scholars mobilized for the shogun's project. The other factor concerned the depletion of the ranks of those scholars and their loss, by Yoshimune's death, of the authority he had provided.

THE *HŌRYAKUREKI:* JAPAN'S SECOND CALENDRICAL REFORM

Harumi died in 1715, the year before Yoshimune became shogun. It was Harumi's disciple, Ikai Toyojirō(?–1741), who was first called upon when the shogun asked the Tenmongata to account for the defects of the Jōkyō system.⁵⁴ Because he was unable to do so satisfactorily, Yoshimune in 1716 turned to the respected samurai mathematician Takebe Katahiro, who drew in his own disciple Nakane Genkei as a consultant. In the following year Genkei reported that the Jesuits in China had translated books on astronomy that were excellent but unavailable because of the import restrictions (Genkei had probably already surreptitiously examined one or more of them). Nishikawa Joken's explanation of the value of Western astronomy, requested in 1719, perhaps helped to persuade Yoshimune to relax the ban on book imports in 1720. In 1726 the *Li-suan ch'üan-shu* (Comprehensive collection of works on astronomic computation; compiled in 1723 but written much earlier), a collection of treatises by the great Mei Wen-ting (1633–1721), who included some material on Western astronomy, was officially received in Japan. By 1733 Genkei, working under Yoshimune's orders, had affixed the necessary notations for a Japanese reading of this text—three years after Joken's son Seikyū had done the same for the better-known but greatly inferior work, *T'ien-ching huo-wen* (p. 261). But in 1733 Genkei also died, as did his master Takebe in 1739.

These setbacks only strengthened Yoshimune's determination. Already he had extracted gifts of astronomical instruments, such as a telescope and astrolabe, from the Dutch company head; and Tenmongata specialists had been ordered to obtain through interpreters explanations of their use and of other related matters. With Genkei gone, Yoshimune in 1736 asked the Dutch to secure from Holland someone to help with the calendrical reform; but this was not realized.⁵⁵ In 1738 the shogun was presented with some Dutch astronomical books. Needing someone to read them was probably one reason for

---

⁵⁴Shibukawa Harumi's son Sekii was appointed to the Tenmongata in 1711 but died in April 1715; Keii, the younger brother of Sekii, replaced him at the Tenmongata. With Harumi's death in October 1715, Ikai Toyojirō became the guardian of Keii. Cf. the chronology in Nihon Gakushiin (5).
⁵⁵Numata (2), p. 42.

assigning Aoki Kon'yō and Noro Genjō to Dutch-language study in 1740.

After his retirement as shogun in 1745 Yoshimune continued to promote calendar revision. To supplement the poor ability of the Tenmongata staff, Nishikawa Seikyū in 1747 was assigned to assist them, as was also in the following year Yamaji Nushizumi (1704–1772), a famed mathematician and third Master (Sōtō) of the Seki school. In 1750 Seikyū, Nushizumi, and a colleague went to Kyoto to discuss their plan with the incumbent court astronomer. Joint observations were begun, but Yoshimune died in 1751, and the Tsuchimikado household quickly sought to regain the upper hand. Opinions clashed and Seikyū was powerless without Yoshimune; he was recalled to Edo in 1752 and dismissed from the Tenmongata in 1755. A new computational system was completed in 1754 according to the Tsuchimikado's traditional methods; named the *Hōryakureki*, it was used officially from 1755. When Seikyū died the next year, the Tenmongata's relative weight in calendrical affairs was seriously diminished.

The *Hōryaku* system was in reality only a slightly revised version of the *Jōkyō* system. It might have been better if imported Western instruments like the telescope and astronomical clock had been effectively used. It might have been still better if a fair competence in Western astronomical theory and modern data had been applied. Here again the possibilities were rather limited. Even if Seikyū had been permitted to continue the revision project, his resources were not the best. His knowledge of Western astronomy was largely that of the *T'ien-ching huo-wen*, an introductory cosmological treatise which Seikyū himself recognized as an insufficient basis for adequate calendrical calculations (though too technical, he thought, for philosophers and physicians). Moreover, his grasp of the *Li-suan ch'üan-shu* was shallow. Trigonometrical functions useful in calculations were available in the latter text, and if Genkei had lived longer, he might have brought the revised calendar at least up to the level of the early eighteenth-century

Jesuits in China, which the latter book reflected. Instead, the *Hōryaku* system was completed with only minor borrowings from the Jesuit work and was in fact inferior to its predecessor. Data listed as observed were actually only computed to maintain consistency with established theories, and thereby to bolster the Tsuchimikado position.[56] When ephemerides based on the *Hōryaku* system (used from 1755) failed to predict a lunar eclipse in September 1763, the fault was detected far in advance by Asada Gōryū and other amateur astronomers, and even by an employee of the Tsuchimikado household. Both the Tenmongata and Tsuchimikado officials were made the laughingstock of the scholarly world.

NAGASAKI INTERPRETERS AND WESTERN ASTRONOMY

Although the interpreter Motoki Ryōei (1735–1794) was born after relaxation of the book-import ban, his working conditions were less than ideal. Both the Tenmongata and the Tsuchimikado household were inactive and uncreative. Ryōei had no access to their counsel anyway and was not in touch with the amateurs in Osaka who were to provide the impetus to the next calendrical reform. His only stimuli were the interpreters' tradition of cosmological studies—unsuccessfully championed in official circles by Seikyū—and his own resolve. It was no small achievement that he single-handedly produced his first translation (*Oranda chikyū zusetsu*, Dutchmen's illustration of the earth) in 1772 and his second (*Tenchi nikyū yōhō*, The use of celestial and terrestrial globes) in 1774, simultaneously with the *Kaitai shinsho*. Ryōei's initial efforts were not as significant as they might have been; the Dutch original on which the *Tenchi nikyū yōhō* was based[57] consisted of two parts: the first explained the inadequacy of the Ptolemaic system, and the second expounded the Copernican system as the true hypoth-

[56] Nakayama (6), pp. 129, 133.
[57] The Dutch original of *Tenchi nikyū yōhō* was *Tweevondigh Onderwiss van de hemelshe en aardische Globen* (Amsterdam, 1666; first edition, 1620), written by Willem Janszoon Blaeu (or Blaaw, 1572–1638), edited and published by his son Johan. Nakayama (6), p. 175.

esis. Ryōei's translation ended three-fourths of the way through the first part.[58]

A later translation of Ryōei's was more important, the *Seijutsu hongen taiyō kyūri ryōkai shinsei tenchi nikyū yōhōki* (The ground of astronomy, newly edited and illustrated; on the use of celestial and terrestrial globes according to the heliocentric system; 1792–1793). Translating fully 90 percent of the original,[59] this was the first text in Japanese to give a description and explanation of the solar system, with the apparent and true courses of the planets explained according to the heliocentric system. Elliptic orbits were introduced, but not associated with Kepler. Kepler's first law was given, although Newtonian concepts were not a part of this work. Ryōei did mention Newton in brief summaries of other materials but only for comparison with the systems of Ptoloemy, Tycho, and Copernicus.[60]

In sum, it can be said that, although the earth was here treated for the first time in a Japanese book as a member of the solar system, Ryōei's own understanding of cosmology was incomplete. As noted earlier, he was chiefly concerned to translate, not advocate. He was not a professional astronomer and knew little of the traditional calendrical scheme. Filed away, neither printed nor made public, his translations had little chance to affect immediately the course of traditional astronomy, though his work was known among Nagasaki intellectuals and influenced men like Shiba Kōkan and Yamagata Bantō.

If there was no greater support for Ryōei's pupil Shizuki Tadao (1760–1806), he had at least the fine example of his master to follow. He spent twenty years in translating a commentary on Newton's mechanics. The original on which he worked was Johan Lulof's Dutch translation (Amsterdam,

---

[58] Ibid., p. 176.
[59] The original was the Dutch translation (Amsterdam, 1770) of George Adams the elder's (d. 1773) *Treatise Describing and Explaining the Construction and Use of New Celestial and Terrestrial Globes* (London, 1776). Ibid., p. 177.
[60] Ibid., pp. 177–178.

1741) of John Keill's *Introductiones ad veram Physicam et veram Astronomiam. Quibus accendunt, Trigonometris, de viribus Centralibus. De legibus Attractiones*, editio novissima (London, 1739).[61] Completed by 1802 in three parts under the title *Rekishō shinsho* (New treatise on calendrical phenomena), this work was a digest of Tadao's notes and not a literal translation. The important ideas introduced were: the relativity of motion depending upon the point of reference; the key Newtonian concept of universal gravitation; the laws of motion; explanations of centrifugal and centripetal forces; and certain properties of the ellipse.

How well Tadao understood Newtonian concepts such as mass, force, acceleration, momentum, and gravitation is open to question. There was in the Japanese tradition nothing resembling mechanistic laws of nature, and Japanese science was not far enough advanced to require Newtonian mechanics as a working tool.[62] Tadao was more concerned to reconcile the geocentric and heliocentric systems than to propagate the latter.

Time and again Japanese science benefited greatly from the industrious and creative work of amateurs. But to influence a wider sector of the Japanese intellectual community it was helpful to win a forum in Edo, the intellectual capital. To induce reform in the calendar, it was important to have access to government circles that controlled its computation. Ryōei and Tadao never had an opportunity themselves to do either. Their translations might have gone altogether unnoticed apart from the popularizers, who of course were even less specialized. Not astronomers, neither man had much hope of being called into the Tenmongata. Hence, there was no direct line of impact of their work on calendrical reform. The next developments took place among private citizens who were practicing astronomers, and who were able to form and sustain a scholarly

---

[61] John Keill (1671–1721) was professor of astronomy at Oxford University; his book was a commentary on the *Principia* by Isaac Newton (1642–1727). Nihon Gakushiin (5), p. 193.

[62] Nakayama (6), pp. 184, 186.

tradition, to spot their own inadequacies and correct them or acquire new sources. Later some of them were invited into government service. The founder of this group, Asada Gōryū, did not even know about Tadao's *Rekishō shinsho*.

That is not to say that there were no repercussions in the intellectual community that derived from the interpreters' translations. Popularization of the heliocentric theory by Shiba Kōkan, Yamagata Bantō, and others posed an implicit challenge to all traditional thought systems—Shinto, Buddhist, and Confucian. Only the Buddhists had a cosmic view which, despite its historical unimportance, seemed a good issue on which to challenge the new Western influence. It was from this quarter that vigorous refutations were mounted. They came from individual enthusiasts, however, not from organized Buddhist institutions (which cared little about cosmology), and never really amounted to a movement. In 1756 the monk Monnō of Ryōrenji temple in Kyoto had reacted to the general threat of Western cosmology in his "Against the *T'ien-ching huo-wen*" (*Hi tenkyō wakumon*). Specific resistance to the heliocentric theory was most strongly manifested by the monk Entsū (1754–1834), who in 1810 wrote *Bukkoku rekishōhen* (On the calendrical astronomy of Buddha's country), which argued the validity of Indian astronomy to defend the pre-Buddhist cosmological theory that placed a mythical mountain, Mt. Sumeru (Shumisen), and not the sun or earth, at the center of the universe.[63] Entsū actually knew much about Western astronomy, and while he attracted a sizable following, he could not get permission from his Buddhist superiors to publish his 1810 work. The best astronomers ignored him. His efforts, in fact, later served to stimulate interest in Western astronomy.

ASADA GŌRYŪ'S SCHOOL AND THE KANSEI CALENDRICAL SYSTEM

Asada Gōryū (1734–1799) was born Ayabe Yasuaki, the fourth son of a Confucian teacher in the Kizuki fief of northern Kyushu. Educated in the Ancient Practice school of medicine, Yasuaki studied astronomy as an avocation from an early age.

[63] For Entsū's argument, see Ibid., pp. 211–212.

In 1767 he was appointed physician to the lord of Kizuki fief but two years later petitioned for release from this obligation in order to devote full time to his avocational interest. Permission was not granted, but he left illegally and went to Osaka, changing his name to Asada Gōryū. There he practiced medicine to support himself while engaging seriously in his favorite subject. He later established a private academy called Senjikan in Osaka and attracted many able students. Several of his top students, whose names we have already encountered, figured prominently in the advance of astronomical knowledge —notably Takahashi Yoshitoki and Hazama Shigetomi.

Gōryū first mastered the *Li-suan ch'üan-shu* and in 1793 devised his own private calendrical system. To refine it, he then mastered the *Li-hsiang k'ao-ch'eng* (Compendium of calendrical phenomena; compiled by imperial order, 1737), based on pre-Newtonian models but with some post-Newtonian data, and particularly the sequel to this, the *Li-hsiang k'ao-ch'eng hou-pien* (1742), edited by the German missionary Ignatius Kögler, which uses Kepler's first and second laws but without reference to the heliocentric system.[64]

Gōryū's interest focused particularly on the problem of variation in the length of the solar year. Known as *shōchō* (in Chinese, *hsiao-ch'ang*; the secular diminution of tropical-year length), the problem had been recognized earlier in the *T'ung-t'ien li* and the *Shou-shih li* calendrical systems of the Sung and Yüan eras respectively. Gōryū's concern was to reconcile inherited sources with the new information from the West and with his own observations. He succeeded in working out an algebraic representation of all the observational data available to him. He also recognized that all parameters of calendrical calculation—lengths of the tropical year, and of the synodic, nodical, and anomalistic months, as well as the inclination of the elliptic—were subject to cyclic variation, and he corrected all parameters every ten years according to his own observations. His pupil Yoshitoki was to continue this research,

[64] Ibid., p. 189; Sivin (2), p. 92.

seeking to give Gōryū's law of variation (*shōchōhō*) the rigorous theoretical foundation it lacked. All of this effort was ultimately wasted, however, for the peculiar variations Gōryū identified were not the minute values of modern astronomy but larger values which arose from the imprecision of very ancient observations.[65]

By Gōryū's time a wider range of observational instruments was available than in earlier eras, and he and his followers gave considerable attention to the acquisition, use, production, and even improvement of such instruments. The total range covered traditional Chinese-style instruments, those borrowed from the Jesuits in China, and new ones imported directly from Europe through the Dutch traders in Japan. The generosity of Shigetomi, an Osaka merchant's son and himself a wealthy proprietor of a large pawnshop, was of great help in this cause; in addition, he himself had great talent for inventing and improving instruments and for making precise observations. Notable among new instruments used for astronomical observations were the pendulum clock, the quadrant, the transit, the eclipse meter, a more elaborate gnomon, and the telescope —most of which Japanese craftsmen sooner or later learned to make. A few telescopes, as well as smoked glasses, octants, and sextants, were imported through the Dutch, though Japanese artisans gradually became accomplished at lens-grinding and telescope construction.

The Asada group carried out sustained observations in order to test and refine theories directly related to calendrical science. Using a telescope, Gōryū was the first Japanese to measure the period of the sun's revolution by observing sun spots and to make detailed observations of the moon's surface, the movements of Jupiter's four satellites, Saturn's rings, and Uranus.

In 1795 the shogunate ordered the Tenmongata staff to make a calendrical revision using Western knowledge. As they were unable to do so, the private group under Asada Gōryū was given the chance to try. Gōryū himself declined, pleading

[65] Nakayama (6), pp. 188–194; also Nakayama (3).

old age but presumably reluctant to embarrass his former lord by coming into public view; so he recommended Yoshitoki and Shigetomi for the job. These two went to Edo, where Yoshitoki the samurai was made a Tenmongata official, but the merchant Shigetomi was given a lower position as "assistant."

A new computational system was completed in 1797 for use in calendars from 1798. Called the *Kansei* calendar, it was based upon the *Li-hsiang k'ao-ch'eng hou-pien*, and Gōryū's law of variation was used to adjust parameters. It was the first Japanese system to incorporate Western methods of computation, though it was still constructed according to the geocentric theories retained by the Jesuits in China. Yoshitoki and Shigetomi were prevented from making full use of their ideas by the incumbent Shibukawa head of the Tenmongata and his subordinates. Not entirely satisfied, therefore, with the *Kansei* system, Yoshitoki and Shigetomi were moved further to study Dutch astronomical materials.

In 1803 Yoshitoki began his translation of Lalande's book.[66] Personally very limited in Dutch, he secured the assistance of Ōtsuki Gentaku and was able, in about six months, to translate portions of the calendrical section of this book, which he recorded in his *Rarande rekisho kanken* (A private review of Lalande's *Astronomie*; 1803). Yoshitoki died in the beginning of the next year at age forty-one.[67] His eldest son Kageyasu was assigned to the Tenmongata at age twenty, keeping Shigetomi as his assistant. His second son, Kagesuke, was adopted into the Shibukawa family in 1808 and entered the Tenmongata the following year. Kageyasu remained active, acquiring considerable facility in the Dutch language, and in 1811 organized the Translation Office of the Tenmongata and served as its first director—until his injudicious exchange of maps with

---

[66] The original was in French, *Astronomie* (Paris, 1771) by Joseph J. L. de Lalande (1732–1807). The Dutch version was published in 1775 in Amsterdam. Nakayama (6), p. 198, n. 28.

[67] Before the end of 1803 by the *Kansei* calendar; after the beginning of 1804 by the Gregorian calendar.

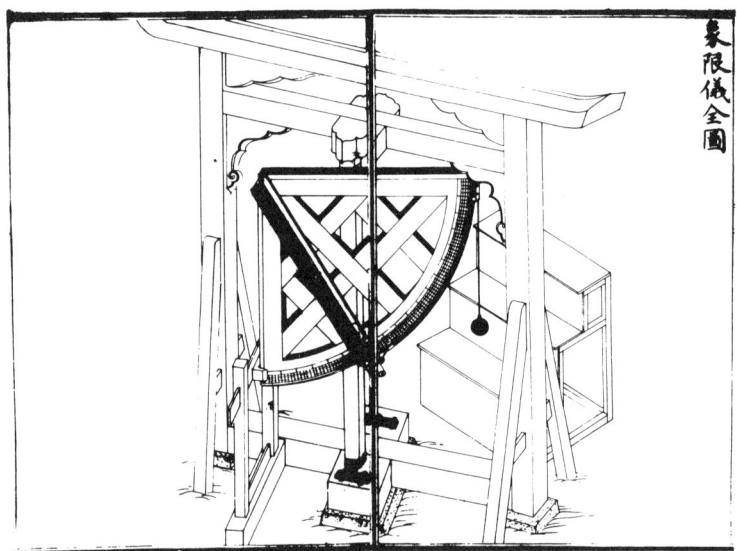

Fig. 40. Quadrant for measuring the altitude of celestial bodies; from the *Kansei rekisho*. (The National Archives, Japan)

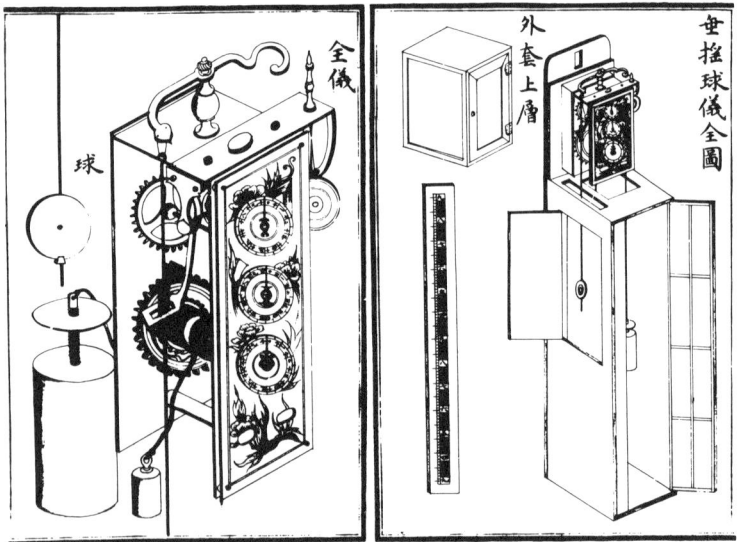

Fig. 41. Diagram of surveyor's clock, taken from the *Kansei rekisho*. (The National Archives, Japan)

Siebold led to his disgrace and decease. Kagesuke meantime persisted in the translation of Lalande, which, with Shigetomi's aid, was completed and published in 1836 as *Shinkō rekisho* (Astronomy by the new technique).[68] In the following year Yamaji Tomotaka, another Tenmongata official, completed a translation, *Seireki shinsho* (A new treatise on Western astronomy; 1837) of Pybo Steenstra's *Gronbeginsels der Steerekunde* (Amsterdam, two volumes, 1711 and 1722). These two translated works were made the basis of the next and last calendrical reform in Japan before modern times.

THE *TENPŌREKI*: LAST REFORM OF THE TRADITIONAL CALENDRICAL SYSTEM

The calendrical revision that produced the *Tenpōreki* was ordered by the shogunate in 1841 and completed the following year for use from 1843. Full use was made of the new knowledge available in the translations of Lalande and Steenstra, and the accuracy of the traditional calendar was greatly improved. The new system was compiled in the treatise *Shinpō rekisho* (Calendrical treatise by the new method), completed in 1846. Generally regarded as the closest to perfection of all traditional systems, its construction followed the lunisolar form, and of course the hemerological annotations were retained.

The Tenmongata regained ascendancy in calendrical affairs by incorporating the ablest civilian specialists, their instruments, and observational data. But in the rapidly changing scene of the next three decades, the relative place of calendrics steadily declined. The Translation Office was separated from the Tenmongata and expanded into the shogunate's Center for Western Learning (Kaiseisho). The Tenmongata, along with all shogunate institutions, was abolished in 1868. Although the Tsuchimikado household retained temporary authority over the public calendar, the actual work of calendrical preparation was later located in the Daigaku nankō (University, south campus), a predecessor of Tokyo University.

---

[68] Some sources attribute a major role in producing this translation to Kageyasu. Cf. Hirose (2), p. 186; and Nōda (3), p. 128.

Tsukamoto Akitaka (1833–1885), chief of the geography section in the new Meiji government, in October 1872 proposed abandoning the traditional lunisolar calendar for the Western solar one, contending that the old calendar's requirement of an intercalary month every two or three years was a bother and that the hemerological annotations retarded human knowledge. On the ninth day of the eleventh lunar month of 1872 (by the Tenpō calendar) the Meiji government decided to adopt a purely Western calendar. Uchida Itsumi, a *wasan* scholar (pp. 345–346) now head of the Meiji calendrical office, hurriedly prepared one, using the British nautical calendar. The third day of the twelfth month, 1872, of the Tenpō calendar was made January 1, 1873, of the Gregorian calendar to correct for differences in the lunisolar and solar new years, and thereafter Japan has used the same calendar as the Western world. It is said that the reason for such haste in making the shift was because the *Tenpōreki* would have required an intercalary month in 1873, and government officials (paid monthly in the new Meiji system) would have to be paid for an extra month, putting an unnecessary budgetary strain on the new government.

MATHEMATICS

In the reintroduction of Western learning into Japan from 1720 and its later development as Dutch studies, activities in calendrical astronomy and medicine intertwined considerably. Scholars often developed competence in more than one field, and translated texts often touched on more than one discipline. Moreover, interaction between Dutch studies and the problems arising within traditional science not infrequently led to revision or abandonment of traditional ideas. Some mathematicians made significant contributions, as we have seen, to calendrical astronomy; but on the whole such boundary-crossings did not affect the content or methods of Japanese mathematics. *Wasan* was not influenced in any essential way by Western learning in general or by its mathematics in particular; *wasan* did not borrow from, mount a challenge to, or vanish immediately because of, Western mathematics. Indeed, the counterpart of the best of *wasan*—the pure mathematics of Europe—was not even introduced into Japan until after isolation ended in 1854, and not systematically until after the Meiji Restoration of 1868.

The pure mathematics of the West was not wholly unavailable in Japan: the partial translation of Euclid's *Elements* into Chinese by Matteo Ricci and his collaborators was brought to Japan after the book ban was eased; there were mathematical portions in the later Jesuit astronomical writings imported from China; and Shizuki Tadao's translation of Keill's book (pp. 352–353) included some differential calculus. But the Japanese did not see these as part of a logical system. To the extent that they took cognizance of Europe's pure mathematics at all, *wasan* practitioners thought its methods clumsy and unnecessarily complicated, and all of it certainly inferior to *wasan*.

Practical mathematics for commercial and technical uses had made no essential progress beyond the seventeenth-century level represented by the *Jinkōki*. As European trigonometry and logarithms became known to Japanese mathematicians from the middle of the eighteenth century, they were introduced primarily as practical skills for use in navigation, land-surveying, and calendrical computation, and hence were regarded as of no importance to the leisured, nonutilitarian mainstream of *wasan*.

Wasan was not, of course, static throughout the century and a half considered here. The achievements of men like Ajima Naonobu (1739–1796) and Wada Yasushi (1787–1840) of the dominant Seki school were comparable to Western integral calculus. Late in the eighteenth century methodological criticism aimed at systematizing and simplifying *wasan*, and there were also efforts to make public the highest achievements of *wasan*. During the first half of the nineteenth century it became more popular. But throughout the process the central interest of its practitioners was the determination of the length, area, and volume of increasingly complicated figures. The problems had nothing to do with other sciences and technology. By mid-nineteenth century *wasan* was in a blind alley and could not by itself find its way out.

THE DEVELOPMENT OF CIRCULAR PRINCIPLES (*ENRI*)

Seki Takakazu, it will be recalled, had calculated the value of π by the method of reiterated approximation, and Takebe Katahiro, one of his disciples, succeeded in expressing this value in the form of an infinite series. Katahiro's work was expounded by Matsunaga Yoshisuke (? –1744) in a book entitled *Enri kenkon no maki* (Monograph on circular principles; n.d.). This book was held in the highest regard by Yamaji Nushizumi (1704–1772), who formally organized the Seki school of *wasan*, and hence most deserving of secrecy.

Ajima Naonobu was one of the most original in the Seki line. He discovered a method for calculating the area of a circle as the limit of the sum of the areas of infinitesimal rectangles stacked within it.[69] By this method one could calculate the area surrounded by any curve and thus the surface area and volume of any closed body. This is equivalent to the method of definite integrals in Western mathematics. Naonobu did not produce any books in his lifetime, but some of his papers were compiled in handwritten copy after his death for limited circulation under the title *Fukyū sanpō* (Immortal mathematical method; 1798).

The last major figure in the quest for circular principles was Wada Yasushi, who constructed tables of definite integrals, precisely equivalent to Western results, for some functions.[70] Using his tables, some of the very difficult problems of previous times became mere exercises—for example, calculation of an ellipse. Many *wasan* scholars became his students, at least long enough to obtain, for a fee, copies of his "Tables of circular principles" (*Enri shohyō*).[71]

---

[69] For Ajima Naonobu's new method of calculating the area and volume of various figures, see Katō (6), 2:90–94; also Ch. 10 on Ajima Chokuyen (alternate reading for Naonobu) in Smith and Mikami, pp. 195–205.
[70] For examples of Wada's tables, see Fujiwara, pp. 242–244.
[71] Nihon Gakushiin (4), p. 39.

CRITICAL DEVELOPMENTS WITHIN *WASAN*

The concerns of *wasan* devotees diverged somewhat in the second half of the eighteenth century. One trend that emerged was the pursuit of highly complicated problems, using consciously acrobatic and technically complex methods of solution. Problems were solved for their own sake, with no interest in possible general propositions implicit in specific problems. In reaction to such a tendency a few specialists advocated the need to systematize, generalize, and simplify *wasan*. Ajima Naonobu, for example, tried to change the descriptions in *wasan* texts from loose and inexact forms common to traditional Chinese-style mathematics to more terse and exact expressions. Fujita Sadatsugu (1734–1807), who shared with Naonobu and Toita Yasusuke (1708–1784) the title of fourth Master (Sōtō) of the Seki school, advanced this attempt to simplify and generalize *wasan* to a considerable degree in his *Seiyō sanpō* (Detailed key to mathematical methods; 1781).[72]

Aida Yasuaki (1747–1817), annoyed by what seemed to him a growing authoritarianism among disciples of the Seki school, in 1785 wrote *Kaisei sanpō* (Reformed mathematical methods), criticizing the content and expressions of Fujita's work. Fujita replied in turn with a refutation of Aida's criticisms. The resulting dispute between Aida and Fujita continued for seventeen years, producing a total of fourteen books, nine of which were published.[73] Their arguments centered mostly on disagreements over proper expressions and other mutual fault-finding. The dispute had little value for the essentials of *wasan*, but the exchange of accusations was so vivid and exciting that it did help arouse public interest in *wasan*. In fact the expressions in *wasan* became more precise and easier to understand.[74]

Aida also advocated the view that *wasan* specialists should search for "generally valid methods" (*tsūjutsu*), rather than settle for those applicable only in particular cases. He founded his own school, called Saijō-ryū[75], to propound his views; and

[72] Ogura (3), p. 79.
[73] On this dispute, see Smith and Mikami, pp. 188–190.
[74] Ogura (3), pp. 84–85.
[75] Saijō-ryū can mean, literally, "the very best school" but can also be read "the flow (or style) of Mogami," the name of a river in Aida's native district.

he is said to have read almost all of the secretly transmitted texts of the Seki school, so its secrecy was no longer easily maintained.

Although *wasan* leaders tried to establish ideals of generality, systematization, and exactitude, the reality of *wasan* practice was, generally speaking, far below these ideals. Along with methodological criticism, there emerged also a self-appraisal of *wasan* within the larger intellectual scheme. When it first sprang from practical mathematics early in the Edo period, *wasan* was not self-conscious about its justification. As it became separated from all practical concerns by the middle of that period, it was purposely developed as "pure" mathematics-for-leisure. Voluntarily divested of all utility, *wasan* was not then regarded as a kind of learning to be institutionalized by either the shogunate or the domains. It was a leisured art for spare time like the tea ceremony or flower-arranging, and its texts were listed in book catalogues along with those for these cultured hobbies.

Part of the problem was that so many of the leading *wasan* scholars were samurai (or *rōnin*), members of a class that despised all calculation and the abacus as tools of townsmen. The judgment on *wasan* by Confucian scholars was almost always negative; and even pure mathematics as a leisured art was not viewed as a hobby in the best of taste. Matsunaga Yoshisuke frankly admitted the propriety of many criticisms against *wasan* in a letter to Takebe Katahiro in 1743 and appealed to its devotees to forego the pursuit of scattered unrelated problems and try to forge mathematics into a system.[76] In this task, no help could be expected from Confucian quarters; for unlike astronomy and medicine, there were no prominent Confucian scholars who were also *wasan* enthusiasts—in great contrast to the early modern Western experience which had several distinguished philosopher-mathematicians, such as Descartes (1596–1650), Pascal (1623–1662), and Leibniz (1646–1716).

In the introduction to his 1781 *Seiyō sanpō*, Fujita Sadatsugu

---

[76] Hirayama Akira (2), p. 57.

classified mathematics into three categories: mathematics "useful for some use" (*yō no yō*), mathematics "useful for no use" (*muyō no yō*), and mathematics "not useful for no use" (*muyō no muyō*). By the first he meant practical mathematics, and by the third he intended the random pursuit of problems using unnecessarily complex and particularistic methods of solution. He favored the second category as the proper approach to *wasan*, which would be highly systematized and internally consistent, of no direct utility, but of some value in life and society. While he vigorously rejected the third category, much of his own work, judged by present-day standards, fell within its bounds. Fujita's second definition of *wasan* as useful-for-no-use most aptly reflected the character of *wasan* as a leisured art. But he was a controversial figure, hardly capable of forestalling long-term trends that tended to disperse rather than unify views on *wasan*.

PUBLICIZING *WASAN* ACHIEVEMENTS

The practice of secrecy was not uniformly applied to all *wasan* knowledge. Introductory problems and methods were increasingly made public in printed books; it was generally only the higher accomplishments that were restricted to transmission in handwritten manuals distributed among chosen disciples. The shared copies included only problems and processes of solution and were almost totally lacking in theoretical explanations. Until the middle of the eighteenth century there were no efforts among *wasan* scholars to make their highest achievements easily accessible in textbooks which systematically explained problems and methods.

In 1767, however, Arima Yoriyuki (1714–1783), lord of the Kurume domain and a disciple of Yamaji Nushizumi, wrote a book entitled *Shūki sanpō* (Selected jewels of mathematical method; 1767) in which he described 150 problems chosen from all fields of *wasan* research in the Seki tradition. Included were problem-solving procedures—though he gave no theoretical explanations of them. This was the first printed book in which one could find, for instance, a detailed description of Seki's written algebra, called *tenzanjutsu*. Thenceforth *tenzanjutsu* became a standard working tool of *wasan* students in general—almost a full century after its invention by Seki[77] (p. 266).

Fujita Sadatsugu's *Seiyō sanpō* (1781) consisted of three parts, the first two of which dealt with the more practical aspects of *wasan*; the third treated the more theoretical aspects, or what might be called the "core" of *wasan*. Rather easy to comprehend, this book was warmly received as the best mathematical text since the *Jinkōki*. Hasegawa Hiroshi (1782–1838) in his *Sanpō shinsho* (New text on mathmathical method; 1830) not only provided an extensive coverage of everything from practical mathematics using the abacus to the circular principles, but also included some of the Seki school's prime secrets. The book's format was very similar to Western texts, first giving theoretical explanations, then problems and their methods of solution, but with no exercise problems. It was easily understood and widely welcomed by *wasan* students. For his trouble, though, Hasegawa was "excommunicated" from the Seki school by its fifth Master, Kusaka Makoto (1764–1839).[78] It is said that Kusaka took this action because Hasegawa, by publishing a self-explanatory text, had deprived *wasan* teachers of their main source of livelihood—fees collected from students upon completion of a course.

WASAN SCHOOLS

Throughout most of the Edo era there were no official institutes solely for *wasan* research and education. These tasks were carried out in four main patterns:

1. Mathematicians of samurai status did research and taught in their spare time from official duties. This pattern was quite common among leading figures in the Seki tradition.
2. Mathematicians who were *rōnin* or townsmen settled in cities and towns where they opened private academies and lived off fees received for instruction, finding time when they could for research.
3. Mathematicians who were *rōnin* or townsmen wandered the provinces, staying to teach for some time in the homes of rich merchants and landowners. This pattern was particularly com-

[77] Ibid., p. 74.
[78] Ogura (3), p. 125.

mon in the nineteenth century and contributed much, along with printed texts, to the spread of *wasan*.
4. Mathematicians were hired to teach *wasan* in the domain schools (*hankō*). This pattern appeared only very late in the Edo period when *wasan* came to be included in the curriculum of some domain schools.

In contrast to medical academies, no *wasan* academy (case 2) lasted for several generations, sustained by a single household. The term "school" understood as "sectarian" not "academy" is applicable to all patterns, though most consistently to the first two, in which particular traditions were maintained through master-disciple relationships. It is in this sense that we speak of the Seki school, the Saijō school, and others. Most of the prominent mathematicians belonged to one of the various schools in this sense. It was entirely in keeping with the characteristics *wasan* shared with other refined pastimes like the tea ceremony and flower-arranging that the term for sectarian *wasan* schools (*ryūha*) should apply also to the sects of these arts. Among the *wasan* schools the greatest in achievements and number of disciples, as well as the longest-lasting, was that which traced its origin to Seki Takakazu.

Some of Seki's works were compiled and collected by his early disciples Araki Murahide (1640–1718) and Matsunaga Yoshisuke. Yoshisuke's disciple Yamaji Nushizumi assigned his own disciple Toita Yasusuke to the task of gathering (though not very systematically) all the achievements of Seki and his followers. Yamaji then classified the books representing these achievements into several ranks according to degree of difficulty (e.g., problems solved arithmetically, those solved by algebraic equations with one unknown, with two or more unknowns, etc.). A system was created that designated the books to be learned and the license to be awarded for each rank. The Seki school thus became the first formally organized *wasan* school with a license-ranking system.[79]

---

[79] There were five ranks with a license for each, beginning with elementary arithmetic and advancing through algebraic problems to the most difficult qualifications for the highest rank. Fujiwara, pp. 185–186.

Yamaji also instituted the post of *Sōtō* (Master) to designate the highest authority in the Seki school and retroactively determined those qualified to stand in the proper line of transmission after Seki. Araki Murahide, who had compiled some of Seki's manuscripts posthumously in a book titled *Katsuyō sanpō* (Essentials of mathematical method; 1709, printed in 1712) was made retroactively the "*Sōtō* of the First Transmission," and Yamaji proclaimed himself as "*Sōtō* of the Third Transmission" (see Table 25).

Some other schools lasted for a few generations and made some achievements in *wasan*.[80] These schools imitated the organization and customs of the Seki school to some extent. A few of them were regionally prominent, such as the Saijō school which took deep root in the Tōhoku (northeastern) district, where some of its followers continued to enjoy *wasan* late into the nineteenth century after it had virtually died out in the main centers of Japan.[81]

WASAN AND WESTERN MATHEMATICS

From the very beginning of Western Wave II, in Yoshimune's calendrical reform project, *wasan* specialists were exposed to aspects of Western mathematics. The most important work available at the time was the collected writings of the great Chinese mathematician Mei Wen-ting, the *Li-suan ch'üan shu* (p. 349), which contained trigonometrical functions, though in a thoroughly traditional setting. Nakane Genkei supplied the marks (*kunten*) for a Japanese reading of this book and wrote a brief commentary on the use of trigonometrical functions based on it. His master Takebe Katahiro composed a simple table of sine, cosine, and inverse sine functions for each degree (to twelve digits).[82] The *Ch'ung-chen li-shu* (Astronomical treatises of the Ch'ung-chen era; completed in 1634) by Jacob Rho, Johann Adam Schall von Bell (1591–1666), and other

---

[80] All names listed here (except that of the Saijō school) come from those of their founders. For a list of nineteen schools, see Smith and Mikami, p. 207, n. 4.
[81] Hirayama Akira (2), p. 156.
[82] Fujiwara, p. 164; Hirayama Akira (2), p. 52; Katō (6), p. 21.

Developments in the Sciences   369

Table 25. Prominent Figures in the Seki School of Japanese Mathematics (*Wasan*) and the Succession of Its Masters (*Sōtō*)

	Seki Takakazu (?–1708) ——————————————————————————————————————— **Takabe Katahiro** (1664–1739)
Sōtō I	ARAKI MURAHIDE (1640–1718) — Nakane Genkei (1662–1733)
Sōtō II	MATSUNAGA YOSHISUKE (?–1744) — Honda Toshiaki (1744–1821)
Sōtō III	┬── YAMAJI NUSHIZUMI (1704–1772) ── Naitō Masaki (1703–1766) ── Arima Yoriyuki (1714–1783)
Sōtō IV	FUJITA SADATSUGU (1734–1807)
Sōtō V	└── AJIMA NAONOBU (1732–1769) ── TOITA YASUSUKE (1708–1784)
Sōtō VI	**WADA YASUSHI** (1789–1840) ── KUSAKA MAKOTO (1764–1839) ── Hasegawa Hiroshi (1782–1838)
	UCHIDA ITSUMI (1805–1886)
Sōtō VII	KAWAKITA CHŌRIN (From 1882) (1841–1919)
Sōtō VIII	HAYASHI TSURUICHI (From 1918) (1873–1935)

Key
Capital letters: Sōtō (Master of First, Second, ... etc., Transmission; not necessarily limited to one. Cf. Sōtō IV and VI)
Bold line: Main line of succession (broken line: retroactive)
Light line: Strong influence on disciple
Bold type: Those with most prominent achievements
Source: Hirayama Akira (2), pp. 145–148.

Jesuits in China, also came to Japan (in 1733), as did Matteo Ricci's partial translation of Euclid's *Elements*, after relaxation of the book-import ban. To *wasan* scholars, as well as to calendrical specialists, were available in these sources portions of European mathematics as developed by the late seventeenth century (i.e., arithmetic, geometry, algebra, and trigonometry, but not Descartes' analytic geometry).

Japanese mathematicians first learned of logarithms from a Chinese work, the *Shu-li ching-yun* (Essential treasury of mathematical principles), printed in 1713 in Peking. Ajima Naonobu was one of the first to use logarithms in practical calculations. The first extensive logarithmic table printed in Japan came very late, in 1844; and not until 1854 did any published work include much in the way of theoretical explanation.[83] In any case, trigonometry and logarithms were accepted as tools for astronomy, navigation, and land-surveying; they had no attraction for *wasan* scholars in their pursuit of abstract mathematics as a leisured art. As for Euclid's *Elements*, they failed utterly to see that it was an exact logical system, thinking it much too elementary and inelegant to define with so many words things intuitively evident or to prove self-evident propositions with such labored demonstrations. In astronomy the West excels, ran the *wasan* conviction, but Japanese mathematics surpasses that of China and the West, from whom *wasan* had nothing to learn.[84]

Japanese demand for the pure mathematics of the West arose only after Perry's coming, that is, only after it was recognized as indispensable to the mastery of Western military technology, because it was integral to the basic sciences upon which that technology depended. *Wasan* had not developed any essential relationships with other sciences and techniques in Japan, so when *wasan* scholars were confronted with Western science and technology, they could not engage in the study of

[83] Smith and Mikami, pp. 268–270.
[84] Ogura (3), p. 129. Even Uchida Itsumi (or Yatarō), sixth Master of the Seki school, said in 1855 that Japanese mathematics is the best in the world. Ibid., p. 132.

Western mathematics in any meaningful way. Those who pioneered in the introduction of Western mathematics, though on an elementary level, were not *wasan* specialists but amateurs with some knowledge of the Dutch language.[85]

Prior to Perry's coming the points of greatest exposure (through Dutch sources, not Jesuit writings in Chinese) to advanced European mathematics were Shizuki Tadao's translation of Newtonian mechanics and the complete translation of Lalande's work by Shibukawa Kagesuke. From translations of Western books on physics available by mid-nineteenth century, *wasan* scholars picked up the idea of center of gravity,[86] which they adopted into their repertory for calculations of various complex figures. But their calculations were purely for intellectual play, not for relating to other sciences or techniques.

The main pre-Perry response in Japan to Western mathematics, then, was thoroughly practical, even when the response came from someone who was himself a *wasan* practitioner, as for example, Honda Toshiaki (1744–1821).[87] A famed but not very creative figure in the Takebe wing of the Seki school (see Table 25), Honda was one of the enlightened thinkers who first advocated the value of learning Western mathematics, along with astronomy, land-surveying, and navigation, in order for Japan to develop its political interests abroad. He not only published a book, *Keisei hisaku* (Secret plan for governing the country; 1798) advocating these views but also translated a Dutch navigational table, under the title *Daisokuhyō* (Tables of large measurements; n.d.), that included trigonometrical and logarithmic tables. He also wrote a manual on Western-style navigation. But his interests did not remain long in these fields.

---

[85] The only book published prior to the Meiji era that explained Western mathematics more or less systematically was the *Yōzan yōhō* (Methods of Western mathematics; 1857) by Yanagawa Shunzō (1832–1870), who was not a *wasan* scholar but a student of Western learning (*yōgaku*) interested in Western mathematics. See Table 23.

[86] Problems of the center of gravity appeared in the early stage of Japanese studies of geometry; see Smith and Mikami, p. 216f.

[87] On Honda Toshiaki, see Keene.

After the crisis provoked by Perry, the center of intellectual activities shifted rapidly from traditional to Western science and technology, but the mainstream of *wasan* scholars had nothing to do with this trend, personally, intellectually, or socially. The logic of decisions made in response to forced opening of the country led directly and unambiguously to the adoption of Western mathematics, but this necessity was more readily appreciated by those whose thinking was closer to the tradition of practical mathematics than to the esoteric *wasan* tradition.

MEDICINE

Publication of the *Kaitai shinsho* in 1774 was the crucial turning point in the eventual shift from traditional to Western medicine in Japan. But it was not an isolated event. It was rather one into which several lines of development converged: the critical stance of the Ancient Practice School (*Kohōha*) within traditional medicine; the sidestream of Dutch (*Kōmō*) surgery cultivated by the Nagasaki interpreters; and the new interest in Dutch studies prefigured in Arai Hakuseki, officially sanctioned by Yoshimune, and served by Aoki Kon'yō's pioneering language study. Having once converged, these lines did not, however, disappear or forfeit their independent development —except for Kōmō surgery, which was completely absorbed into the broader scope of the new Dutch medical practice (*Ranpō*).

As Dutch studies in general are not our immediate concern, this section must focus initially on developments in *Kohō* medicine, picking up the story from the 1750s. The account centers on two men, Yoshimasu Tōdō and Yamawaki Tōyō, whose works led to entirely different results. In Tōdō's case we find inflexibility provoking reaction, and ultimately eclecticism— all within the scope of traditional medicine. Tōyō's greater flexibility and empiricism led to the awakening of interest in Dutch medical studies that flowered first in the publication of the *Kaitai shinsho*. For about half a century not enough was known about Dutch medicine as a whole to rely solely upon it in practice. Though a general mania for things Dutch (*Ranpeki*) affected many intellectuals, actual practice in the field of medicine was eclectic, using traditional medicine where Dutch medicine was not yet known or accepted.

The situation changed after Siebold came and provided instruction in clinical procedures so that practitioners of Dutch medicine had an opportunity to see for themselves how the new medicine was applied in practice. New translations appeared more and more frequently, rapidly expanding the scope and depth of understanding. *Ranpō* doctors became more confident and began to abandon the eclectic Dutch-traditional approach during the 1830s and 1840s when, ironically, the Dutch fever was suddenly chilled by the Siebold and *Bansha* incidents (pp. 338–347). In medicine as in other fields, these issues were eventually resolved in favor of the Western side, less by academic debate than by reversal of the isolation policy due to the urgency of national defense.

FROM ANCIENT PRACTICE TO ECLECTICISM

By mid-eighteenth century the school of Ancient Practice had almost completely annihilated the "metaphysical nonsense" of the Li and Chu (*Goseihō*) and the Liu and Chang (*Goseihō beppa*) schools of medicine, most recently by the attack of Kagawa Shūan (1683–1755). Final demolition of the remaining authority of the "modernist" schools awaited the masterful assaults of Yoshimasu Tōdō and Yamawaki Tōyō.

Yoshimasu Tōdō (1702–1773) was not a direct disciple of Kagawa Shūan, yet he exceeded his vehemence in attacking the phase energetics theories of the Li-Chu and Liu-Chang traditions and in rejecting all references to the *yin-yang* and Five Phases concepts in relation to the human body. The only correspondence he was concerned with was that between sets of symptoms and medicinal prescriptions, though in his time there was no rigorous methodology for testing clinical experiences. In 1764 he published a book titled *Ruijūhō* (Classified prescriptions), listing 220 prescriptions which he considered confirmed by experience.[88] The initial printing of 1764 was 10,000 copies, all of which sold out in a month, about 5,000 each in the Edo and Kyoto-Osaka districts; reprints were issued many times. This book is regarded as a best-seller in Japan without precedent in works on medicine or Confucianism.[89]

An unprecedented amount of discussion, pro and con, among medical scholars was provoked by a published collection of Tōdō's theories, compiled in 1759 by one of his disciples, Tsuru Gen'itsu. This work, called *Idan* (New perspective on medicine), included an elaboration of a pathological theory that attributed all diseases to a single toxic principle (*manbyō ichidoku setsu*). Tōdō's speculative position was attacked from many quarters, not only from the Li-chu and Liu-Chang schools but from within the Ancient Practice school as well. For example, Gotō Boan, grandson of Gotō Gonzan, one of the founders of the *Kohō* school, argued strongly against the "single poison" theory.[90] Hata Kōzan (1721–1804), another *Kohō* advocate, in 1762 published a vigorous chapter-by-chapter rebuttal of Tōdō's theory, entitled *Seki "idan"* (Refutation of "Idan").[91] Mochizuki Rokumon (1697–1769), a founder of the Eclectic School of medicine, criticized Tōdō's frequent use of strong drugs and pointed out that "although Tōdō claims to respect Chang Chung-ching [author of the *Shang-han lun*], in his therapy he actually follows the way of Chang Ts'ung-cheng [founder of the 'modernist' Chang school, marked by use of strong drugs]."[92] The heated dispute over *Idan* served at least to stimulate Japanese medical thought. Eventually Tōdō's position gained wide acceptance, overshadowing both *Goseihō* and *Goseihō beppa* medicine.

[88] These 220 prescriptions were taken from the *Shang-han lun* (On Cold Damage Disorders) and *Chin-kuei yao-lueh* (Essential [prescriptions] of the Golden Casket), both by Chang Chung-ching (third century).
[89] Fujikawa Yū, p. 362.
[90] Ibid., pp. 360–361.
[91] Ibid., pp. 361–362.
[92] Ibid., p. 361.

Perhaps it was unavoidable for Tōdō to go to extremes in trying to eradicate the persistent influence of the unnecessarily complicated and therapeutically irrelevant theories and practices of the Li-Chu and Liu-Chang traditions. But Tōdō's defense of his own one-sided standpoint in time became too dogmatic. Many doctors realized that it could not be fully accepted any more than the theories he attacked, and it was only natural that there emerged an Eclectic School that sought to syncretize the various positions. Started in Edo by Mochizuki Rokumon, eclecticism was made the mainstream of Japanese medicine early in the nineteenth century through the efforts of Taki Keizan (1755–1810) and his descendants in Edo. Government policy on medicine became almost identical with views of the Taki household.

The Taki household's lineage goes back to Taki Mototaka (or Genkō; 1695–1766). He was a descendant of Tanba no Yasuyori, author of the *Ishinpō*, whose twenty-ninth descendant Tanba Motoyasu (or Gentai) had taken the surname Kaneyasu. Motoyasu and his descendants, including Mototaka, served the shogunate, specializing in oral hygiene. Mototaka, who first altered the surname to Taki, also practiced internal medicine; in 1747 he was appointed physician to the shogun.[93] In 1765 Mototaka founded in Edo a private medical academy named Seijukan. In 1791, under his son Motonori (1732–1801), this academy became, as we have seen, the shogunate's official Institute of Medicine (Igakkan, pp. 302–303), where shogunate-related doctors were required to train. It also exercised control over appointees to government medical service.

Taki Motoyasu (or Genkan; 1755–1810), son of Motonori and more widely known as Taki Keizan, was tutored in the Eclectic School of Confucianism by Inoue Kinga (1739–1784).[94] He applied the methods of textual criticism to many of the medical classics and developed a wider tradition of eclectic textual criticism. He and his descendants promoted

---

[93] Ibid., pp. 424–425.
[94] A medical doctor serving the Kasama domain in the province of Hitachi; he later resigned his post, moved to Edo, and advocated eclecticism in Confucianism, along with Katayama Kenzan (p. 298).

this line in the Institute of Medicine; and as they also monopolized the directorate of the Institute, the Eclectic School of medicine was made the predominant authority all over Japan. Though failing to evolve any new theories or therapies, the eclectics produced many critical studies of medical classics and published many versions of them.

One notable effort was the printing of the *Ishinpō*, completed in 984 by Tanba no Yasuyori. A manuscript in perfect condition was in the possession of a descendant of the Wake household living in Kyoto. In 1854 Taki Motokata (or Genken; 1796–1857), then head of the Igakkan, persuaded the government to order this person to submit his copy of the *Ishinpō* to the Igakkan. This was actually a last desperate attempt to find good prescriptions to compete with the growing influence of Western medicine. Printing of the *Ishinpō* was not completed until 1860, by which time the shogunate was already taking steps toward adoption of Western medicine. The most general effect of the textual studies of this school was to make reliable medical texts available to medical men throughout the country. Though this too was done partly to counter the spread of Western medicine, it was far from enough to stem the tide of excitement and changing loyalties as translation followed translation in the rapidly rising *Ranpō* movement.

Still, the Eclectic School was the main counterforce to Dutch medicine. In the period immediately following publication of the *Kaitai shinsho*, the eclectics were not necessarily antagonistic to Dutch medicine. For example, Taki Keizan wrote a foreword to Udagawa Genzui's translation *Seisetsu naika sen'yō* (1793). This drew adverse criticism because of his official position, and in the enlarged, revised edition (1822), Keizan's foreword was omitted. As the sense of jeopardy intensified, the Eclectic School tried every means to check the rise of Western medicine, taking advantage of popular anti-Western prejudices in general and of popular revulsion to human dissections in particular. The Siebold and *Bansha* incidents provided additional political leverage, and soon after the issuance of the government directives requiring translators to obtain publishing permission from town commissioners (1842) and then from the Tenmongata (1845), the Igakkan in 1849 acquired control over printing of all medical books, traditional and Western. In 1854 Ōtsuki Shunsai (1806–1862) compiled some translated excerpts from a Dutch book on the treatment of gunshot wounds in a book entitled *Jūsō sagen* (Trivial words on gun wounds), which the shogunate allowed to be printed without the Igakkan's permission. When the thirteenth shogun, Tokugawa Iesada (ruled 1853–1858), fell seriously ill and the shogunate's traditionalist doctors could not cure him, two doctors of Western medicine, Itō Genboku and another student of Siebold, were appointed as shogunate doctors; a few days later, four more were added. Though Iesada died that very day, the shogunate's policy was definitely shifting toward recognition of Western medicine.

Finally, smallpox reached epidemic proportions in Japan in the nineteenth century. In China during the Sung period a smallpox vaccine had been discovered that used powdered scabs from human patients, taken by nasal inhalation. Though this vaccine had at least as much possibility of inducing death as of immunizing effectively, it was brought to Japan, and in 1798 a smallpox section was established in the Igakkan. But use of this method of vaccination never spread widely in Japan.

The British physician Edward Jenner (1749–1823) in the same year perfected a smallpox vaccine that used active lymph taken from cows, not humans. Its high rate of success, with no direct possibility of death, became known to the Japanese through various channels. One of these channels was Siebold, who had tried to introduce this vaccine into Japan; the batch of immunizing lymph he brought from Batavia had lost its potency, however, and no vaccinia resulted. Though he offered to procure active lymph, the Japanese regarded it as too risky. Russia was an earlier source of information on smallpox vaccination. A Japanese fisherman living in Siberia after being rescued when shipwrecked, learned of this method of vaccination and in 1812 brought home to Japan a Russian book on the subject. This proved a good incentive for Baba Sadayoshi of the Tenmongata's Translation Office to study

Russian seriously, and in 1820 his Japanese version of this book appeared.[95]

In 1849 Otto Mohnike, a Dejima doctor in Japan from 1848 to 1850, had a cowpox scab sent to Japan from Batavia; with it he infected three children, from one of whom he obtained an effective lymph. Its use in vaccination spread quickly to cities like Nagasaki, Kyoto, Osaka, and Fukui in western Japan. Due to strong opposition from the traditionalists in the Igakkan, its use in Edo was delayed until 1858. It was in that year that eighty-two doctors, under the leadership of Itō Genboku, privately gathered funds and opened the Shutōjo (Center for Vaccination), which was destined in 1861 to become the shogunate's own Center for Western Medicine and later, after the Meiji Restoration, the medical faculty of Tokyo University.

FROM ANCIENT PRACTICE TO DUTCH MEDICINE

Yamawaki Tōyō (1705–1762) studied Ancient Practice medicine under both Gotō Gonzan and Kagawa Shūan and knew all the arguments against the Li-Chu and Liu-Chang theories. But Tōyō was more interested in what could be achieved on an empirical basis. He was dissatisfied with simply ascertaining the correspondence between sets of symptoms and prescriptions oriented to the *Shang-han lun* and wanted to know more about the structure of the human body and the functions of its organs. Early in his career he developed a healthy skepticism toward the theories of the internal organs sometimes appended to discussions of the functional systems in Chinese medical books. He was told by Gonzan that the internal organs of the otter resemble those of the human body, so he performed several dissections on otters; but he could not find the distinctions, for example, between large and small intestines which appeared in some of the Chinese illustrations of the body's interior.[96]

In ancient China human dissections for medical purposes were performed a few times, and the results appear in several late classical texts. They were mostly rough sketches, not pro-

---

[95] This is regarded as the first translation of a Russian book into Japanese. Ogawa (1), p. 145.
[96] Nihon Gakushiin (1), 1: 85.

ducts of thorough dissections and detailed observations. They also differ in minor details; the large and small intestines are distinguished in some but not in others, although the functional systems with which they are associated are always carefully distinguished in the texts. Some of these sketches show an organ to correspond to the functional system called by an untranslatable name (Chinese, *san-chiao*; Japanese, *sanshō*; see p. 95). The Chinese medical tradition was thoroughly functional, not structural, in its approach and thus did not regard anatomy as relevant to pathology or therapy. From dissections it was sufficient—and certainly not essential—to acquire a rough idea of the existence of the five viscera and the six bowels, as they were merely the anatomical substrata of more broadly defined functional orbs. The first known record of a human dissection for medical purposes, occurring in A.D. 16, is found in the official history of the Han era (*Han shu*). The next extant record is of an observed vivisection of fifty-six political prisoners in the Sung period, executed as a warning to would-be rebels. Sketches made at this observation appear in the Japanese text *Ton'ishō* by Kajiwara Shōzen (p. 143).

The Confucian tradition, however, viewed the body as a gift of one's parents, and keeping it intact was the beginning of filial piety. To damage it was disrespectful, and dissections were thought inhuman. There were strong social pressures against the use of dissections for medical research, and after the Sung period there is no record of human dissections so used until late in the nineteenth century. Even the great Chinese physician Wang Ch'ing-jen (1768–1831), who in China began the empirical criticism of the traditional ideas about the organs, relied on observations of exposed corpses supplemented by a few animal dissections.

During Chinese Wave I human dissections were prohibited in Japan by the *ritsuryō* laws.[97] After the *ritsuryō* system disintegrated and there no longer existed laws against dissections, it

---

[97] The crime of mutilation of corpses was clearly defined in the criminal code of the *ritsuryō* system; see Yamazaki Tasuku (1), pp. 328–329.

nevertheless remained the custom not to perform them. Executions of criminals in the Tokugawa period were done by the *eta* (one of two outcast groups), and it seems they made use of the dead bodies to collect human bile which was made into a medication for various ailments, such as abdominal cramps. The *eta* were unlearned, but because they were made to engage in butchery, tanning, and other tasks regarded by the Japanese as impure, their services were crucial to medical scholars wanting to witness dissections, as the learned doctors could not—nor did they wish to—wield the dissector's knife. Only men trained in Chinese-style medicine, though, could confirm or deny the accuracy of traditional illustrations of the internal organs.

Tōyō's opportunity came in 1754 when five thieves were executed in Kyoto, and three medical men from Obama fief asked for and received, for the first time, official permission to dissect one corpse. This permission was granted by the lord of Obama fief, Sakai Tadamochi, who was at that time magistrate (*shoshidai*) of Kyoto. The three petitioners were Kosugi Genteki, Hara Shōan, and Itō Tomonobu;[98] it was Kosugi who later first inspired Sugita Genpaku to take an interest in anatomy. Tōyō was invited to join this group for the observation. Their interests focused mainly on the internal organs; for example, they blew through a bamboo tube inserted into the trachea and saw the lungs expand, and pressure applied on the urinary bladder produced an emission.

Tōyō recorded his observations in 1759 in his *Zōshi* (On the viscera) which consisted of six pages of explanations and four pages of sketches (*Zōzu*) drawn by his pupil Asanuma Sukemitsu (pp. 317–319).[99] The sketches were reproduced by woodblock printing; in a revised edition five years later they were multicolored. The content was simple, elementary, and limited.

[98] Both Genteki and Tomonobu were students of Tōyō. Most probably it was Tōyō who initiated this project, though he pretended to have been invited by the other three to witness the dissection. See Nihon Gakushiin (1), 1: 86.
[99] The *Zōshi*, with annotations by Saigusa Hiroto and Kodama Reizō, is included in Saigusa (1), 8: 133–203.

Nothing of the head, muscles, nerves, or skeleton (except the spine) was recorded—all these had only minor places in Chinese medicine. Still, as the first record of a human dissection observed by Japanese medical scholars, it was an outstanding achievement. The impact on the Japanese medical world was extensive.

This event occurred more than two centuries after Andreas Vesalius' publication in 1543 of his *De fabrica corporis humani* which accurately described human anatomy and corrected certain errors in the Galenic tradition caused by excessive reliance on animal dissections. Vesalius had been a professor of anatomy at the University of Padua, thereafter the European center for anatomical studies. At the time of the 1754 dissection Tōyō had in his possession a Dutch version of the Latin text *Syntagma anatomicum*, published in 1633 (and translated into many European languages) by Johann Vesling (1598–1649), a professor of anatomy at Padua. Unable to read the text, Tōyō compared the copperplate prints of Vesling's book with what he observed and was shocked by the accuracy and precision of its charts.

Publication of *Zōshi* stirred up a vigorous debate over the propriety of dissections. Notable among negative views was that of Tōyō's own colleague, Yoshimasu Tōdō, who claimed in his *Idan*, published in the same year (1759), that anatomical knowledge is of no use in curing diseases.[100] Sano Yasusada, in his 1760 publication *Hi "zōshi"* (Against "Zōshi"), insisted that it was useless to observe dead organs that have ceased to function.[101]

Despite the criticisms of more conservative doctors, the implications of Tōyō's work were clear, namely, that the few Chinese crude illustrations of the organs were unreliable, that there was much to be learned from further research based upon dissections, and that anatomical charts in Dutch medical books merited careful study due to their accuracy and precision.

[100] Ibid., p. 94.
[101] Ibid., pp. 92–93.

Moreover, as the permission granted in 1754 had set a precedent, others were now able to gain official permission for dissections with less difficulty. In 1758, after seeing the manuscript of *Zōshi* before its publication, Kuriyama Kōan (1728–1791), one of Tōyō's disciples, arranged for a post-mortem dissection at Hagi in Chōshū. Many others were made in various places in Japan, usually performed by outcasts on executed criminals. They were generally done rather hastily, within the day of execution and on the execution ground, not under more controlled conditions in a house or some medical facility.[102] The doctors just watched and gave instructions to the dissecting outcasts. The second book on anatomy printed in Japan was *Kaishihen* (On dissection of corpses), published in 1772 by Kawaguchi Nobutō (1736–1811), a Kōmō surgeon who was able to perform the dissection himself. Kawaguchi in 1770 had obtained permission to dissect two bodies, again in Kyoto, and the content of this book was much advanced over *Zōshi*.[103]

There is also a strange incident connected with *Zōshi* that reveals the difficulty experienced in Japan in this age in evolving a new attitude and receptivity toward foreign learning. One of the early Nagasaki interpreters, Motoki Ryōi (1628–1697) had translated a Western anatomical book (the second edition of Johann Remmlin's *Pinax microcosmographicus*, translated from German into Dutch by Justus Gratianus and published in Amsterdam in 1667) under the title "Oranda zenku naigai bungōzu" (Anatomical chart of the human body).[104] Possibly known to a limited circle of *Kōmō* surgeons in Nagasaki, it had no discernible effect on Japanese medicine. It was brought to light and published in 1772 by Suzuki Sōden of the province of Suō, who recognized the accuracy of Ryōi's translation when he first saw *Zōshi*.[105] This publication preceded the *Kaitai shinsho* by two years, and even its preview,

[102] Ibid., p. 114.
[103] Ibid., p. 110.
[104] Ibid., p. 131.
[105] Ibid., p. 136.

the *Kaitai yakuzu*, by one year—but it came into print seventy-five years after Ryōi's death. Strictly speaking, it was the first published translation of a Western anatomical book; but it was completely overshadowed by the sensation created by the *Kaitai shinsho*.[106]

The details surrounding the translation and publication of the *Kaitai shinsho* have already been given. Here it is important to trace the main lines of its impact on the Japanese medical community, remembering that it also administered a profound shock to intellectuals and especially to Nagasaki interpreters. Japanese medical doctors saw readily that anatomy is an indispensable part of medicine, whether Western or traditional. There was no need to get bogged down in a disabling theoretical debate. The impact, therefore, was electric and immediate, precipitating a second dissection boom, though dissections were still done rather unmethodically. They did serve, however, to confirm again and again what was learned by continued study of Western anatomical charts. They also stimulated activity among those whose work remained largely oriented to traditional medicine (explained below as Dutch-Chinese eclectic medicine).

As no political repercussions followed publication of the *Kaitai shinsho*, many medical men came forward eager to learn how to read Dutch books. The Nagasaki interpreters' monopoly of the Dutch language had been broken, and now many qualified medical doctors could read, study, translate, and publish without the help of interpreters. Edo was the center of this new Dutch study boom. The interpreters did make an indispensable contribution to this boom, of course, by helping produce the *Edo haruma* (1796), the *Yakken* (1810), and the *Nagasaki haruma* (1833) dictionaries. The importance of the private Dutch academies as training grounds for eager aspirants has also been emphasized earlier.

Here must be stressed the difficulty Japanese medical schol-

---

[106] Neither the *Kaitai shinsho* nor Sugita's record of its production, *Rangaku kotohajime*, makes any mention of this book. Ibid., p. 134.

ars experienced late in the eighteenth and early in the nineteenth centuries in achieving a correct and comprehensive grasp of Western medicine as a whole. Their understanding depended primarily on increasing the number of translations and of the medical fields covered. The usual pattern involved the appearance in a particular field of an initial translation of rather poor quality due to the strangeness of the Western ideas and technical terms to doctors trained in traditional medicine. These first translations were frequently short and circulated only in handwritten copies. With the second or third translations in the same field, the quality of translation improved measurably, the manuscripts were longer, and they were often printed. These improved translations generally exercised a wide and crucial influence, immediately inviting many further translations in the field. Not all such translations were printed, but the range of choice as to the best in the specific field certainly widened. Once a field had extensive materials, there often emerged some doctors who claimed to practice that specialty in a purely Western manner. Examples drawn from several fields may serve to illustrate concretely the rapid development of Dutch medical studies in Japan.

After the *Kaitai shinsho*, about a hundred manuscripts on Western anatomy were translated or compiled before the Meiji era.[107] The best anatomical atlas, Udagawa Genshin's *Naishō dōbanzu* (Copperplate charts of anatomy), was published in 1808 and was the first in Japan to use copper plates for printing anatomical charts. It was printed to accompany his earlier *Oranda naikei ihan teikō* (Outline of a Dutch anatomical text; 1805).

Though fundamental to modern medicine, books on physiology were translated rather late, possibly because of difficulty in understanding in this field before Siebold came. The first translation to be printed (although two were completed during the preceding year) was volume one of Takano Chōei's twelve-volume *Igen sūyō* (Principles of physiology) completed in 1832. Ozeki San'ei, fellow member of the *Shōshikai* (p. 345) with

[107] Ishihara (2), p. 149.

Chōei, wrote a commentary, also in 1832, on the technical terms of Western physiology, entitled *Seii genbyō ryaku* (Outline of Western physiology). In that same year Ogata Kōan rendered "physiology" as *jinshin kyūrigaku*, borrowing the Confucian term *kyūri* then in vogue, until an 1866 translation, *Seiri hatsumō* (Enlightenment on physiology) by Shimamura Teiho (1830–1881), introduced the term *seiri*, which has been used since.[108]

Translations in the field of pathology came even later and were few in number. A four-volume translation entitled *Genbyō yakusetsu* (Summary of pathology) by Ogata Kōan was never printed. The first pathology book to be printed was the same author's 1847 translation, *Byōgaku tsūron* (General theory of pathology), in twelve books, the first three of which were printed in 1849.

In pharmacology, many descriptive books appeared from an early stage; the more theoretical works came quite late. Katsuragawa Hoshū's unpublished *Oranda yakusen* (Selected drugs of Holland) was completed in 1785 and was followed by many translations of "Dutch" pharmacopeia to make known the specifications of drugs used in Dutch medical practice. The first translation of pharmacological theory appeared in 1850, the *Wātoru yakuseiron* (Wātoru's pharmacology) by Hayashi Dōkai (1813–1895). His application to the Igakkan for permission to publish was not granted until 1856.

Turning to clinical medicine, the first translation of internal medicine was Udagawa Genzui's *Seisetsu naika sen'yō* (Essentials of Western internal medicine), completed in 1792 and printed a few volumes at a time from 1793.[109] Many other translations appeared in this field, but the most influential one before the

---

[108] In 1851 Benjamin Hobson (1816–1873), a British missionary medical doctor, wrote a book on physiology titled *Chüan-t'i hsin-lun* (New treatise on the human body) with the help of his Chinese students. Along with many Japanese books on physiology, Hobson's book was widely read. Ishihara (2), p. 150.

[109] The original was Johannes de Gorter's *Medicinae compendium* (Leiden, 1731); the Dutch version was titled *Gezuiverde Geneeskonst, of Kort Onderwys der Meeste Inwendige Ziekten; Ten nutte van Chirurgyns* (Amsterdam, 1744, 1762). Asahi, p. 188; Bowers, p. 99.

Meiji era was *Fushi keiken ikun* (Lessons from Mr. Fu's [C. W. Hufeland] experience) by Ogata Kōan. It was printed in thirty volumes between 1842 and 1857. By this stage the time lag between the appearance of the original European texts and completion of the Japanese translations was decreasing. The original was C. W. Hufeland's *Enchiridon medicum*, published in German in 1836; translated into many languages, the Dutch version with which Kōan worked was printed in 1838.[110]

The most influential text on pediatrics, *Yōyō seigi* (Details of pediatrics), was translated by Horiuchi Sodō (1801–1854) and published in 1843. Traditional medicine had advanced considerably in obstetrics and gynecology under the leadership of Kagawa Gen'etsu (1700–1777), so there was less pressing demand in this field (though some appeared[111]).

Long before the *Kaitai shinsho*, Narabayashi Chinzan had partially translated Paré's book on surgery (p. 284). Though incomplete, this unprinted work was copied and circulated widely in various forms among *Kōmō* surgeons. The first translation on surgery to appear as part of the *Rangaku* movement was *Yōi shinsho* (New text on surgery),[112] started by Sugita Genpaku and completed in 1792 by Ōtsuki Gentaku and others. The 1825 printing of this work included only very limited portions of the original translation. The crucial work on surgery was *Yōka shinsen* (New selections on surgery) by Sugita Ryūkei (1786–1845; Genpaku's son by a mistress), which was printed in 1832. Afterward, about thirty trans-

---

[110] Ishihara (2), pp. 154–155.
[111] The first (unpublished and without much influence) was *Karin sankasho* (Obstetrics by [John van] Hoorn; 1823) by Aochi Rinsō. The Chinese reading for *karin* is *ho-lun*, a near approximation of Hoorn, suggesting Chinese influence in alerting the Japanese to this work. The most important (but unpublished) was *Sanshi sanron* (Mr. San's [Salomon] obstetrics; 1845) by Yatabe Keiun (1819–1857), of which the original was by Gottlieb Salomon (1774–1864). Nihon Gakushiin (1), 4: 205.
[112] Laurens (or Lorenz) Heister (1683–1758) wrote *Chirurgie* in German (Nürnberg, 1718); *Yōi shinsho* was translated from the Dutch version (*Heelkundige onderwyzingen*; Amsterdam, 1776) of the original. Numata (2), p. 91; also Asahi, p. 190.

lations were finished before the Meiji era, not all of which were printed.[113]

ECLECTIC DUTCH-CHINESE MEDICINE

Though it is clear from the above that many translations appeared in the decades following the *Kaitai shinsho* (1774), in the treatment of patients it was impossible to rely solely on Western medicine. Most medical practice was based on traditional medicine, though practitioners were eager to absorb any available Western knowledge of anatomy, drugs, instruments, and clinical and surgical procedures. Actual practice for at least half a century after the *Kaitai shinsho* consisted of a mixture of Dutch and traditional Chinese-style medicine. Not only were more diseases cured, there were a number of remarkable academic achievements that reveal the continuing productivity of traditional medicine in the hands of creative men. It was only after Siebold's introduction of direct clinical instruction during his first stay in Japan (1823–1829) that responsible doctors gained enough confidence in their knowledge and skill to dare practice "pure" Western medicine. Then it became clear that two traditions so different in conceptual foundations and methods could not be syncretized. Dutch-Chinese eclecticism in medicine was destined to vanish sooner or later, depending upon the specific branch.

A few examples may suffice to indicate the kinds of scholarly activity pursued in the Dutch-Chinese eclectic interim. One of the most creative men was Hanaoka Seishū (1760–1835), who was trained in both *Kohō* medicine and *Kōmō* surgery. He invented a number of surgical techniques and was adept at syncretizing technical aspects of Western and traditional medicine. His major academic achievements were the rediscovery of general anesthesia and the invention of techniques for major surgery. In China, in the Three Kingdoms period (third to fourth centuries), a legendary doctor named Hua T'o is said to have done major surgical operations under anesthesia using

[113] Ishihara (2), p. 157.

a drug named *Ma-fei-t'ang* (in Japanese, *mafutsutō*; the story probably came from India). Seishū used this name for his own prescription. Pain-killing drugs had been used in China since the Yüan era to set compound fractures and must have often produced total anesthesia; in fact, strong doses were specified to put patients to sleep if light doses did not deaden the pain. The main ingredients of Seishū's compound were *mandarage* (*Datura alba*, i.e., white datura, a relative of jimsonweed) and the root of *torikabuto* (*Aconitum japonicum*, an aconite).[114]

With a drug for general anesthesia, Seishū could perform major surgical operations previously not possible. In 1805 he performed an operation for the removal of breast cancer, the first operation in the world done under total anesthesia for anything more serious than compound fractures. (General anesthesia was not discovered in the West until 1846.) He also developed surgery for the removal of stones in the bladder and for amputation at the ankle joint. He attracted a great many disciples, about 1,130 from all over Japan, who further incorporated knowledge and techniques from Dutch sources. The most prominent was Honma Sōken (1804–1872), who was the first to develop rib surgery and amputations at the knee and lower thigh. He is also said to have violated his master's wish to keep secret these new methods and drugs and therefore to have been dismissed from the Hanaoka school of surgery.[115]

Kagawa Gen'etsu, though unlearned, was skillful in technical matters and a genius at inventing medical instruments. His main interest was obstetrics, and the school he founded in Kyoto attracted many followers; it was unchallenged until early in the Meiji era, diminishing incentives to explore Western sources in this field. His claim to fame was the discovery of the normal position of the fetus in the womb. For generations it had been thought that throughout pregnancy the fetus stood upright, then turned around at birth. In his only book, *Sanron* (Obstetrics; 1766), he derided this notion, explaining that

[114] Miyashita Saburō, pp. 273–278.
[115] Ogawa (1), p. 153.

Fig. 42. From Hanaoka Seishū's record, *Kikanzu* (Illustrated report of a rare disease), of his successful breast surgery under general anesthesia in 1805. In the multicolored original, the spot on the enlarged breast is red, to indicate a cancerous growth. (University of Tokyo General Library)

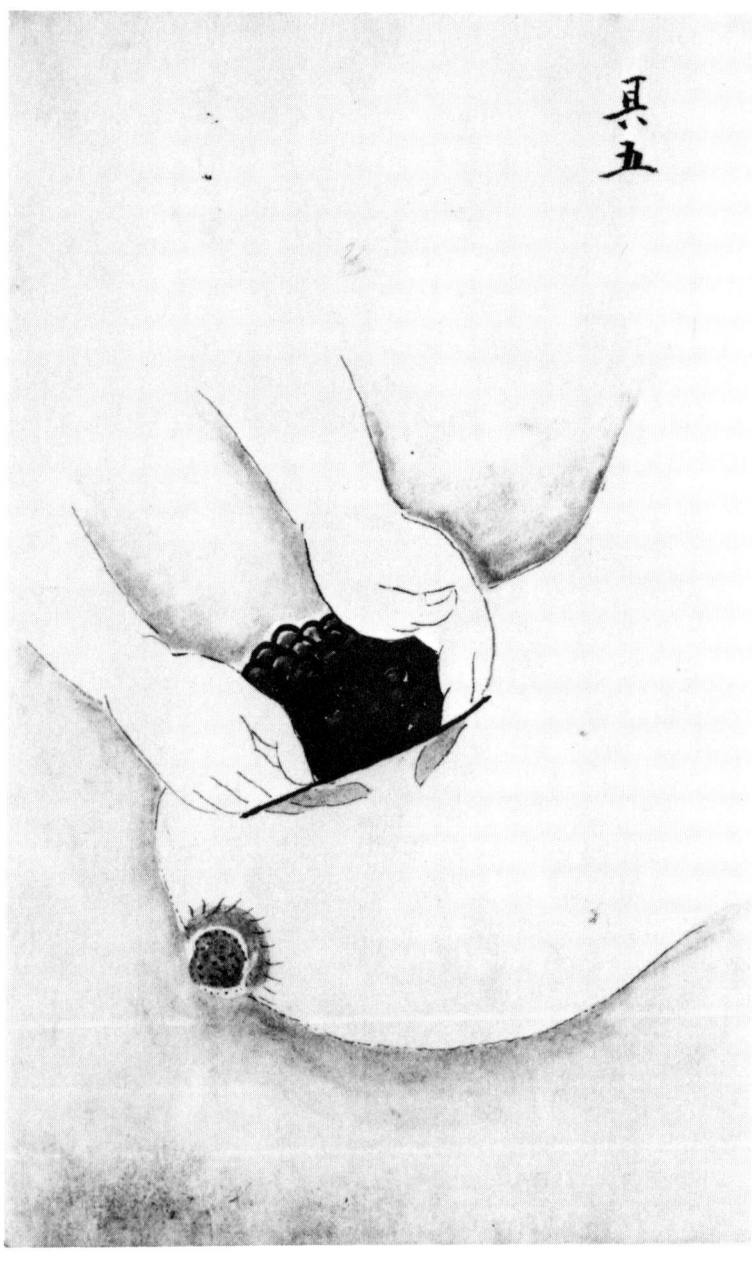

Fig. 43. Close-up from Hanaoka Seishū's record of breast surgery, showing the cancerous growth being removed; this is one of a series of illustrations showing the entire process of the surgical operation. From the *Kikanzu*. (University of Tokyo General Library)

usually by the fifth month of pregnancy the fetus is about the size of a small gourd (*uri*) and is positioned head down, with the forehead near the upper part of the mother's pelvis.[116] It is generally thought that he arrived at this conclusion independently of the discovery of normal fetal position in the West by William Smellie (1697–1763) in 1751. The writing of the *Sanron* was done by his friend Minagawa Kien (1734–1807), a Confucian scholar, as Gen'etsu was unlearned, if not illiterate.

Fuseya Soteki (1747–1811) studied Li-Chu and then *Kohō* medicine before becoming interested in Dutch medicine, though he could not read Dutch. He too was caught up in the dissection boom. In his *Oranda iwa* (Medical tales from Holland), written in 1803 and printed in 1805, he records an experiment he made by putting ink into the kidney to discover its filtering function. This discounted the traditional Chinese theory that the kidney system functions to produce an alimentary essence (sperm) and to store it as well as other essences produced by other functional orbs. The Western discovery of the kidney's function was made in 1842 by William Beaumont (1785–1853), an American army surgeon.[117]

These discoveries and inventions, remarkable as they were, occurred when it was still thought that Dutch and traditional medicine could, at least on more technical levels, be syncretized. Concurrently in the West, and particularly in the early nineteenth century, discoveries were being made one after another on the digestive system, the nervous system, and the function and structure of the brain. These Western developments involved a considerable amount of chemical and biological research; that is, they were related to an overall system of science that was to be more systematically introduced into Japan from the middle of the nineteenth century and was to make most difficult, if not impossible, any further eclectic or syncretistic approach. The contribution of Siebold was to speed up the process by which Japanese doctors became capable of relying almost exclusively on Western medical practice.

[116] Nihon Gakushiin (1), 4: 134.
[117] Ishihara (2), pp. 164–167.

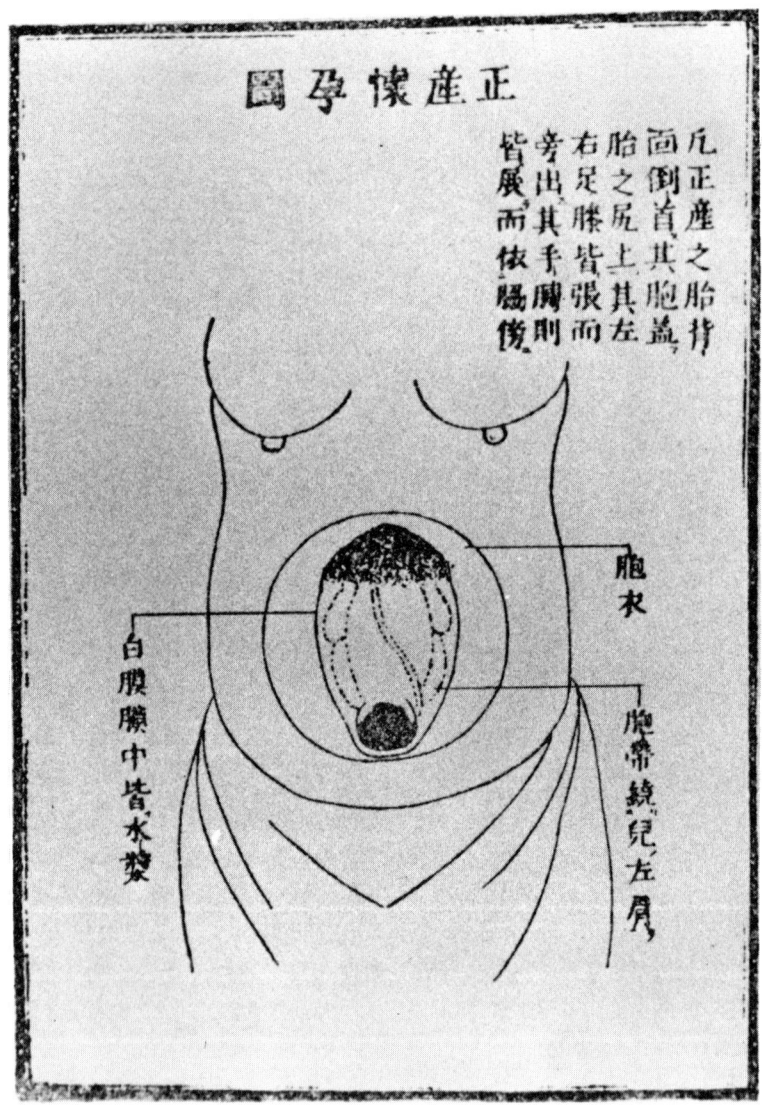

Fig. 44. Sketch showing the inverse position of the fetus in normal pregnancy; from *Sanronyoku* (Obstetrics, Enlarged, vol. 2; 1775), an expanded edition of Kagawa Gen'etsu's *Sanron* (Obstetrics; 1766) by his son Genteki. (Courtesy of Professor Teizō Ogawa, Department of Medical History, School of Medicine, Juntendō University, Tokyo)

WESTERN MEDICINE AFTER SIEBOLD

Siebold's willingness to help, with all of his qualifications, would have meant little had there not been many Japanese doctors and would-be doctors with some background in the study of the Dutch language and of Western medical books, and with an eagerness to learn Western medicine. By the same token, the Japanese aspirants definitely needed the kind of disciplined, detailed, and direct instruction in clinical procedures that Siebold provided. The Nagasaki magistrate, by tolerating Siebold's uprecedented freedom to treat patients and teach students, helped make the time ripe for both sides. Siebold's teaching covered internal medicine, and many other practices as well, but he is said to have been at his best in surgery, ophthalmology, and gynecology.[118]

Relating Siebold's teaching to the flow of new translations, some of his students gradually grasped the overall spectrum of Western medicine, and this undoubtedly guided decisions as to what kinds of additional translations were needed. Not only did the level of Dutch medical studies rise rapidly, but practitioners also gained the necessary confidence to claim that they were now practicing Western medicine as such, and not an eclectic Dutch-Chinese medicine.

For instance, one of the earliest translations was on internal medicine (Udagawa Genzui's *Seisetsu naika sen'yō*; completed in 1792, published from 1839 on). But only after related knowledge of physiology, pathology, and the proper use of drugs was fairly well understood, and the drugs were made available, could a doctor conscientiously claim to practice Western internal medicine. Yoshida Chōshuku (1779–1824) in Edo was the first to make that claim. The translator of the key work on pediatrics (*Yōyō seigi*; 1843), Horiuchi Sodō, was the first to make a similar claim in that field. By way of contrast, the longer history of *Kōmō* surgery, the relative advantage of making on-the-spot observations of Siebold's operations, and the many published translations on surgery made it possible for a multi-

[118] Itazawa (2), p. 14.

plicity of claims to be advanced with respect to the practice of Western surgery.

In the meantime, Dutch studies were being expanded to many fields beyond medicine proper, and the shogunate increasingly coopted the ablest medical scholars into institutions oriented to national defense. The advance of medicine in Japan might have been greatly delayed had not programs initiated after Perry to adopt Western learning and science also included medical education, as in the Naval Training Center in Nagasaki, and later in the Center for Western Medicine (see below).

In the changes already begun or soon to occur, more was involved than merely exchanging established Oriental techniques in medicine for newly imported Occidental alternatives. A significant shift in intellectual ethos was taking place. The long-established tradition of Chinese-style medicine was characterized—variously, according to particular schools—by a physiological understanding of the total human organism as integrated in and correlated with the total natural environment; a pathological understanding of functional correspondence between cosmos and organism that placed as much, if not more, emphasis on external, environmental causes of diseases and contextual diagnoses in relation to them; and a therapeutical approach focused on the whole organism, including the emotions, an approach disinclined to strong manipulation of the body and its organs and certainly reluctant to sacrifice parts of the body through the radical therapy of surgery. The system of Western medicine being rapidly adopted in the late Tokugawa years (and soon to become the only legally sanctioned system under the Meiji government) was physiologically, pathologically, and therapeutically organ-centered, with a more highly developed surgical skill and willingness to use it. By no means heedless of the relations of bodily functions to the natural environment, modern Western medicine relied less on organismic philosophy to define those relations than on sciences like biology and chemistry, verified by rigorous empirical methods. Holistic environmental concerns were to emerge once again in Japanese health care, not so much because some Japanese persisted (as they do today) unofficially to use Chinese-style medicines (*kanpōyaku*) but rather because of the high incidence of pollution-caused diseases in the post–World War II era of rapid and often reckless industrial development. Treatment of the whole person, not just a body, is just beginning to make its way into the medical thought, practice, and education of the modern world, and in this cause it may be hoped that Japanese medicine will once again pioneer.

Aftermath of Chinese Wave II and Western Wave II

Japan's isolation ended with the signing of treaties of amity with the United States, England, and Russia in 1854, and with Holland in 1855. Ratified trade agreements, demanded by the United States in 1857 and negotiated and signed in 1858, were exchanged in 1860. Similar agreements with other Western states followed. Japan did not, however, accept passively this sudden exposure to the imperialist pressures of world politics; there was an immediate acceleration in defense preparations. Indeed, without waiting for Perry's promised return after his first entry into Edo Bay in 1853, the shogunate had ordered from Holland warships, guns and small arms, and books on military tactics. Though confusion and debate reigned in officialdom and the public was torn between strong anti-foreign sentiments and excitement over new access to world culture, the inescapable reality confronted by all parties was the overwhelming superiority of Western military technology.

Consensus quickly formed on one point: the only effective way to defend the nation, and thus avoid possible colonization, was to learn to produce and use Western weapons and strategy. To meet ships with ships, and guns with guns, no traditional system, whether Confucian, Buddhist, or National Learning, was of any immediate help. Only Western learning could provide the necessary information of the real world situation, and only Western science and technology could supply the means to prevent national calamity.

Given this premise, the national priorities fell clearly into place. Military needs were primary, and these alone were quite extensive: Western armaments, including skills in iron production, gun casting, battery construction and ballistics; modern warships, requiring know-how in shipbuilding, navigation, maneuvers, and steam engine manufacture and maintenance; and knowledge of Western military organization, strategy, tactics, logistics, and training. There was also urgent need for reliable information about Western social, political, and economic systems, and world geography. Moreover, effective build-up of a Western-style defense system meant that studies in the basic sciences of mathematics, physics, chemistry, biology, botany, etc., had to be promoted. Such studies enhanced, of course, the continued study of Western medicine as a whole system and in its various branches.

Though the decisions that followed from these priorities were no less than revolutionary, the requisite personnel and facilities were not automatically available. The ablest *Rangaku* scholars were rapidly mobilized to begin research on Western military technology, using Dutch books. Few samurai, however, were immediately qualified for high-level service in any of the areas of top priority, and thus thousands from the shogunate and domain ranks flocked to the private *Rangaku* academies. (Hardly a commoner figured prominently in the frenetic 1854–1868 defense-oriented efforts to meet the new national priorities.)

Traditional institutions of learning were, of course, woefully inadequate for the task ahead. One agency in particular, though, had potential—the Translation Office (Bansho wage goyō) of the Bureau of Astronomy (Tenmongata). In 1855 the shogunate began laying plans to develop it as an independent center for the introduction of Western learning. Established the following year as the Bansho shirabesho (Office for investigating barbarian documents), it was renamed several times until finally designated in 1863 as the Kaiseisho (lit., "Center for [national] development," but, given the purpose and curriculum, best rendered Center for Western Learning). Its functions were twofold: for diplomatic necessities, to compose and study documents exchanged with other countries, and to gather intelligence about their social, political, economic, and military conditions; and secondly, for defense needs, to engage in research and education related to the advanced military technology of the West and to the basic sciences necessary for that technology.

At first students were drawn exclusively from the shogunate's own samurai—about 350 of them when lectures began in 1857—but from 1858 samurai from the domains (those from Saga, Satsuma, and Chōshū were most numerous) were also admitted. Most of those trained in the Center assumed leadership roles in the early Meiji reforms based on the introduction of Western science and technology, though they were replaced in time by a generation who received a more systematic Western-style education in the new Meiji institutions of higher learning.

Japan needed, of course, a broader base from which to acquire scientific and technological knowledge than the Kaiseisho alone provided. In the fourteen postisolation years prior to the Meiji era, a number of shogunate research and training institutes employing Dutch specialists were founded (though they can only be mentioned here, leaving fuller treatment to the second volume on *Science and Culture in Modern Japan*). In Nagasaki a new naval training center (Nagasaki kaigun denshūsho) was established in 1855. In addition to on-board and ground drills, it offered instruction in shipbuilding, steam engines, gunnery, naval techniques, and basic physics, chemistry, and mathematics. It also had an auxiliary program for medical training initiated in 1857 by Pompe van Meerdervoort (1829–1908) in collaboration with a shogunate-appointed physician, Matsumoto Ryōjun (1832–1907). Pompe's successor A. F. Bauduin in 1864 developed a further subsidiary, a laboratory for physics and chemistry (Bunseki kyūrisho). Two years later the full-time services of W. K. Gratama were secured for this laboratory, and in 1867 both he and the laboratory's research equipment were moved to Edo to become part of the Kaiseisho.

Other shogunate agencies providing further access to Western science and technology included a Center for Western Medicine (Seiyō igakusho), an 1861 reorganization of the earlier private Vaccination Center (Shutōjo; est. 1858). In 1863 the term "Western" was dropped, as the official shift to Western medicine made it unnecessary to be so specific. More directly related to defense studies were a military training center (Kōbusho), founded in 1856 to train shogunate samurai and renamed the Military Academy (Rikugunsho) in 1866, and the navy's "warship training center" (Gunkan kyōjusho), started in 1857 and renamed the Naval Academy (Kaigunsho) in 1866.

In these various centers for research and education, the linguistic base for Western studies was rapidly and systematically extended to embrace English, German, French, and Russian, in addition to the Dutch language; and the academic coverage was also broadened toward inclusion of all basic disciplines. The great advantage over all activities prior to opening the country was that studies of Western science and technology were now based upon the latest knowledge, taught often by Dutch personnel using more systematic methods that tied theory to practice, and supported by the basic modern sciences. Compared to the Meiji institutions of higher learning soon to be established, however, all these agencies together (including all similar but lesser domain institutes) were far less adequate as a system for the comprehensive introduction of Western science and technology.

In 1866 the shogunate rescinded its ban on overseas travel by Japanese citizens so that dozens of persons and a number of official missions could be sent abroad to study and observe Western developments—after more than two centuries' interdiction of this means of advancing learning. Steamship travel greatly facilitated overseas travel for study, which in the Meiji years rapidly escalated. Translating and printing expanded as restrictions on them were removed; and with the Meiji reforms came new communications systems—electric telegraphy and a modern postal system—that helped the advancement of learning.

In retrospect it is clear that, insofar as the modern transformation of Japan depended upon the rapid utilization of modern Western science and technology, it was already under way when the Meiji Restoration occurred in 1868. At least, the institutional initiatives taken by the shogunate and domains from 1854 to 1868 constituted a decisive transition toward that end. In the aggregate, though, these initiatives were only an ad hoc, uncoordinated effort oriented to the hurried acquisition of ships, guns, small arms, and the like. Had Japan's modern transformation remained at this initial preparatory level, it may well have ended in a dead end. The early shogunate initiatives were, however, soon escalated to a second, highly concerted level by the new Meiji government (particularly after the complete abolition of the feudal domains in 1871). Though political power in the new government was dominated by the antishogunate samurai of the former Satsuma, Chōshū, Tosa, and Hizen domains, the Western-oriented shogunate facilities (far superior to those of the domains) and their personnel formed the core of the new Meiji institutions of higher learning which produced the professionals so essential to the modernizing designs of the Meiji planners and politicians.

In some respects the transitional period from 1854 to 1868 was the natural evolution of Western Wave II. But it was much more than that. The end of isolation and the steps taken for national defense made this transition the initial phase of Western Wave III—and thus the starting point for the second volume on *Science and Culture in Modern Japan*.

Appendix: Chronological Charts

Table A.1. Brief Outline of the Sociocultural History of Japan, 8,000 B.C. to 600 A.D.

	−1,500	−1000	−500	−400	−300	−200	−100	0	100	200	300	400	500	600
Period	−16th −11th	−722 ← Western | Shang | Chou	Eastern Chou → −481 −403 Spring and Autumn	Warring States	−221 −206	Former Han	−8 25 Later Han		222 280 Three Kingdoms	Six Dynasties → 589 420 Chin | Northern and Southern Courts			
China	−5000/−3000 Han people settled in Yellow River basin		Use of bronze	Small Han states	Use of iron		Political unification of all China; Han expansion into surrounding areas	Confucianism as orthodoxy Books compiled in many fields Buddhism to China		Confusion under contending states				
Period Culture Korea				Chinese settlers and influx of Chinese culture	−206 −108 Han colony in north	North: direct Chinese rule −57 −37 −18 South: small states of Korean clans				313 Silla Koguryo Paekche Mimana → 562 Buddhism to Korea				

	Period	c. −8000	c. −250	Mid-3rd century		552
		Pre-Jōmon (Paleolithic age)	Jōmon (Neolithic age)	Yayoi		Kofun
Japan	Chinese Cultural Influx		Isolated from continental culture Hunting, fishing, plant-gathering No agriculture, no use of metal	Influx of Chinese material culture: Agriculture and domestic techniques (Mainly by way of Korean peninsula) Irrigated farming Use of bronze, iron No native writing Primitive Shinto		Peak of large tomb construction Large-scale irrigation More advanced crafts Initial influx of scholarship
	Political system				Village states in North Kyushu (Tribute sent to China)	First unified state under Yamato court	413‥‥‥‥502 Envoys to China: 13 times
	Top intellectuals					Learned immigrants: mostly Koreans (some Chinese)
	Institutions of learning					
	Western Cultural Influx					

Table A.1 (continued). Brief Outline of the Sociocultural History of Japan, 600 A.D.–1912 A.D.

		600	700	800	900	1000	1100	1200	1300	1400	1500	1600	1700	1800	1900
China	Period	589 618			907 960		1126-Chin-1234 Sung 1127	1279 Yüan	1368		1644			1912
		Sui	T'ang		Five Dynasties	(North)	(South)	(Mongols)	Ming		Ch'ing (Manchus)			Republic
	Culture	Reunification of all China		Buddhism as state religion	Disunion	Classical renaissance; new secular culture		Rise of Neo-Confucianism		Rise of Wang Yang-ming Confucianism			Opium War 1840–1842	
Korea	Period			Silla		935		Koryŏ	1392			Yi		1910
		Koguryŏ 668 Paekche 663												
	Culture		Unification of all Korea						Promotion of Neo-Confucianism				Colonized by Japan	
Japan	Period	552	710–794		894		1192	1336	1401 1467		1603— Tokugawa —1868 1912		1854	
		Asuka	Nara		Heian	Early phase	Kamakura	Muromachi	Sengoku		(Edo) 1639 Isolation 1854 Country opened		Meiji	
						Semiseclusion Era	Late phase							
	Chinese Cultural Influx	ca. 600 Chinese Cultural Wave I Sophisticated culture imported through envoys: Bureaucratic system				Decline of learning and science Growth of native literature, art, institutions, etc. "New Buddhism" flourishes				Chinese Cultural Wave II Renewed influx and importation of most sophisticated Chinese culture (esp. Confucianism and science)				

Japan

	Buddhist complex Learning and science				17th century Intellectual outburst
	646 Taika Reform	1192		1868	
Political system	Ritsuryō System: Centralized government under emperor →	Disintegration: rise of large manors	Feudalism: Civil wars, Tally Trade, "Red Seal Ships"		Centralization by Tokugawa shogunate
Top intellectuals	Ritsuryō officials	Buddhist priests, court nobility	Wakō (pirates); Zen priests		Secular Confucian scholars, scientists (esp. in medicine)
Institutions of learning	Ritsuryō Institutes: University, provincial colleges, institutes of divination and medicine	"House Learning" of court nobility	Gozan temples		Shogunate Confucian academy, Bureau of Astronomy, Institute of Medicine; private academies (mathematics and medicine)
Western Cultural Influx			1543 — Western Wave I: Catholic schools and hospitals, Techniques, Trade	1639–Isolation–1854: 1720 Western Wave II: Dutch on Dejima (medical doctors), Limited trade in Nagasaki (Dutch and Chinese)	1854 Western Wave III → All-out importation of Western science and techniques

Table A.2. Four Major Traditions of Learning in Traditional Japan

	Chinese Wave I	Semiseclusion Era	Western Wave I	Chinese Wave II / Isolation Policy	Western Wave II
Periods	ca. 600	894 — 1401	1543 — 1639	1639 — 1720 — 1854	1720 — 1854
Chinese	T'ang-style learning and institutions: *University*: Confucianism, law, literature, history, mathematics. *Institute of Divination*: calendrics, divination. *Institute of Medicine*: medicine, acupuncture, massage, incantations, herb cultivation.	Learning of former era preserved by lower court nobility (House Learning), but gradually declined. Decline of University; rise and fall of private academies of nobility.	Study of Neo-Confucianism and Chinese classics by Zen priests.	Tokugawa shogunate promoted Neo-Confucianism as official ideology. Late 17th and 18th century critical movements; eclecticism. By 19th century Confucianism as system no longer accepted. 17th-century Intellectual Outburst in all fields, including sciences. Rise of professional scholars.	
Buddhist	Imperial household and wealthy nobility promoted temple building (not mass movement; but as "guardian of the state"). Sutras in Chinese.	Learned Buddhist priests served as political advisors to military leaders. Some studies in astrology, medicine, and Neo-Confucianism (esp. Zen priests).	Buddhism lost its huge political and economic powers, and esp. its intellectual leadership.	Criticized by both Confucianists and "National Studies" scholars. Occasional reactions to new Western ideas, esp. in cosmology.	

Japanese	Earliest extant documents: 712: *Kojiki* (Record of Ancient Matters) 720: *Nihon shoki* (Chronicles of Japan)	Early 11th century: *Genji monogatari* (Tale of Genji) Study of Japanese poetry by court nobles.			Development of "National Studies" (*Kokugaku*), using Confucian critical methods, but increasingly stressing Shinto ideology.
Western			Catholic mission (esp. Jesuits) taught Christian doctrine and Scholasticism in church schools, seminaries, and college. Western medicine, esp. surgery, taught in Catholic hospitals. Techniques, astronomy, mathematics adopted by some Japanese.	Only traces of Western cosmology and surgery remain. All else vanished.	Late 18th century dissection "boom" in medicine led to study of Western anatomy, then medicine, through Dutch books; spread to other sciences and languages. Early 19th century: upsurge of study of Western military techniques. Preparation for all-out importation after 1868.

Table A.3. Brief Outline of Astrology and Calendrical Astronomy in Traditional Japan, 600 A.D.–1639 A.D.

```
ca. 600         Chinese Wave I
                                              Semiseclusion Era              Chinese Wave II
                                         894                          1401              Western Wave I
                                                                                 1543                  1639
```

Chinese scheme

Calendrical systems:
7th Century   764   858   862
Yüan-chia li   Ta-yen li   Wu-chi li

├──────────────── Senmyōreki ────────────────┤
                 (Hsüan-ming li)

Institute of Divination (Onmyōryō)
  Astrology
  Calendrical astronomy
  Divination
  Timekeeping

675: Astronomical observatory

"*House Learning*"
Astrology: Abe House ────────→ Tsuchimikado House ─ ─ ─ ─ ─→ (Meiji)
                                (both astrology and calendrical
Cal. astron.: Kamo House   astronomy)
                                      ↓
                                   extinct
Some court mathematicians   (c. 1560)   Spread of
competed in predicting eclipses.   calendrical
                                                 knowledge to
                                                 civilians.

Buddhism:
Sukuyōdō
(based on Sukuyōkyō canon)
eclipse prediction; horoscopes
```
```

Confucianism: Anti-Jesuit critique of "spherical earth" by Hayashi Razan, relying on Neo-Confucian ideology.

Western scheme

Jesuit sources: Ptolemaic universe spherical earth church calendar

(In China: translations into Chinese language)

Mariners: Instruments and techniques

1618: *Genna kōkaisho* (Navigation manual)

Table A.3 (continued). Brief Outline of Astrology and Calendrical Astronomy in Traditional Japan, 1639 A.D.–1854 A.D.

1639	---------- Isolation Period ----------	1854
	Chinese Wave II	
	1720	
	Western Wave II	1854
		1854

Chinese scheme

1685:
Jōkyōreki
first Japanese
revision;
unsatisfactory

1755:
Hōryakureki
minor
corrections of
Jōkyōreki; less
satisfactory

1798:
Kanseireki
better, but
still inadequate

1843:
Tenpōreki
based on
Lalande;
excellent

1684:
Tenmongata (Bureau of Astronomy)
established by shogunate

1720:
Yoshimune's calendrical
revision project;
book ban relaxed

1811
Translation Office
established in
Tenmongata

(1873:
Gregorian
calendar
adopted)

Shibukawa Harumi   Asada Gōryū   Takahashi Yoshitoki
(1639–1715)   (1734–1799)   (1764–1804)
                                   Hazama Shigetomi
                                   (1756–1816)

Outburst of civilians' calendrical studies:
criticism of *Senmyōreki*, study of *Shou-shih li*;
rise of professionals outside Tsuchimikado
House.

Bonreki movement to defend Buddhist cosmology
against Western astronomy:

1756:   1810:
Monno's Hi-"Tenkyō wakumon"   Entsū's (1754–1834)
                               Bukkoku rekishō hen

Hirata Atsutane (1776–1843) and other Shinto
ideologues support spherical earth and heliocentric universe vs. Confucianism and Buddhism

Translation of Dutch books by amateurs: Dutch books translated by experts

1774:   1802:   1836:
Motoki Ryōei   Shizuki Tadao   Calendrical sections
(1735–1794) first   (1760–1806)   of Lalande
introduction of   introduction of   translated
heliocentric theory   Newtonian
                         mechanics

Imports of Jesuits'
Chinese translation

Western scheme
Residual Western
astronomy studied
by amateurs:
Sawano Chūan,
Kobayashi Kentei,
et al.

Table A.4. Brief Outline of Mathematics in Traditional Japan, 600 A.D.–1639 A.D.

	Chinese Wave I		Semiseclusion Era		Chinese Wave II		Western Wave I
ca. 600		894		1401		1543	1639

Chinese scheme

Ritsuryō institute:
University
  Mathematics course

Practical mathematics for administrative use

"*House Learning*" of lower court nobility:
  Ozuki House
  Miyoshi House

Decline of interest and ability in higher mathematics; games and superstitions flourish

Some Buddhist priests use mathematics in Tally Trade

Increased need for practical mathematics in commerce, civil engineering, surveying

1620s: Commercial printing of books on practical mathematics; use of abacus spread

Early 17th century: No intellectual bias against mathematics; leading mathematicians studied computational system of calendars, esp. *Shou-shih li*.

Western scheme

Potential in Catholic missions and Western navigation, gunnery not realized.

Table A.4 (continued). Brief Outline of Mathematics in Traditional Japan, 1639 A.D.–1854 A.D.

```
1639 ---------------------------- Isolation Period ---------------------------- 1854
                                   Chinese Wave II
     ----------------------------|---------------------------------------------- 1854
                               1720              Western Wave II
     ------------------------------------------------------------------------- 1854
```

Chinese scheme

1641: Problem Succession begins; problems rapidly become more difficult

Understanding of Chinese algebra (*tengenjutsu*)	Seki Takakazu (?–1708) invents written notations (*tenzanjutsu*) for algebra Theory of determinants (1683)			
	Circular Principles Seki Takakazu; Takebe Katahiro (1664–1739)		Calculation of area and volume by method of definite integrals Ajima Naonobu (1739–1796)	Tables of definite integrals Wada Yasushi (1787–1840)
		Emergence of various schools: Seki, Takuma, Saijō, etc.		
		Itinerant mathematics teachers; rudimentary education for ordinary citizens in *terakoya* includes abacus instruction		
		Votive plaques (*sangaku*) popular for displaying mathematical problems and answers		
		Growing contempt for mathematics among Confucian ideologues of samurai class; e.g., Ogyū Sorai's criticism of mathematics as useless acrobatics.	Fujita Sadatsugu's (1734–1807) defense of mathematics as intellectually "useful for no use" (not profit-oriented)	

Western scheme

Following 1720 relaxation of book ban, Jesuit calendrical books in Chinese are introduced; some include mathematics (e.g., trigonometry)

Trigonometry and logarithms for use in navigation, from Dutch sources.

Western mathematics rejected by Japanese as inferior to *wasan* until after Meiji Reformation (1868)

Table A.5. Brief Outline of Medicine in Traditional Japan, 600 A.D.–1639 A.D.

ca. 600 Chinese Wave I	894 Semiseclusion Era	1401 Chinese Wave II	1543 Western Wave I	1639

Chinese scheme

Ritsuryō institutes:
  Institute of Medicine
  medicine, acupuncture,
  massage, incantations,
  herb cultivation
  Provincial Colleges
  Similar but limited
  curriculum.

Buddhist medical activities:
  Temple clinics

"*House Learning*" of
lower court nobility:
  Tanba House
  Wake House

Specialization:
e.g., eye diseases, gynecology, "combat" medicine

Priests-practioners to China
for medical study

1259–1333: Gokurakuji hospital
  in Kamakura

Li-Chu Schools: mainstream of *Japanese medicine*
  forerunner: main figure:
  Tashiro Sanki   Manase Dōsan
  (1465–1537)   (1507–1594)

Kohō School
  forerunner:
  Nagata Tokuhon
  (ca. 1600)

Western scheme

Catholic medical activities:
Hospitals
Leprosaria   Nanban surgery

Luis de Almeida
(1525–1584)
founder of principal
Jesuit hospital in Funai

Table A.5 (continued). Brief Outline of Medicine in Traditional Japan, 1639 A.D.–1854 A.D.

1639 ---------------------------------- Isolation Period ---------------------------------- 1854

| Chinese Wave II | 1720 | Western Wave II | 1854 |

Chinese scheme   Early 19th century: Wahō School

Furubayashi Kengi
(1597–1657)

(Li-Chu Schools)   c. 1750:   c. 1800: absorbed
    Aeba Tōan   lost leadership   by Electic School
Liu-Chang Schools: (1615–1673)
                          Okamoto Ippō   c. 1800: absorbed
                          (1686–1754)   by Eclectic School

    Nagoya Gen'i   Gotō Gonzan   Yoshimasu Tōdō   To Chinese-Dutch Eclectic,
    (1629–1696)   (1659–1733)   (1702–1773)   or Dutch-Western Schools

(Kohō School)   Yamawaki Tōyō
                             (1703–1762)
                              Dissection boom
                          Mochizuki Rokumon
                        Eclectic School: (1697–1769)   Taki Keizan
                        (c. 1800: mainstream)   (1755–1810)

                                  1765: Seijukan   1791: Igakkan
                                  (private)   (shogunate)

	Arai Hakuseki (1657–1725) corrected views of the West	Chinese-Dutch Eclectic School:	Fuseya Soteki (1747–1811)	Hanaoka Seishū (1760–1835)

1720: book ban ⟶ Dutch-Western School:
relaxed

Sugita Genpaku (1733–1817)

1823–1829: Siebold in Japan

Western Scheme

Kōmō surgery: Dutch interpreters' avocation

Maeno Ryōtaku (1723–1803)

Dictionaries: ⟶ Dutch Learning academies:
Edo haruma (1796)   Shiradō (est. 1786) and
Nagasaki haruma   others in Edo, Osaka, Kyoto
(1833)

Motoki Ryōi   Narabayashi Chinzan
(1618–1697)   (1643–1711)

1858: Shutōjo

Shogunate orders Dutch
language study:
Noro Genjō   Aoki Kon'yō
(1693–1761) (1698–1769)

1857:
Nagasaki
Naval
Center

Bibliography

Japanese-language series of which two or more volumes or numbers appear are listed here with publisher, series title, Chinese characters, and English equivalents.

Series
Chūōkōronsha 中央公論社
Nihon no rekishi 日本の歴史 (History of Japan) series.
Shinsho 新書 (New books) series.
Fuji Tanki Daigaku Shuppanbu 富士短期大學出版部
Tōzai sūgaku 東西數學 (East-West mathematics) series.
Heibonsha 平凡社
Tōyō bunko 東洋文庫 (East Asian library) series.
Iwanami Shoten 岩波書店
Shinsho 新書 (New books) series.
Tōyō shichō 東洋思潮 (East Asian thought) series.
Kadokawa Shoten 角川書店
Shinsho 新書 (New books) series.
Shibundō 至文堂
Nihon rekishi shinsho 日本歴史新書 (New books on Japanese history) series.
Yamakawa Shuppansha 山川出版社
Taikei Nihonshi sōsho 體系日本史叢書 (Systematic history of Japan) series.
Yoshikawa Kōbunkan 吉川弘文館
Jinbutsu sōsho 人物叢書 (Biography) series.
Nihon rekishi sōsho 日本歴史叢書 (Japanese history) series.

Books and Articles
Akiyama Kenzō 秋山謙三.
Nisshi kōshōshi kenkyū 日支交渉史研究 (Studies of the history of Sino-Japanese relations). Tokyo: Iwanami Shoten, 1939.

Alston, W. G., Jr., tr.
Nihongi: Chronicles of Japan from the Earliest Times to A.D. 697. London: George Allen and Unwin, 1956.

Anesaki Masaharu 姉崎正治.
Kirishitan dendō no kōhai 切支丹傳道の興廢 (Rise and fall of Catholic missions in feudal Japan). Tokyo: Dōbunkan, 1930.

Aoki Kazuo 青木和夫.
Nara no miyako 奈良の都 (Nara, capital city). Nihon no rekishi series, no. 3. Tokyo: Chūōkōronsha, 1965.

Araki Toshima 荒木俊馬.
Nihon rekigakushi gaisetsu 日本曆學史概説 (Outline history of Japanese calendrics). Tokyo: Kōseisha Kōseikaku, 1960.

Arima Seiho (1) 有馬成甫.
Ikkansai Kunitomo Tōbei den 一貫齋國友藤兵衞傳 (Biography of Ikkansai Kunitomo Tōbei). Tokyo: Musashino Shoin, 1932.

Arima Seiho (2).
Takashima Shūhan 高島秋帆 (Biography of Takashima Shūhan). Jinbutsu sōsho series. Tokyo: Yoshikawa Kōbunkan, 1958.

Arima Seiho (3).
Kahō no kigen to sono denryū 火砲の起源とその傳流 (Origin of firearms and their transmission [to Japan]). Tokyo: Yoshikawa Kōbunkan, 1962.

Arima Seiho (4).
"The Western Influence on Japanese Military Science, Shipbuilding, and Navigation." *Monumenta Nipponica*, vol. 19 (1964), nos. 3–4, pp. 118–145.

Asahi Shinbunsha 朝日新聞社
Nihon kagaku gijutsu shi 日本科學技術史 (History of sciences and techniques in Japan). Tokyo: Asahi Shinbunsha, 1962.

Bock, Felicia Gressitt, tr., with introduction and notes.
Engi-shiki: Procedures of the Engi Era, Books I–V, Books VI–X (2 vols.). *Monumenta Nipponica* monographs. Tokyo: Sophia University Press, 1970, 1972.

Bowers, John Z.
Western Medical Pioneers in Feudal Japan. Baltimore and London: The Johns Hopkins Press, 1970.

Boxer, C. R.
The Christian Century in Japan, 1549–1650. Berkeley and Los Angeles: University of California Press, 1967.

Carter, T. F.
The Invention of Printing in China and Its Spread Westward (1925). 2nd ed., revised by L. Carrington Goodrich. New York: Ronald Press, 1955.

Ching, Julia.
"Chu Shun-shui, 1600–82: A Chinese Confucian Scholar in Tokugawa Japan." *Monumenta Nipponica*, vol. 30, no. 2 (Summer 1975), pp. 177–191.

Cieslik, Hubert.
"Overseas Studies in the Early *Kirishitan* Period." Translated by John Timmer. *Japan Christian Quarterly*, vol. 38, no. 1 (Winter 1972), pp. 3–10.

Copper, Michael, S. J.
Rodrigues the Interpreter: An Early Jesuit in Japan and China. New York and Tokyo: John Weatherhill, Inc., 1974.

Craig, Albert M.
"Science and Confucianism in Tokugawa Japan." In *Changing Japanese Attitudes toward Modernization*, ed. Marius Jansen. Princeton, N.J.: Princeton University Press, 1965.

Croizier, Ralph C.
Traditional Medicine in Modern China. Harvard East Asian series, no. 34. Cambridge, Mass.: Harvard University Press, 1968.

Dore, Ronald P.
Education in Tokugawa Japan. London: Routledge & Kegan Paul, 1965.

Ebisawa Arimichi (1) 海老澤有道.
Kirishitan no shakai katsudō oyobi nanban igaku 切支丹の社會活動及南蠻醫學

(Social activities of Catholics and medicine of the "Southern Barbarians"). Tokyo: Fuzanbō, 1944.

Ebisawa Arimichi (2).
Nanban gakutō no kenkyū 南蠻學統の研究 (Studies of the learning tradition of the "Southern Barbarians"). Tokyo: Sōbunsha, 1958.

Ebisawa Arimichi (3).
Nanban bunka 南蠻文化 (Culture of the "Southern Barbarians"). Nihon rekishi shinsho series. Tokyo: Shibundō, 1962.

Endō Toshisada 遠藤利貞
Zōshū: Nihon sūgakushi 増修日本數學史 (History of Japanese mathematics; enlarged and revised). Tokyo: Kōseisha Kōseikaku, 1960.

Forke, Alfred
Shina shizen kagaku shisōshi 支那自然科學思想史 (Translation of *The world-conception of the Chinese; their astronomical, cosmological and physicophilosophical speculations*. London: A. Probsthain, 1925). Translated by Kowada Takenori 小和田武紀. Tokyo: Seikatsusha, 1939.

Fujii Naohisa 藤井尚久
Igaku bunkashi nenpyō 醫學文化史年表 (Chronology of cultural history of medicine). Tokyo: Nisshin Shoin, 1942.

Fujikawa Hideo, ed. 富士川英郎
Fujikawa Yū kagaku zuihitsu ishi sōdan 富士川游科學隨筆醫史叢談 (Fujikawa Yū's scientific essays and discourses on the history of medicine). Tokyo: Shomotsu Tenbōsha, 1942.

Fujikawa Yū 富士川游
Nihon igakushi 日本醫學史 (History of Japanese medicine). Tokyo: Nisshin Shoin, 1941.

Fujinami Gōichi 藤浪剛一
Ika sentetsu shōzōshū 醫家先哲肖像集 (Collected portraits of medical pioneers). Tokyo: Tōkō Shoin, 1936.

Fujiwara Matsusaburō 藤原松三郎
Nihon sūgaku shiyō 日本數學史要 (Short history of Japanese mathematics). Tokyo: Hōbunkan, 1952.

Fukui Kyūzō 福井久藏
Shodaimyō no gakujutsu to bungei no kenkyū 諸大名の學術と文藝の研究 (Studies of learning and literature [sponsored] by various daimyo). Tokyo: Kōseisha Kōseikaku, 1937.

Goodman, Grant Kohn
The Dutch Impact on Japan (1640–1853). Leiden: E. J. Brill, 1967.

Goodrich, L. Carrington
A Short History of the Chinese People. Harper Torchbooks/The University Library. New York: Harper and Row Publishers, 1963.

Hagino Kōgō (1) 萩野公剛
Nihon sūgakushi kenkyū benran 日本數學史研究便覽 (Handbook for research

in the history of Japanese mathematics). Tokyo: Fuji Tanki Daigaku Shuppanbu, 1961.

Hagino Kōgō (2)
Wasanka jinmei jiten 和算家人名辞典 (Biographical dictionary of *wasan* practitioners). Kenkyū sōsho series 研究叢書 (Research series) vol. 6, no. 1. Tokyo: Fuji Tanki Daigaku Shuppanbu, 1964.

Han Woo-keun.
The History of Korea. Tr. Lee Kyung-shik; ed. Grafton K. Mintz. Honolulu: East-West Center Press, 1970.

Hatada Takashi (1)
A History of Korea. Tr. and ed. Warren W. Smith, Jr., and Benjamin H. Hazard. Santa Barbara, Cal.: American Bibliographical Center, Clio Press, 1969.

Hatada Takashi (2) 旗田巍
Chōsenshi 朝鮮史 (History of Korea).
Zensho series 全書 (Monograph series), no. 154. Tokyo: Iwanami Shoten, 1966.

Hattori Toshiyoshi (1) 服部敏良
Nara jidai igaku no kenkyū 奈良時代醫學の研究 (Studies of medicine in the Nara period). Tokyo: Tōkyōdō, 1945.

Hattori Toshiyoshi (2)
Heian jidai igaku no kenkyū 平安時代醫學の研究 (Studies of medicine in the Heian period). Tokyo: Kuwana Bunseidō, 1955.

Hattori Toshiyoshi (3)
Kamakura jidai igaku no kenkyū 鎌倉時代醫學の研究 (Studies of medicine in the Kamakura period). Tokyo: Yoshikawa Kōbunkan, 1964.

Hattori Toshiyoshi (4)
Shaka no igaku 釋迦の醫學 (Buddhist medicine). Nagoya: Reimei Shobō, 1968.

Hattori Toshiyoshi (5)
Muromachi Azuchi Momoyama jidai igakushi no kenkyū 室町安土桃山時代醫學史の研究 (Studies of the history of medicine in the Muromachi, Azuchi, and Momoyama periods). Tokyo: Yoshikawa Kōbunkan, 1971.

Hayashi Tsuruichi 林鶴一
Wasan kenkyū shūroku 和算研究集録 (Collected studies on *wasan*). Vols. 1–2. Tokyo: Tokyo Kaiseikan, 1937.

Hildreth, Richard
Japan As It Was and Is (1855). Reprint ed. Wilmington, Del.: Scholarly Resources Inc., 1973.

Hirayama Akira (1) 平山諦
Seki Takakazu 關孝和 (Biography of Seki Takakazu). Tokyo: Kōseisha Kōseikaku, 1959.

Hirayama Akira (2)
Wasan no rekishi 和算の歴史 (History of *wasan*). Nihon rekishi shinsho series. Tokyo: Shibundō, 1961.

Hirayama Akira (3)
Wasanshijō no hitobito 和算史上の人々 (Personalities in the history of *wasan*). Tōzai sūgaku series, no. 9. Tokyo: Fuji Tanki Daigaku Shuppanbu, 1965.

Hirayama Seiji 平山清次
Rekihō oyobi jihō 暦法及時法 (Calendrical and timekeeping methods). Tokyo: Kōseisha Kōseikaku, 1933.

Hirose Hideo (1)
"The European Influence on Japanese Astronomy." *Monumenta Nipponica*, vol. 19 (1964), nos. 3–4, 61–80.

Hirose Hideo (2) 廣瀬秀雄
Nihonjin no tenmonkan 日本人の天文観 (Japanese view of astrology and astronomy). NHK Books, no. 167. Tokyo: Nihon Hōsō Shuppan Kyōkai, 1972.

Hisaki Yukio 久木幸男
Daigakuryō to kodai jukyō 大學寮と古代儒教 (National University and Confucianism in antiquity). Tokyo: Saimaru Shuppankai, 1968.

Hoashi Kinen Toshokan 帆足記念圖書館, ed.
Hoashi Banri zenshū 帆足萬里全集 (Complete works of Hoashi Banri). Vols. 1–2. Hidemachi, Oita Prefecture: Hoashi Kinen Toshokan, 1926.

Hong I-sŏp 洪以燮 (Japanese reading: Kō Ishō)
Chōsen kagakushi 朝鮮科學史 (History of Korean science). Tokyo: Sanseidō Shuppan Sōritsu Jimusho, 1944.

Hora Tomio 洞富雄
Tanegashima jū: denrai to sono eikyō 種子島銃,傳來とその影響 (Tanegashima arquebus: its transmission and influence). Tokyo: Awaji Shobō Shinsha, 1958.

Hori Isao (1) 堀勇雄
Yamaga Sokō 山鹿素行 (Biography of Yamaga Sokō). Jinbutsu sōsho series, no. 33. Tokyo: Yoshikawa Kōbunkan, 1959.

Hori Isao (2)
Hayashi Razan 林羅山 (Biography of Hayashi Razan). Jinbutsu sōsho series, no. 118. Tokyo: Yoshiawa Kōbunkan, 1964.

Horiuchi Gōji 堀内剛二
Kindai kagaku shisō no keifu 近代科學思想の系譜 (Genealogy of modern scientific thought). Nihon rekishi shinsho series. Tokyo: Shibundō, 1964.

Hosoi Sō (1) 細井綜
Nihon kagaku no tokushitsu: sūgaku 日本科學の特質（數學）(Characteristics of Japanese science: mathematics). Tōyō shichō series. Tokyo: Iwanami Shoten, 1935.

Hosoi Sō (2)
Wasan shisō no tokushitsu 和算思想の特質 (Characteristics of *wasan* thought). Tokyo: Kyōritsusha, 1941.

Hübotter, Franz
Die Chinesische Medizin zu Beginn des XX Jahrhunderts und ihr historischer Entwicklungsgang. Leipzig: Verlag Der Asia Major, 1929.

Ienaga Saburō (1) 家永三郎
Nihon bunkashi 日本文化史 (Cultural history of Japan). Shinsho series, no. 367. Tokyo: Iwanami Shoten, 1959.

Ienaga Saburō (2)
Jōdai Bukkyō shisōshi kenkyū; shinteiban 上代佛教思想史研究, 新訂版 (Studies of Buddhist thought in antiquity; rev. ed.). Kyoto: Hōzōkan, 1966.

Ienaga Saburō (3)
Chūsei Bukkyō shisōshi kenkyū 中世佛教思想史研究 (Studies of Buddhist thought in medieval times). Kyoto: Hōzōkan, 1966.

Iijima Tadao (1) 飯島忠雄
Tenmon rekihō to inyō gogyō setsu 天文暦法と陰陽五行説(Astrology and calendrics, and the Yin-Yang and Five Phases theory). Tokyo: Kōseisha Kōseikaku, 1939.

Iijima Tadao (2)
Shina kodaishi to tenmongaku 支那古代史と天文學 (Astrology and astronomy in ancient Chinese history). Tokyo: Kōseisha Kōseikaku, 1942.

Imaizumi Genkichi 今泉源吉
Rangaku no ie: Katsuragawa-ke no hitobito 蘭學の家，桂川家の人々 (The Katsuragawa family: a Dutch learning household). Vols. 1–2. Tokyo: Shinozaki Shorin, 1965, 1968.

Inoue Kiyoshi 井上清
Nihon no rekishi 日本の歴史 (History of Japan). Vols. 1–3. Shinsho series, nos. 500a, 500b, 500c. Tokyo: Iwanami Shoten, 1963–1966.

Inoue Mitsusada (1) 井上光貞
Nihon kokka no kigen 日本國家の起源 (Origins of the Japanese state). Shinsho series, no. 380. Tokyo: Iwanami Shoten, 1960.

Inoue Mitsusada (2)
Shinwa kara rekishi e 神話から歴史へ (From myth to history). Nihon no rekishi series, no. 1. Tokyo: Chūōkōronsha, 1965.

Inoue Tadashi 井上忠
Kaibara Ekken 貝原益軒 (Biography of Kaibara Ekken). Jinbutsu sōsho series, no. 103. Tokyo: Yoshikawa Kōbunkan, 1963.

Ishida Eiichirō 石田英一郎, ed.
Nihon kokka no kigen 日本國家の起源 (Origins of the Japanese state). Shinsho series, no. 220. Tokyo: Kadokawa Shoten, 1967.

Ishida Eiichirō and Izumi Seiichi 石田英一郎，泉成一, eds.
Nihon nōkō bunka no kigen 日本農耕文化の起源 (Origins of Japanese agriculture). Shinsho series, no. 234. Tokyo: Kadokawa Shoten, 1968.

Ishida Ichirō 石田一良
Itō Jinsai 伊藤仁齋 (Biography of Itō Jinsai). Jinbutsu sōsho series, no. 39. Tokyo: Yoshikawa Kōbunkan, 1960.

Ishihara Akira (1) 石原明
Kanpō 漢方 (Chinese medicine). Shinsho series, no. 26. Tokyo: Chūōkōronsha, 1963.

Ishihara Akira (2)
Nihon no igaku 日本の醫學 (Medicine in Japan). Nihon rekishi shinsho series. Tokyo: Shibundō, 1963.

Ishii Susumu 石井進
Kamakura bakufu 鎌倉幕府 (Kamakura shogunate). Nihon no rekishi series, no. 7. Tokyo: Chūōkōronsha, 1965.

Ishikawa Ken 石川謙
Terakoya 寺小屋 (Temple schools). Nihon rekishi shinsho series. Tokyo: Shibundō, 1963.

Itazawa Takeo (1) 板澤武雄
Nichiran bunka kōshōshi no kenkyū 日蘭文化交渉史の研究 (Studies in the history of cultural intercourse between Japan and Holland). Nihon shigaku kenkyū sōsho series 日本史學研究叢書 (Series on Japanese historical research). Tokyo: Yoshikawa Kōbunkan, 1959.

Itazawa Takeo (2)
Shīboruto シーボルト (Siebold). Jinbutsu sōsho series, no. 45. Tokyo: Yoshikawa Kōbunkan, 1960.

Itazawa Takeo (3)
Nihon to Oranda 日本とオランダ (Japan and Holland). Nihon rekishi shinsho series. Tokyo: Shibundō, 1962.

Itō Chūta 伊東忠太
Nihon kenchiku no kenkyū 日本建築の研究 (Studies on Japanese architecture). Vols. 1–2. Tokyo: Ryūginsha, 1942.

Itō Sakae 伊東榮
Itō Genboku 伊東玄朴 (Biography of Itō Genboku). Tokyo: Genbunsha, 1916.

Iwao Seiichi (1) 岩生成一
Shuinsen bōekishi no kenkyū 朱印船貿易史の研究 (Studies in the history of trade by Red Seal Ships). Tokyo: Kōbundō, 1958

Iwao Seiichi (2)
Shuinsen to Nihonmachi 朱印船と日本町 (Red Seal Ships and Japanese towns [in Southeast Asia]). Nihon rekishi shinsho series. Tokyo: Shibundō, 1963.

Iwao Seiichi (3)
Nanyō Nihonmachi no kenkyū; zōteiban 南洋日本町の研究，増訂版 (Studies of Japanese towns in the southern seas; enlarged and rev. ed.). Tokyo: Iwanami Shoten, 1966.

Iwao Seiichi (4)
Sakoku 鎖國 (National isolation). Nihon no rekishi series, no. 14. Tokyo: Chūōkōronsha, 1966.

Jennes, Joseph, C.I.C.M.
History of the Catholic Church in Japan: From Its Beginnings to the Early Meiji Period (1549–1873). Missionary Bulletin Series, no. 8. Tokyo: The Committee of the Apostolate, 1959.

Jeon Sang-woon
Science and Technology in Korea: Traditional Instruments and Techniques. M.I.T. East Asian Science Series, vol. 4. Cambridge, Mass.: The M.I.T. Press, 1974.

Jugaku Bunshō 壽岳文章
Nihon no kami 日本の紙 (Paper of Japan). Nihon rekishi sōsho series. Tokyo: Yoshikawa Kōbunkan, 1967.

Kamstra, J. H.
Encounter or Syncretism: The Initial Growth of Japanese Buddhism. Leiden: E. J. Brill, 1967.

Kanda Shigeru, ed. (1) 神田茂
Nihon tenmon shiryō sōran 日本天文資料總覽 (General survey of Japanese astronomical records). Tokyo: Maruzen, 1934.

Kanda Shigeru, ed. (2)
Nihon tenmon shiryō 日本天文資料 (Japanese astronomical records). Tokyo: Maruzen, 1935.

Kasahara Kazuo 笠原一男
Nihonshi hyakushō 日本史百章 (One hundred chapters of Japanese history). Vols. 1–3. Tōdai shinsho 東大新書 (University of Tokyo new books). Tokyo: Tokyo Daigaku Shuppankai, 1962–1963.

Katō Heizaemon (1) 加藤平左衞門
Wasan no kenkyū: gyōretsushiki oyobi enri 和算の研究, 行列式及圓理 (*Wasan* studies: determinants and circular principles). Tokyo: Kaiseikan, 1944.

Katō Heizaemon (2)
Wasan no kenkyū: zatsuron 和算の研究, 雑論 (*Wasan* studies: miscellany). Vols. 1–3. Tokyo: Nihon Gakujutsu Shinkōkai, 1954–1956.

Katō Heizaemon (3)
Wasan no kenkyū: hōteishikiron 和算の研究, 方程式論 (*Wasan* studies: equations). Tokyo: Nihon Gakujutsu Shinkōkai, 1957.

Katō Heizaemon (4)
Wasan no kenkyū: seisūron 和算の研究, 整數論 (*Wasan* studies: number theory). Tokyo: Nihon Gakujutsu Shinkōkai, 1964.

Katō Heizaemon (5)
Wada Yasushi no gyōseki 和田寧の業蹟 (Achievements of Wada Yasushi). Nagoya: Meijō Daigaku, Rikōgakubu Sūgaku Kyōshitsu, 1967.

Katō Heizaemon (6)
Nihon sūgakushi 日本數學史 (History of Japanese mathematics). Sūgaku sensho series 數學選書 (Selected works on mathematics), nos. 1–2. Tokyo: Maki Shoten, 1967–1968.

Kattendyke, W. J. C. R. H. van
Nagasaki kaigun denshūjo no hibi 長崎海軍傳習所の日々 (Days at the Nagasaki Naval Training Center). Translated by Mizuta Nobutoshi. Tōyō bunko series, no. 26. Tokyo: Heibonsha, 1964.

Keene, Donald
The Japanese Discovery of Europe: Honda Toshiaki and Other Discoverers, 1720–1798. London: Routledge and Kegan Paul, 1952.

Kidder, Jonathan Edward, Jr.
Japan before Buddhism. Rev. ed. New York: Praeger, 1966.

Kimiya Yasuhiko, 木宮泰彦, ed.
Nihon koinsatsu bunkashi 日本古印刷文化史 (Cultural history of early Japanese printing). Tokyo: Fuzanbō, 1932.

Kimura Yōjirō 木村陽二郎
Nihon shizenshi no seiritsu: Rangaku to honzōgaku 日本自然誌の成立: 蘭學と本草學 (Formation of natural history in Japan: Dutch learning and materia medica). Shizen sensho series 自然選書 (Selected natural science monographs). Tokyo: Chūōkōronsha, 1974.

Kirkwood, Kenneth P.
Renaissance in Japan: A Cultural Survey of the Seventeenth Century (1938). Reprint ed. Tokyo: Charles E. Tuttle Co., 1970.

Kitagawa, Joseph
Religion in Japanese History. New York: Columbia University Press, 1966.

Kitajima Masamoto (1) 北島正元
Edo jidai 江戸時代 (Edo period). Shinsho series, no. 332. Tokyo: Iwanami Shoten, 1958.

Kitajima Masamoto (2)
Bakuhansei no kumon 幕藩制の苦悶 (Troubles in the Tokugawa system). Nihon no rekishi series, no. 18. Tokyo: Chūōkōronsha, 1966.

Kobata Atsushi (1) 小葉田淳
Chūsei Nisshi tsūkō bōeki shi no kenkyū 中世日支通交貿易史の研究 (Studies in the history of intercourse and trade between Japan and China in medieval times). Tokyo: Tōkō Shoin, 1941.

Kobata Atsushi (2)
Kōzan no rekishi 鑛山の歴史 (History of mines). Nihon rekishi shinsho series. Tokyo: Shibundō, 1956.

Kobata Atsushi (3)
Nihon kōzanshi no kenkyū 日本鑛山史の研究 (Studies in the history of Japanese mines). Tokyo: Iwanami Shoten, 1968.

Kobayashi Yukio 小林行雄
Kofun no hanashi 古墳の話 (Ancient burial mound tales). Shinsho series, no. 342. Tokyo: Iwanami Shoten, 1962.

Kodama Kōta, ed. (1) 兒玉幸多
Sangyōshi II 産業史 (Industrial history, vol. 2). Taikei Nihonshi series, no. 11. Tokyo: Yamakawa Shuppansha, 1965.

Kodama Kōta (2)
Genroku jidai 元禄時代 (Genroku era). Nihon no rekishi series, no. 16. Tokyo: Chūōkōronsha, 1966.

Kodama Kōta, ed. (3)
Zuroku: Shokuhō kara bakumatsu 圖錄, 織豊から幕末 (Collected pictures: from the time of Oda Nobunaga to the end of the Tokugawa period). Nihon no rekishi series, appendix no. 3. Tokyo: Chūōkōronsha, 1967.

Koga Jūjirō (1) 古賀十二郎
Seiyō ijutsu denraishi 西洋醫術傳來史 (History of transmission [to Japan] of Western medical arts). Tokyo: Nisshin Shoin, 1942.

Koga Jūjirō (2)
Nagasaki yōgakushi 長崎洋學史 (History of Western learning in Nagasaki). vols. 1–3. Edited for republication by Nagasaki Gakkai. Nagasaki: Nagasaki Bunkensha, 1966–1968.

Koizumi Akira 小泉丹
Nihon kagakushi shikō 日本科學史私攷 (Private view of the history of science in Japan). Tokyo: Iwanami Shoten, 1943.

Konishi Shirō 小西四郎
Kaikoku to jōi 開國と攘夷 (Opening the country and expelling the barbarians). Nihon no rekishi series, no. 19. Tokyo: Chūōkōronsha, 1966.

Kure Shūzō (1) 呉秀三
Mitsukuri Genpo 箕作阮甫 (Biography of Mitsukuri Genpo). Tokyo: Dainihon Tosho, 1914.

Kure Shūzō (2)
Shīboruto sensei, I: sono shōgai oyobi kōgyō シーボルト先生, I: その生涯及功業 (Dr. Siebold, vol. I: his life and work). Tōyō bunko series, no. 103. Tokyo: Heibonsha, 1967.

Kuroda Toshio 黒田俊雄
Mōko shūrai 蒙古襲來 (Mongol invasion). Nihon no rekishi series, no. 8. Tokyo: Chūōkōronsha, 1965.

Laures, Johannes, S.J.
The Catholic Church in Japan: A Short History. Rutland, Vt. and Tokyo: Charles E. Tuttle Co., 1954.

Ledyard, Carl
"Culture and Political Aspects of Traditional Korean Buddhism." *Asia* (Winter, 1968), pp. 46–61.

Levin, Maksim Grigor'evich
Ethnic Origins of the Peoples of Northeastern Asia. Toronto: University of Toronto Press, 1963.

Liao Wen-jen 廖溫仁 (Japanese reading: Ryō Onjin)
Shina chūsei igakushi 支那中世醫學史 (History of medieval Chinese medicine). Kyoto: Kaniya Shoten, 1932.

Libbrecht, Ulrich
Chinese Mathematics in the Thirteenth Century: the Shu-shu Chiu-chang of Ch'in Chiu-shao. M.I.T. East Asian Science Series, vol. 1. Cambridge, Mass.: The M.I.T. Press, 1973.

Maruyama Masao (1) 丸山眞男
Nihon seiji shisōshi kenkyū 日本政治思想史研究 (Studies in the history of Japanese political thought). Tokyo: Tokyo Daigaku Shuppankai, 1952.

Maruyama Masao (2)
Studies in the Intellectual History of Tokugawa Japan. Translated by Mikiso Hane. Tokyo: University of Tokyo Press, 1974.

Matsuda Kiichi (1) 松田毅一
Nippo kōshōshi 日葡交渉史 (History of intercourse between Japan and Portugal). Shinwa bunko series 親和文庫 (Friendship library.), no. 6. Tokyo: Kyōbunkwan, 1963.

Matsuda Kiichi (2)
Nanban shiryō no hakken 南蠻資料の發見 (Discovery of materials on "Southern Barbarians"). Shinsho series, no. 51. Tokyo: Chūōkōronsha, 1964.

Matsuda Kiichi (3)
Tenshō shōnen shisetsu 天正少年使節 (Youth envoys of the Tenshō era). Shinsho series, no. 203. Tokyo: Kadokawa Shoten, 1965.

Matsuda Kiichi (4)
Nanban shiryō no kenkyū 南蠻資料の研究 (Studies of the "Southern Barbarians" materials). Tokyo: Kazama Shobō, 1967.

Meijizen Nihon Kagakushi Kankōkai 明治前日本科學史刊行會
Meijizen Nihon kagakushi: sōsetsu, nenpyō 明治前日本科學史,總説・年表 (Pre-Meiji Japanese science history: general outline and chronology). Tokyo: Nihon Gakujutsu Shinkōkai, 1968.

Mikami Yoshio (1)
The Development of Mathematics in China and Japan. New York: Chelsea Publishing Co., 1913.

Mikami Yoshio (2) 三上義夫
Nihon kagaku no tokushitsu: tenmon 日本科學の特質（天文）(Characteristics of Japanese science: astronomy). Tōyō shichō series. Tokyo: Iwanami Shoten, 1936.

Mikami Yoshio (3)
Bunkashijō yori mitaru Nihon no sūgaku 文化史上より見たる日本の數學 (Japanese mathematics in the perspective of cultural history). Tokyo: Sōgensha, 1947.

Mikami Yoshio (4)
Nihon sūgakushi 日本數學史 (History of mathematics in Japan). Tokyo: Tōkai Shobō, 1947.

Mikami Yoshio (5)
Nihon sokuryōshi no kenkyū 日本測量史の研究 (Studies in the history of land-surveying in Japan). Tokyo: Kōseisha Kōseikaku, 1951.

Miyagi Eishō 宮城榮昌
Engishiki no kenkyū 延喜式の研究 (Studies of "Procedures of the Engi era"). 2 vols. Tokyo: Kōseisha Kōseikaku, 1955–1957.

Miyamoto Chū 宮本仲
Sakuma Shōzan 佐久間象山 (Biography of Sakuma Shōzan). Tokyo: Iwanami Shoten, 1932.

Miyashita Saburō
"A Neglected Source for the Early History of Anesthesia in China and Japan." In *Chinese Science: Explorations of an Ancient Tradition*, ed. Nakayama Shigeru and Nathan Sivin. M.I.T. East Asian Science Series, vol. 2, pp. 273–278. Cambridge, Mass.: The M.I.T. Press, 1973.

Miyazaki Michio 宮崎道生
Arai Hakuseki 新井白石 (Biography of Arai Hakuseki). Nihon rekishi shinsho series. Tokyo: Shibundō, 1957.

Momo Hiroyuki 桃裕行
Jōdai gakusei no kenkyū 上代學制の研究 (Studies of the educational system in antiquity). Tokyo: Meguro Shoten, 1947.

Mori Katsumi (1) 森克己
Nissō bōeki no kenkyū 日宋貿易の研究 (Studies of trade between Japan and Sung China). Tokyo: Kunitachi Shoin, 1948.

Mori Katsumi (2)
Kentōshi 遣唐使 (Envoys to T'ang China). Nihon rekishi shinsho series. Tokyo: Shibundō, 1962.

Moriya Tatsusaburō 森屋辰三郎
Tenka ittō 天下一統 (Unification of Japan). Nihon no rekishi series, no. 12. Tokyo: Chūōkōronsha, 1966.

Mozai Torao 茂在寅男
Kōkaijutsu 航海術 (Navigation). Shinsho series, no. 135. Tokyo: Chūōkōronsha, 1967.

Murakami Naojirō, tr. (1) 村上直次郎
Yasokaishi Nihon tsūshin 耶蘇會士日本通信 (Jesuit dispatches from Japan). Annotated by Watanabe Yosuke 渡邊世祐. Ikoku sōsho 異國叢書 (Foreign country series), vols. 1–2. Tokyo: Sunnansha, 1931–1932.

Murakami Naojirō, tr. (2)
Dejima rankan nisshi 出島蘭館日誌 (Daily records of the Dutch factory on Dejima). Vols. 1–3. Tokyo: Bunmei Kyōkai, 1938–1939.

Nagahama Yoshio 長濱善夫
Tōyō igaku gaisetsu 東洋醫學概説 (Outline of East Asian medicine). Osaka: Sōgensha, 1961.

Nagahara Keiji 永原慶二
Gekokujō no jidai 下克上の時代 (Age of revolt of vassals against their lords) Nihon no rekishi series, no. 10. Tokyo: Chūōkōronsha, 1965.

Nagata Hiroshi 永田廣志
Nihon tetsugaku shisōshi 日本哲學思想史. (Intellectual history of Japanese philosophy). Nagata Hiroshi Nihon shisōshi kenkyū, vol. 1 永田廣志日本思想史研究 (Vol. 1 of Nagata Hiroshi's series of studies of Japanese thought). Tokyo: Hōsei Daigaku Shuppankyoku, 1967.

Nakamura Hajime
Ways of Thinking of Eastern Peoples: India, China, Tibet, Japan. Rev. English translation. Ed. Philip P. Wiener. Honolulu: East-West Center Press, 1964.

Nakano Misao 中野操
Kōkoku iji dainenpyō 皇國醫事大年表 (Comprehensive chronology of medicine in Japan). Tokyo: Nankōdō, 1942.

Nakayama Shigeru (1) 中山茂
"Motoki Ryōei yaku 'Oranda Chikyūsetsu' ni tsuite" 本木良永譯「阿蘭陀地球説」について (On Motoki Ryōei's translation of the 'Dutch theory of the earth'). In *Rangaku shiryō kenkyūkai kenkyū hōkoku* 蘭學資料研究會研究報告 (Reports of the Research Group for Dutch Studies). No. 112 (1962) and no. 162 (1964).

Nakayama Shigeru (2)
"Abhorrence of 'God' in the Introduction of Copernicanism into Japan." *Japanese Studies in the History of Science*, no. 3 (1964), pp. 60–67.

Nakayama Shigeru (3)
"Cyclic Variation of Astronomical Parameters and the Revival of Trepidation in Japan." *Japanese Studies in the History of Science*, no. 3 (1964), pp. 68–80.

Nakayama Shigeru (4)
"Edo jidai ni okeru jusha no kagakukan" 江戸時代に於ける儒者の科學觀 (Confucian scholars' views of science in the Edo period). *Kagakushi kenkyū* 科學史研究 (Journal of history of science, Japan), no. 72 (1964), pp. 157–168.

Nakayama Shigeru (5)
Senseijutsu 占星術 (Astrology). Shinsho series, C-6 新書 (New books, no. 6 of Class C). Tokyo: Kinokuniya Shoten, 1964.

Nakayama Shigeru (6)
A History of Japanese Astronomy: Chinese Background and Western Impact. Harvard-Yenching Institute monograph series, no. 18. Cambridge, Mass.: Harvard University Press, 1969.

Nakayama Shigeru and Sivin, Nathan, eds.
Chinese Science: Explorations of an Ancient Tradition. M.I.T. East Asian Science Series, vol. 2. Cambridge, Mass.: The M.I.T. Press, 1973.

Naoki Kōjirō 直木孝次郎
Kodai kokka no seiritsu 古代國家の成立 (Establishment of the ancient state). Nihon no rekishi series, no. 2. Tokyo: Chūōkōronsha, 1965.

Naramoto Tatsuya 奈良本辰也
Chōnin no jitsuryoku 町人の實力 (Power of townsmen). Nihon no rekishi series, no. 17. Tokyo: Chūōkōronsha, 1966.

Needham, Joseph
Science and Civilisation in China. Vols. 1–5. Cambridge: At the University Press, 1954–1974.

Nihon Gakushiin (1) 日本學士院
Meijizen Nihon igakushi 明治前日本醫學史 (History of pre-Meiji Japanese medicine). Vols. 1–5. Tokyo: Nihon Gakujutsu Shinkōkai, 1955–1957.

Nihon Gakushiin (2)
Meijizen Nihon kenchiku gijutsushi 明治前日本建築技術史 (History of pre-Meiji Japanese architectural techniques). Tokyo: Nihon Gakujutsu Shinkōkai, 1961.

Nihon Gakushiin (3)
Meijizen Nihon nōgyō gijutsushi 明治前日本農業技術史 (History of pre-Meiji Japanese agricultural techniques). Tokyo: Nihon Gakujutsu Shinkōkai, 1964.

Nihon Gakushiin (4)
Meijizen Nihon sūgakushi 明治前日本數學史 (History of pre-Meiji Japanese mathematics). Vols. 1–5. Tokyo: Iwanami Shoten, 1954–1960.

Nihon Gakushiin (5)
Meijizen Nihon tenmongakushi 明治前日本天文學史 (History of pre-Meiji Japanese astrology and astronomy). Tokyo: Nihon Gakujutsu Shinkōkai, 1960.

Nihon Gakushiin (6)
Meijizen Nihon yakubutsugakushi 明治前日本藥物學史 (History of pre-Meiji Japanese pharmacology). Vols. 1–2. Tokyo: Nihon Gakujutsu Shinkōkai, 1957–1958.

Nihon Gakushiin (7)
Meijizen Nihon zōheishi 明治前日本造兵史 (History of pre-Meiji Japanese weapons production). Tokyo: Nihon Gakujutsu Shinkōkai, 1960.

Nihon rekishi daijiten 日本歷史大辭典
(Dictionary of Japanese history). Vols. 1–20; Appendixes, 2 vols. Tokyo: Kawade Shobō Shinsha, 1956–1961.

Nihonshi jiten 日本史辭典
(Historical dictionary of Japan). 2d ed. Tokyo: Kadokawa Shoten, 1974.

Nishi Naiga 西內雅
Shibukawa Harumi no kenkyū 澁川春海の研究 (Studies on Shibukawa Harumi). Tokyo: Shibundō, 1940.

Nōda Chūryō (1) 能田忠亮
Tōyō tenmongakushi ronsō 東洋天文學史論叢 (Discussions of East Asian astrology and astronomy). Tokyo Kōseisha Kōseikaku, 1943.

Nōda Chūryō (2)
Rekigaku shiron 曆學史論 (On the history of calendrics). Tokyo: Seikatsusha, 1948.

Nōda Chūryō (3)
Koyomi 曆 (Calendars). Nihon rekishi shinsho series. Tokyo: Shibundō, 1962.

Numata Jirō (1) 沼田次郎
Bakumatsu yōgakushi 幕末洋學史 (History of Western learning in the late Tokugawa period). Tokyo: Tōkō Shoin, 1961.

Numata Jirō (2)
Yōgaku denrai no rekishi 洋學傳來の歷史 (History of transmission of Western learning [to Japan]). Nihon rekishi shinsho series. Tokyo: Shibundō, 1963.

Numata Jirō (3)
"The Introduction of Dutch Language." *Monumenta Nipponica*, vol. 19 (1964), nos. 3–4, pp. 9–19.

Ogata Hiroyasu 尾形裕康
Nihon kyōiku tsūshi 日本教育通史 (General history of Japanese education). Tokyo: Waseda Daigaku Shuppanbu, 1963.

Ogata Tomio 緒方富雄
Ogata Kōan den 緒方洪庵傳 (Biography of Ogata Kōan). Tokyo: Iwanami Shoten, 1942.

Ogata Tomio et al.
"Monjin ga Shīboruto ni teikyō shitaru Rango ronbun no kenkyū" 門人ガシーボルトに提供したる蘭語論文の研究 (Studies of Dutch-language essays written for Siebold by his disciples). In *Shīboruto kenkyū* シーボルト研究 (Studies on Siebold), compiled by Nichidoku Bunka Kyōkai 日獨文化協會 (Japanese-German Cultural Association). Tokyo: Iwanami Shoten, 1938.

Ogawa Teizō (1) 小川鼎三
Igaku no rekishi 醫學の歷史 (History of medicine). Shinsho series, no. 39. Tokyo: Chūōkōronsha, 1964.

Ogawa Teizō (2)
Kaitai shinsho 解體新書 (New treatise on anatomy). Shinsho series, no. 165. Tokyo: Chūōkōronsha, 1968.

Ogura Kinnosuke (1) 小倉金之助
Sūgaku kyōikushi 數學教育史 (History of mathematical education). Tokyo: Iwanami Shoten, 1932.

Ogura Kinnosuke (2)
Sūgakushi kenkyū 數學史研究 (Studies of mathematical history). Vols. 1–2. Tokyo: Iwanami Shoten, 1935, 1948.

Ogura Kinnosuke (3)
Nihon no sūgaku 日本の數學 (Mathematics in Japan). Shinsho (red) series, no. 61. Tokyo: Iwanami Shoten, 1961.

Ogyū Sorai 荻生徂來
"Gakusoku" (Principles of learning) 學則. In *Nihon tetsugaku shisō zensho: gakumon* 日本哲學思想全書, 學問 (Complete works on Japanese philosophical thought: learning). Tokyo: Heibonsha, 1956.

Ōhira Kimata 大平喜間多
Sakuma Shōzan 佐久間象山 (Biography of Sakuma Shōzan). Jinbutsu sōsho series. Tokyo: Yoshikawa Kōbunkan, 1964.

Oka Senkichi 岡專吉
Nihon sūgaku gaisetsu 日本數學概説 (Outline of mathematics in Japan). Tokyo: Iwanami Shoten, 1933.

Okada Akio 岡田章雄
Kirishitan bateren キリシタン・バテレン (Catholic padres). Nihon rekishi shinsho series. Tokyo: Shibundō, 1962.

Ōkubo Toshiaki and Ebisawa Arimichi 大久保利謙, 海老澤有道 eds., *Nihon shigaku nyūmon* 日本史學入門 (Introduction to study of Japanese history). Tokyo: Kōbunsha, 1965.

Okumura Shōji (1) 奥村正二
Hinawajū kara kurobune made 火繩銃から黒船まで (From arquebuses to black ships). Shinsho series, no. 750. Tokyo: Iwanami Shoten, 1970.

Okumura Shōji (2)
Koban, kiito, watetsu 小判・生絲・和鐵 (Coins, raw silk, Japanese iron). Shinsho series, no. 863. Tokyo: Iwanami Shoten, 1970.

Ōta Hirotarō 太田博太郎
Nihon kenchikushi josetsu: zōho shinpan 日本建築史序説,増補新版 (Introduction to the history of Japanese architecture: enlarged new edition). Tokyo: Shōkokusha, 1969.

Ōta Masao 太田正雄, ed.
Nihon no igaku 日本の醫學 (Medicine in Japan). Tokyo: Minpūsha, 1946.

Ōtani Ryōkichi 大谷亮吉
Inō Tadataka 伊能忠敬 (Biography of Inō Tadataka). Tokyo: Iwanami Shoten, 1917.

Ōtori Ranzaburō
"The Acceptance of Western Medicine in Japan." *Monumenta Nipponica*, 19 (1964), nos. 3–4, pp. 20–40.

Ōtsuka Keisetsu 大塚敬節
Rinshō ōyō "Shōkanron" kaisetsu 臨床応用傷寒論解説 (Explanation of 'On Cold Damage Disorders' for Clinical Use). Tōyō igaku sensho series 東洋醫學選書 (Selected East Asian medical treatises). Osaka: Sōgensha, 1966.

Ōtsuki Nyoden 大槻如電
Nihon yōgaku hennenshi 日本洋學編年史 (Chronology of Western learning in Japan). Enlarged and revised by Satō Eishichi 佐藤榮七. Tokyo: Kinseisha, 1965.

Philippi, Donald L., tr., with introduction and notes.
Kojiki (Record of ancient matters). Tokyo: University of Tokyo Press, 1968.

Pledge, H. T.
Science since 1500: A Short History of Mathematics, Physics, Chemistry, Biology. Harper Torchbooks/Science Library. New York: Harper and Brothers Publishers, 1959.

Porkert, Manfred
The Theoretical Foundations of Chinese Medicine: Systems of Correspondence. M.I.T. East Asian Science Series, vol. 3. Cambridge, Mass.: The M.I.T. Press, 1974.

Reischauer, Edwin O., tr. (1)
Ennin's Dairy: The Record of a Pilgrimage to China in Search of the Law. New York: Ronald Press, 1955.

Reischauer, Edwin O. (2)
Ennin's Travels to T'ang China. New York: Ronald Press, 1955.

Reischauer, Edwin O. and Fairbank, John K.
East Asia: The Great Tradition. Boston: Houghton Mifflin Co., 1958 and 1960. (Tokyo: Charles E. Tuttle Co., 1958 and 1960. Modern Asia Edition).

Rekishigaku Kenkyūkai 歷史學研究會, ed.
Nihonshi nenpyō 日本史年表 (Chronology of Japanese history). Tokyo: Iwanami Shoten, 1966.

Sagara Tōru 相良亨
Kinsei Nihon ni okeru jukyō undō no keifu 近世日本に於ける儒教運動の系譜 (Evolution of Confucian movements in feudal Japan). Tetsugaku zensho series, no. 3 哲學全書 (Series of philosophical works, no. 3). Tokyo: Risōsha, 1965.

Saigusa Hiroto, ed. (1) 三枝博音
Nihon kagaku koten zensho 日本科學古典全書 (Collected classics of Japanese science). Vols. 1, 6, 8–14. Tokyo: Asahi Shinbunsha, 1944, 1942, 1942–1948.

Saigusa Hiroto, ed. (2)
Shūkyō to kagaku no rekishi 宗教と科學の歷史 (History of religion and science). Nihon bunkashi kōza series, no. 5 日本文化史講座 (Japanese cultural history, lecture series, no. 5). Tokyo: Shinhyōronsha, 1955.

Saitō Tokutarō 齋藤悳太郎
Nijūroku taihan no hangaku to shifū 二十六大藩の藩學と士風 (Fief schools and samurai codes of twenty-six major domains). Osaka: Zenkoku Shobō, 1944.

Saitō Tsutomu 齋藤勵
Ōchō jidai no onmyōdō 王朝時代の陰陽道 (*Yin-yang* divination in the age of the nobility). Nihon bunka meichosen series 日本文化名著選 (Selected famous works on Japanese culture). Tokyo: Sōgensha, 1947.

Sasaki Junnosuke 佐々木潤之助
Daimyō to hyakushō 大名と百姓 (Feudal lords and farmers). Nihon no rekishi series, no. 15. Tokyo: Chūōkōronsha, 1966.

Satō Kan'ichi 佐藤貫一
Nihon no tōken 日本の刀劍 (Japanese swords). Nihon rekishi shinsho series. Tokyo: Shibundō, 1961.

Satō Shin'ichi 佐藤進一
Nanbokuchō no dōran 南北朝の動亂 (Turmoil in the age of the Southern and Northern Courts). Nihon no rekishi series, no. 9. Tokyo: Chūōkōronsha, 1965.

Satō Shōsuke 佐藤昌介
Yōgakushi kenkyū josetsu 洋學史研究序説 (Introduction to study of Western learning [in Japan]). Tokyo: Iwanami Shoten, 1964.

Sawada Goichi 澤田吾一
Nihon sūgakushi kōwa 日本數學史講話 (Discourses on the history of Japanese mathematics). Tokyo: Tōkō Shoin, 1928.

Seki Akira 關晃
Kikajin 歸化人 (Naturalized persons). Nihon rekishi shinsho series. Tokyo: Shibundō, 1962.

Shimizu Tōtarō 清水藤太郎
Nihon yakugakushi 日本藥學史 (History of Japanese pharmacology). Tokyo: Nanzandō, 1949.

Shimodaira Kazuo 下平和夫
Wasan no rekishi 和算の歷史 (History of *wasan*). Tōzai sūgaku series, nos. 7, 8, 9. Tokyo: Fuji Tanki Daigaku Shuppanbu, 1965–1967.

Shimode Sekiyo 下出積與
Shinsen shisō 神仙思想 (Thought of religious recluses). Nihon rekishi sōsho series, no. 22. Tokyo: Yoshikawa Kōbunkan, 1968.

Shinjō Shinzō 新城新藏
Tōyō tenmongakushi kenkyū 東洋天文學史研究 (Studies in the history of East Asian astrology and astronomy). Kyoto: Kōbundō, 1928.

Shirai Kōtarō (1) 白井光太郎
Honzōgaku ronkō 本草學論攷 (On materia medica). Vols. 1–4. Tokyo: Shun'yōdō, 1933–1934.

Shirai Kōtarō (2)
Nihon hakubutsugakushi nenpyō: kaitei zōho 日本博物學史年表, 改訂增補 (Chronology of the history of natural history in Japan: revised and enlarged). Tokyo: Ōokayama Shoten, 1934.

Shōji Takeo 庄司武夫
Kahō no hattatsu 火砲の發達 (Development of cannons). Tokyo: Shinkyō Shuppansha, 1945.

Sivin, Nathan (1)
Chinese Alchemy: Preliminary Studies. Cambridge, Mass.: Harvard University Press, 1968.

Sivin, Nathan (2)
"Copernicus in China." In *Colloquia Copernicana*, vol. 2. Ed. Union Internationale d'Histoire des Sciences. Warsaw: The Compilers, 1973.

Sivin, Nathan, tr. (3)
Traditional Medicine in Contemporary China (forthcoming).

Sivin, Nathan (4)
"Shen Kua (1031–1095)." In *Dictionary of Scientific Biography, s.v.*

Smith, David Eugene, and Mikami Yoshio
History of Japanese Mathematics. Chicago: The Open Court Publishing Co., 1914.

Sohn Pow-key, Kim Chol-choon, and Hong Yi-sup
The History of Korea. Seoul: Korean National Commission for UNESCO, 1970.

Sudō Toshiichi 須藤利一, ed.
Fune 船 (Ships). Mono to ningen no bunkashi series 物と人間の文化史 (Cultural history of men and materials). Tokyo: Hōsei Daigaku Shuppankyoku, 1968.

Sugimoto Isao (1) 杉本勳
Kinsei jitsugakushi no kenkyū 近世實學史の研究 (Studies in the history of practical learning in Tokugawa Japan). Tokyo: Yoshikawa Kōbunkan, 1962.

Sugimoto Isao, ed. (2)
Kagakushi 科學史 (History of [Japanese] science). Taikei Nihonshi sōsho series, no. 19. Tokyo: Yamakawa Shuppansha, 1967.

Sugita Genpaku
Dawn of Western Science in Japan. Translation of *Rangaku kotohajime* by Matsumoto Ryōzō and Kiyooka Eichi. Tokyo: Hokuseidō, 1969.

Sugiyama Hiroshi 杉山博
Sengoku daimyō 戰國大名 (Daimyo in the Warring States era). Nihon no rekishi series, no. 11. Tokyo: Chūōkōronsha, 1965.

Sun E-tu Zen and Sun Shiou-chuan, trs.
Chinese Technology in the Seventeenth Century (Translation of *T'ien-kung k'ai-wu*). University Park, Pa.: Pennsylvania State University Press, 1966.

Suzuki Hisao 鈴木久男
Shuzan no rekishi 珠算の歷史 (History of the abacus). Tōzai sūgaku series, no. 11. Tokyo: Fuji Tanki Daigaku Shuppanbu, 1964.

Suzuki Keishin 鈴木敬信
Koyomi to meishin 曆と迷信 (Calendars and superstitions). Tokyo: Kōseisha Kōseikaku, 1969.

Tahara Tsuguo 田原嗣郎
Tokugawa shisōshi kenkyū 德川思想史研究 (Studies of Tokugawa intellectual history). Tokyo: Miraisha, 1967.

Takagi Kikusaburō 高木菊三郎
Nihon ni okeru chizu sokuryō no hattatsu ni kansuru kenkyū 日本に於ける地圖測量の發達に關する研究 (Studies of the development of map measurement in Japan). Tokyo: Kazama Shobō, 1966.

Takahashi Shin'ichi 高橋磌一
Yōgaku shisōshi ron 洋學思想史論 (On the intellectual history of Western learning [in Japan]). Tokyo: Shin Nihon Shuppansha, 1972.

Takano Chōun 高野長運
Takano Chōei den 高野長英傳 (Biography of Takano Chōei). Tokyo: Iwanami Shoten, 1943.

Takeuchi Rizō 竹内理三
Bushi no tōjō 武士の登場 (Advent of the samurai). Nihon no rekishi series, no. 6. Tokyo: Chūōkōronsha, 1965.

Takeuchi Rizō and Nagahara Keiji 竹内理三, 永原慶二, eds.
Zuroku: Kamakura kara Sengoku 圖錄, 鎌倉から戰國 (Collected pictures: from Kamakura period to Sengoku period). Nihon no rekishi series, Appendix no. 2. Tokyo: Chūōkōronsha, 1967.

Tanaka Takeo (1) 田中健夫
Chūsei kaigai kōshōshi no kenkyū 中世海外交渉史の研究 (Studies of overseas contacts in medieval times). Tōdai Jinmon Kagaku Kenkyū Sōsho series 東大人文科學研究叢書 (University of Tokyo Humanities Research series). Tokyo: Tokyo Daigaku Shuppankai, 1959.

Tanaka Takeo (2)
Wakō to kangō bōeki 倭寇と勘合貿易 (Japanese pirate-traders and the Tally Trade). Nihon rekishi shinsho series. Tokyo: Shibundō, 1963.

Tokoro Sōkichi 所荘吉
Hinawajū: zōhoban 火繩銃,増補版 (Arquebus: enlarged edition). Tokyo: Yūzankaku, 1969.

Tokyo Kagaku Hakubutsukan 東京科學博物館
Edo jidai no kagaku 江戸時代の科學 (Science in the Edo period). Tokyo: Hakubunkan, 1934.

Tōyama Shigeki and Satō Shin'ichi 遠山茂樹、佐藤進一 eds.,
Nihonshi kenkyū nyūmon 日本史研究入門 (Introduction to the study of Japanese history). Vols. 1–2. Tokyo: Tokyo Daigaku Shuppankai, 1954, 1962.

Toyama Usaburō 外山卯三郎
Nanban gakkō 南蠻學考 (On the learning of the "Southern Barbarians"). Tokyo: Kokuminsha Sōritsu Jimusho, 1944.

Toyoda Takeshi 豊田武, ed.
Sangyōshi I 産業史 (History of industry, I). Taikei Nihonshi sōsho series, no. 10. Tokyo: Yamakawa Shuppansha, 1964.

Tsuchida Naoshige 土田直鎮
Ōchō no kizoku 王朝の貴族 (Aristocrats of the court). Nihon no rekishi series, no. 5. Tokyo: Chūōkōronsha, 1965.

Tsuda Sōkichi (1) 津田左右吉
Nihon bunka to Shina oyobi Chōsen no bunka to no kōryū 日本文化と支那及び朝鮮の文化との交流 (Japanese cultural intercourse with China and Korea). Tōyō shichō series. Tokyo: Iwanami Shoten, 1936.

Tsuda Sōkichi (2)
Nihon no Shintō 日本の神道 (Shintoism in Japan). Tokyo: Iwanami Shoten, 1949.

Tsuda Sōkichi (3)
Jukyō no kenkyū 儒教の研究 (Studies on Confucianism). Vols. 1–3. Tokyo: Iwanami Shoten, 1950–1956.

Tsuge Hideomi, ed.
Historical Development of Science and Technology in Japan. Series on Japanese Life and Culture, vol. 5. Tokyo: Kokusai Bunka Shinkōkai, 1961.

Tsuji Tatsuya 辻達也
Edo kaifu 江戸開府 (Founding of Edo). Nihon no rekishi series, no. 13. Tokyo: Chūōkōronsha, 1966.

Tsuji Zennosuke (1) 辻善之助
Zōtei: Kaigai kōtsū shiwa 増訂 海外交通史話 (Discussions of the history of overseas transportation; revised and enlarged). Tokyo: Naigai Shoseki, 1930.

Tsuji Zennosuke (2)
Nihon Bukkyōshi 日本佛教史 (History of Japanese Buddhism). Vols. 1–3. Tokyo: Iwanami Shoten, 1944–1955.

Tsunoda Ryusaku et al.
Sources of the Japanese Tradition. New York: Columbia University Press, 1958.

Tsurumi Shunsuke 鶴見俊輔
Takano Chōei 高野長英 (Biography of Takano Chōei). Tokyo: Asahi Shinbunsha, 1975.

"Tung-i ch'uan" 東夷傳
(Record of eastern barbarians). In *Hou Han shu* 後漢書 (History of Later Han). Compiled in the fifth century A.D.

Ueda Masaaki 上田正昭
Kikajin 歸化人 (Naturalized persons). Shinsho series, no. 70. Tokyo: Chūōkōronsha, 1965.

Ueda Sanpei 上田三平
Nihon yakuenshi no kenkyū 日本藥園史の研究 (Studies in the history of herbal gardens in Japan). Tokyo: Maruzen, 1930.

Ueno Masuzō (1) 上野益三
Nihon seibutsugaku no rekishi 日本生物學の歴史 (History of biology in Japan). Tokyo: Kōbundō, 1939.

Ueno Masuzō (2)
Nihon hakubutsugakushi 日本博物學史 (History of natural history in Japan). Tokyo: Heibonsha, 1973.

Veith, Ilza, tr.
Huang ti nei ching su wen: The Yellow Emperor's Classic of Internal Medicine. Berkeley, Los Angeles, London: University of California Press, 1970.

Wada Shinjirō 和田信二郎
Nakagawa Jun'an sensei 中川淳庵先生 (Biography of Nakagawa Jun'an, teacher). Kyoto: Ritsumeikan Shuppanbu, 1941.

Wajima Yoshio (1) 和島芳男
Shōheikō to hangaku 昌平校と藩學 (The Shōheikō and domain schools). Nihon rekishi shinsho series. Tokyo: Shibundō, 1962.

Wajima Yoshio (2)
Chūsei no jugaku 中世の儒學 (Medieval Confucianism). Nihon rekishi sōsho series, no. 11. Tokyo: Yoshikawa Kōbunkan, 1965.

Wang Ling and Needham, Joseph
"Horner's Method in Chinese Mathematics: Its Origins in the Root-Extraction Procedures of the Han Dynasty." In *T'oung Pao (Archives concernant L'Histoire, les Langues, la Géographie, l'Ethnographie et les Arts de l'Asie Orientale).* Leiden, vol. 43 (1955), p.345.

Watanabe Toshio 渡邊敏夫
Hazama Shigetomi to sono ikka 間重富とその一家 (Hazama Shigetomi and his household). Kyoto: Yamaguchi Shoten, 1943.

Watsuji Tetsurō 和辻哲郎
Sakoku 鎖國 (National isolation). Tokyo: Chikuma Shobō, 1964.

Wong K. Chimin and Wu Lien-teh
History of Chinese Medicine. Tientsin: The Tientsin Press, 1932.

Yabuuchi Kiyoshi, ed. (1) 藪内清
Tenkō kaibutsu no kenkyū 天工開物の研究 (Studies of the *T'ien-kung k'ai-wu* [Exploitation of the works of nature]). Tokyo: Kōseisha Kōseikaku, 1953.

Yabuuchi Kiyoshi, ed. (2)
Chūgoku chūsei kagaku gijutsu shi no kenkyū 中國中世科學技術史の研究 (Studies in the history of sciences and techniques in medieval China). Tokyo: Kadokawa Shoten, 1963.

Yabuuchi Kiyoshi (3)
Chūgoku kodai no kagaku 中國古代の科學 (Science in ancient China). Shinsho series, no. 180. Tokyo: Kadokawa Shoten, 1964.

Yabuuchi Kiyoshi, ed. (4)
Sō Gen jidai no kagaku gijutsu shi 宋元時代の科學技術史 (History of Chinese sciences and techniques in the Sung and Yüan periods). Kyoto: Kyoto Daigaku, Jinmon Kagaku Kenkyūjo, 1967.

Yabuuchi Kiyoshi (5)
Chūgoku no tenmon rekihō 中國の天文曆法 (Astrology and calendrics in China). Tokyo: Heibonsha, 1969.

Yabuuchi Kiyoshi (6)
Chūgoku no kagaku bunmei 中國の科學文明 (Scientific civilization in China). Shinsho series, no. 759. Tokyo: Iwanami Shoten, 1970.

Yabuuchi Kiyoshi (7)
Chūgoku no sūgaku 中國の數學 (Mathematics in China). Shinsho series, no. 906. Tokyo: Iwanami Shoten, 1974.

Yabuuchi Kiyoshi and Sōda Hajime 藪内清, 宗田一, eds.
Edo jidai no kagaku kikai 江戸時代の科學器械 (Scientific instruments in the Edo period). Tokyo: Kōseisha Kōseikaku, 1964.

Yabuuchi Kiyoshi and Yoshida Mitsukuni 藪内清, 吉田光邦, eds.
Min Shin jidai no kagaku gijutsu shi 明清時代の科學技術史 (History of Chinese sciences and techniques in the Ming and Ch'ing periods). Kyoto: Kyoto Daigaku, Jinmon Kagaku Kenkyūjo, 1970.

Yamaguchi Ryūji 山口隆二
Nihon no tokei 日本の時計 (Clocks in Japan). Tokyo: Nihon Hyōronsha, 1950.

Yamawaki Teijirō 山脇悌二郎
Nagasaki no tōjin bōeki 長崎の唐人貿易 (Chinese trading in Nagasaki). Tokyo: Yoshikawa Kōbunkan, 1964.

Yamazaki Tasuku (1) 山崎佐
Nihon ekishi oyobi bōekishi 日本疫史及防疫史 (History of epidemics and their prevention in Japan). Tokyo: Kokuseidō, 1931.

Yamazaki Tasuku (2)
Edokizen Nihon iji hōsei no kenkyū 江戶期前日本醫事法制の研究 (Studies of legal provisions for medical care in Japan before the Edo period). Tokyo: Chūgai Igakusha, 1953.

Yamazaki Tasuku (3)
Kakuhan igaku kyōiku no tenbō 各藩醫學教育の展望 (Survey of medical education by domains). Tokyo: Kokudosha, 1955.

Yamazaki Yoemon (1) 山崎與右衛門
Tōzai soroban bunkenshū 東西算盤文献集 (Collected Oriental and Occidental references on the abacus). Vols. 1-2. Tokyo: Morikita Shuppansha, 1956, 1962.

Yamazaki Yoemon (2)
Jinkōki no kenkyū 塵却記の研究 (Studies on the *Jinkōki*). Tokyo: Morikita Shuppansha, 1966.

Yazaki Takeo
Social Change and the City in Japan: From Earliest Times through the Industrial Revolution. Translated by David L. Swain. Tokyo: Japan Publications, 1968.

Yoshida Mitsukuni (1) 吉田光邦
Nihon kagakushi 日本科學史 (History of Japanese science). Tokyo: Asakura Shoten, 1955.

Yoshida Mitsukuni (2)
Edo no kagakushatachi 江戶の科學者たち (Scientists of the Edo period). Gendai kyōyō bunko series, no. 664 現代教養文庫 (Contemporary education library series). Tokyo: Shakai Shisōsha, 1969.

Yuasa Akira 湯淺明
Nihon shokubutsugakushi 日本植物學史 (History of botany in Japan). Tokyo: Kenkyūsha, 1948.

Yuasa Mitsutomo 湯淺光朝
Kaisetsu kagaku bunkashi nenpyō 解説科學文化史年表 (Annotated chronology of the history of science and culture). Revised and enlarged. Tokyo: Chūōkōronsha, 1966.

Index

Explanatory notes:
Persons mainly active in specific scientific fields are identified by one or more of the following abbreviations: astron. = astronomy (not directly related to calendars); astrol. = astrology; cal. = calendrical astronomy (cf. Introduction, xxvii); med. = medicine.

Persons mainly active in one of the main traditions of learning are identified by the following: Conf. = Confucian learning; National Learning; Dutch Learning; Western Learning.

Others are identified variously by status, occupation, or activity in some field other than those specified above.

Chinese characters follow respective names or terms.

Abacus (*soroban* 算盤), 79–80, 132, 180, 204, 205, 208, 264, 266, 274, 276, 364, 366
  as inhibiting factor in the development of abstract in mathematics, 275
Abdominal cramps, 380
Abe 安倍 family, 123, 124, 199
Abe no Manao 安倍眞直 (med., fl. 808), 88, 309
Abe no Seimei 安倍淸明 (astrol., 921–1005), 123, 124, 253
Academic institutions, 30, 234
  challenge of Buddhist learning to, 41–42
  for leadership selection, 28
  in the *ritsuryō* system, 31–40, 41, 45
Acceleration, concept of, 173, 353
Aconitum japonicum, 388
Acupuncture, 36, 45, 85, 87, 98, 99, 100, 217, 287, 289. *See also* Hari
  loci for application of, 96
Adams, William (1564–1620), 177
Addition, 74, 75, 79
  on the abacus, 205
Aeba Tōan 饗場東庵 (med., 1615–1673), 139, 238, 280
Aesop's Fables, 185
Agatai monjinroku 縣居門人錄 (Directory of disciples of Kamo no Mabuchi), 307n.19
Agriculture, xxiv, 2n.1, 6, 7, 24, 25, 103, 112, 223, 229–230, 293
  in China, 107
  student-monks and, 42
Aida Yasuaki 會田安明 (math., 1747–1817), 363–364
Aizu 會津 district, calendar, 199
Ajima Naonobu 安島直圓 (math., 1739–1796), 361, 362, 363, 369, 370
Alchemy, 9, 78, 87, 90, 127
Algebra, 74, 76, 131–132, 173, 180, 370
  "heaven-origin." *See* Heaven-

origin algebraic method written, 365
Algebraic equations, 81, 82, 264
Algebraic notations. *See* Notations
Algebraic unknowns, coefficients for solving, 129
Almanac (*guchūreki* 具注暦), 56ff., 67, 128, 198
Almeida, Luis de (Jesuit, med., 1525–1584), 217–218
Amakusa 天草 islands, 161, 189
Amoghavajra (Pu-k'ung 不空, Indian astrol., fl. 759), 69, 253
Amputations, 98, 221, 388
Anatomical charts, 317, 322, 323, 381, 384
  Western, 383
Anatomical illustrations, 322, 378, 379, 380, 381
Anatomical observations, 96, 379, 380
Anatomische Tabellen (*Tabulae anatomicae*). *See Tafel anatomia*
Anatomy, 99, 221, 317, 332, 379, 380, 381, 383
  Dutch books on, 322
  Western, 387
"Ancient Learning" school (*Kogakuha* 古學派), 235, 238, 241, 246–247, 277, 282, 283, 297, 298, 299, 302, 303, 304
"Ancient Practice School" (*Kohōha* 古方派) of medicine, 247, 277, 278, 279, 280, 281, 282, 283, 289, 298, 302, 317, 354, 372, 373, 374, 378, 387, 381, 391
Andō Yūeki 安藤有益 (cal., fl. 1663), 253
Anesthesia, 98, 221, 222, 387–388
Angeria gorin taisei 暗厄利亞語林大成 (English-Japanese dictionary), 333
Animal names, 315
Animal substances, used in drugs, 85

Anma 按摩 (massage) course, 37
Anomalistic months, 355
Annotated calendar. *See* Almanac
Anti-Catholic prejudices and fears, 237, 245, 313
Anti-Western prejudices and fears, 237, 247, 310, 347, 376, 395
Aochi Rinsō 青地林宗 (med., fl. 1823), 330
Aoki Kon'yō 青木昆陽 (Conf., 1698–1769), 315, 316, 317, 350, 372
Apotropaic medicine, 36
Arabic numerals, 180, 208
Arabs, 6, 104, 170, 175
Arai Hakuseki 新井白石 (Conf., 1657–1725), 238, 244, 299–300, 304, 310–311, 316, 372
Arakan, Burma, 153
Araki Murahide 荒木村秀 (math., 1640–1718), 369
Arashiyama 嵐山 family, 286
Arashiyama Hoan 嵐山甫安 (med., 1633–1693), 286
Arashiyama 嵐山 school, 286
Archimedan screw, 182, 183, 233n.18
Architecture, 26, 43, 168, 206
Areas, measurement of, 75, 82, 205, 206, 361
Arihara 在原 household, 116
Arima 有馬, 160
  seminary at, 188, 189
Arima Yoriyuki 有馬賴僮 (math., 1714–1783), 365, 369
Aristotelian elements. *See* Four elements
Aristotle, and Aristotelianism, xv, xvi, 194, 201
Arithmetic, 44, 75, 76, 77, 78, 82, 134, 135, 203, 276, 370
Arithmetical progression. *See* Progressions
Armies, tactics and equipment of, 172
Armillary sphere, 54, 122, 312

Index   443

Armor, 168
Arms production, 174
Arquebus, 171–173, 232
Artificial gold, 90
Artisans, 83, 112, 138, 164, 166, 230, 356
Asada Gōryū 麻田剛立 (b. Ayabe Yasuaki 綾部妥章; astron., 1734–1799), 305, 314, 351, 354–356
Asai Sōzui 阿佐井宗瑞 (med., ?–1532), 214
Asakusa 淺草, astronomical observatory at, 312
Asanuma Sukemitsu 淺沼佐盈 (med., fl. 1750s), 317, 380
Ashikaga School (Ashikaga Gakkō 足利學校), 119, 127–128, 215
Ashikaga 足利 shogunate, 105, 119, 141, 150, 151, 152, 161, 214, 216
Ashikaga Yoshimitsu 足利義滿 (shogun, 1358–1408), 19, 150, 151, 198
Astrolabe, 122, 176, 179, 180, 181, 203, 349
Astrology, xxvii, 7, 8, 31, 33, 43, 45, 46, 53, 88, 117, 123, 133, 198, 199, 200, 202, 240, 251, 252
　Chinese. *See* Chinese astrology
　first book by a Japanese related to, 88
　Indian. *See* Indian astrology
　portent, 52, 54, 55, 128
　private initiative and, 73
　the state and, 47, 52, 73
　and Taoism, 50
Astronomical abnormalities, 36, 121
Astronomical books
　in the Dutch language, 300, 349
　Jesuit translations of, 349
Astronomical clock, 350
Astronomical observational instruments, 122, 179, 195, 202, 258, 349, 356, 359
Astronomical observations, 180, 312, 350, 355–356
Astronomical observatories, 50, 312
Astronomical predictions, xxvii, 122, 252
Astronomical tables, 48, 199, 252
Astronomie (de Lalande), 314, 337
Astronomy and astronomers, xvi, xxvii, xxix, 43, 125–126, 128, 133, 176, 255, 261, 305, 311, 370, 371
　algebraic techniques, 258
　calendrical. *See* Calendrical science
　Chinese. *See* Chinese astronomy
　circumpolar, 62
　Copernican. *See* Copernicus
　first Japanese serious attempt to study, 45
　Greek, 69
　Indian. *See* Indian astronomy
　Islamic. *See* Islamic astronomy
　Jesuits' teaching of, 200–203
　Korean. *See* Korean astronomy
　low level of Japanese interest in, 77, 202
　observational, 252
　Ptolmaic. *See* Ptolemy
　Taoism and. *See* Taoism
　Western influence on, 313
Asuka 飛鳥 period (552–645), 11, 26, 45
Augustinian missionaries, 157, 219
Aurifaction, 90
Authoritarianism, 229
Ayabe Yasuaki. *See* Asada Gōryū
Ayutthaya, Siam, 155
Azabu 麻布, 188
Azuchi 安土, 188

Baba Sadayoshi 馬場貞由 (or Sajūrō 佐十郎, interpreter, 1787–1822), 337, 341, 377
Babylonian calendar, 47, 70
Bacteriology, 220
Bakufu 幕府 (shogunate), 164
Ballistics, 173, 396
Bangaku shachū 蕃學社中 (Circle for barbarian learning), 346

Bankoku chihō ruijūteki den 蠻國治方類聚的傳 (Classified records of the barbarians' therapy), 286
Bansha 蕃社 incident. *See* Barbarian circle incident
Bansha no goku 蕃社の獄 ("Prosecution of the barbarian circle"), 346
Bansho shirabesho. *See* Office for investigating barbarian documents
Bansho wage goyō. *See* Office for the Translation of Barbarian Texts
Barbarian circle incident, 344–347, 373, 376
Batavia, 338, 377, 378
  Governor of, 292
Battery construction, 396
Bauduin, A. F., (med., fl. 1864), 397
Bear's gall, 282
Beaumont, William (med., 1785–1853), 391
Bile, used for medication, 380
Biological research, 391
Biology, xxviii, 394, 396
  terminology, 306
Biwa 琵琶, Lake, 245
"Black Ships," xvi
Blaeu, Johan (cartographer), 311
Blockades, 173
Blood circulation, 95, 99, 221
Bodhisattvas, 24
Boils, 98, 214
Bone setting, 36, 85, 87, 221
Bone structures, 328
Bonin Islands, 346
Book of Changes. *See I-ching*
Book import ban, 161, 162, 165, 202, 208, 237, 349
  ban relaxed, 224, 262, 291, 294, 312–313, 349, 351, 361, 370
Books imported from China, 168, 169. *See also* Medical books
Bōsō 房總 peninsula, 177
Botanikakyō 菩多尼訶經 (botany text), 330
Botany, 305, 340, 341, 396
  terminology, 306
Brahe, Tycho, xv, xvi, 201, 352
Brain, function and structure of, 391
Brawn, Hans Wolfgang (German gunner, 1609–1660), 232
Breast cancer, 388, 389, 390
Breathing exercises, 85, 98
Britain and the British, 292, 338
British Museum, 287
Buddha, offerings to, 276–277
Buddhas, 24
Buddhism, 5, 10, 11, 14, 16, 18, 19–24, 26, 27, 28, 40–42, 46, 87, 97, 100, 110, 157, 159, 187, 241, 295, 307
  and astrology, 69, 74, 124
  and calendrical divination, 59
  in China, 9–11, 20, 26, 30, 91, 108, 109, 187, 231, 395
  Confucianism and, 241, 242
  decline of, in Japan, 157–158
  economic strength of, 10
  Indian, 10, 49, 91
  in Korea, 21, 25, 26, 110
  medicine and, 94, 100, 102, 121, 142–146, 216–217
  National Learning and, 307, 308
  new forms of. *See* New Buddhism
  reaction against novel ideas from the West, 231
  religious communities, 10
  Roman Catholicism and, 231
  science and, 41, 121
  and the secularization of Japanese culture, 167
  Shinto and, 23–24
  sociopolitical role of, 21
  teaching of, 44
  and Western astronomy, 301
  Zen, 10n.12, 113
Buddhist art and architecture, 26
Buddhist canons, 135, 140
Buddhist compassion, 142, 216n.72
Buddhist cosmology, 301, 354
Buddhist learning, 21, 41–42, 120–

Index   445

121
Buddhist priests and monks, 19, 42,
    74, 94, 113, 120, 125, 136,
    170, 184, 187, 195, 199, 236,
    307, 308, 329. *See also* New
    Buddhism, priests; Zen
    priests
  Confucianism and, 235
  and mathematics, 135, 137, 203
  and medicine, 121, 139, 141, 146,
    147
  and the samurai, 119
  and trade relations with China,
    148
  visits to and from China, 106,
    119, 121, 122, 126, 142, 147,
    156
Buddhist scholars, ruling class
    patronage of, 41
Buddhist secular power, 158, 160
Buddhist statues and bells, casting
    of, 173
Buddhist temples, 184, 185
  building of, 5, 21, 22, 25–26, 29,
    105, 113, 120, 203
  burning of, 160
  clinics in, 101, 141
  compulsory registration of
    citizens at, 231
  economic and political influence
    of, 21–22
  New Buddhist, 120
  rituals to ward off evils, 126
  secular power, 158, 160
  tax-exempt landholdings of, 27,
    111
Buddhist titles, awarded to
    physicians, 279, 283–284,
    286, 288
Bukkoku rekishōhen 佛國曆象篇 (On
    the calendrical astronomy of
    Buddha's country), 354
Bunkyū 文久 era (1861–1864), 64
Bunseki kyūrisho. *See* Laboratory
    for physics and chemistry
Bureau of Astronomy. *See*
    Tenmongata
Bureau of Divination. *See* T'ai-pu

shu
Bureaucracy, 6, 107, 109
  and Confucianism, 11
  and mathematics, 275
Buredan (Dutch ship), 232
Burial mounds, 5, 24, 25. *See also*
    Kofun period
Burma, 154
Butchery, 380
Butsumetsu 佛滅, 66
Butsurigaku 物理學 (physics), 306
Byōgaku tsūron 病學通論 (General
    theory of pathology), 385
Byōrigaku 病理學 (pathology), 306

Calculation
  with abacus or counting board,
    275
  skills in, 25, 43
  written, 267, 275
Calculations of complex figures,
    261, 371
Calculus, 173
  differential, 361
  integral, 361
Calendars, 44, 45, 46–52, 202
  annotated (*guchūreki*), 67, 128
  annual, 38, 72
  Babylonian, 47, 70
  Chinese. *See* Chinese calendars
  differences between local and
    Kyoto calculations, 199
  Fu-t'ien li, 125
  Greek, 70
  Gregorian, 69, 123, 360
  Hsüan-ming li, 49, 51, 72, 128
  Indian, 70
  made by the Institute of
    Divination specialists, 128
  Islamic, 70
  Japanese utilitarian attitude
    toward, 52, 72, 77
  Julian, 47
  liturgical, 203
  lunar, 70
  lunisolar, 69, 71, 202, 203, 360
  nautical, 176, 178, 179, 180, 203,
    360

Calendars (continued)
  official, 47, 48
  printed and calculated independently of Kyoto, 199
  proposal to adopt Chinese calendar, 198
  Roman, 71n.78
  San-t'ung li, 48
  solar, 202, 203, 360
  Ssu-fen li, 47, 48, 70
  Ta-ming li, 76
  Ta-t'ung li, 123, 151, 256
  Ta-yen li, 49, 51
  T'ai-chu li, 48
  Ts'ui Hao's, 50
  Western, Japanese adoption of, 360
  Wu-chi li, 51
  Yüan-chia li, 51
Calendrical astronomy course (*rekidō* 暦道), 35, 38
Calendrical parameters, 258
Calendrical reforms, 48, 52, 73, 223, 224, 251–252, 254–258, 262, 275, 291, 295, 311, 313, 348, 349, 350–351, 353–360, 368
  Kansei, 337
  Tenpō, 337
  using Western knowledge, 356, 357
  Western methods in, 313, 314, 315
Calendrical research, 251
Calendrical science, xxvii, xxviii, 7, 8, 31, 35, 38, 43, 46, 53, 73, 74, 88, 117, 123, 126, 133, 198–200, 251, 252, 255, 275, 286, 298, 299, 300, 311, 348–351, 360
  Japanese failure to engage in research in, 73, 198
  and mathematics, 263, 275, 370
Callippus (fl. 370–332 B.C.), 47
Camphor, 226
Cannonball wounds, 221
Cannons, 107, 169, 172–173, 174, 232

breech-loading, 232
muzzle-loading, 173
Cape of Good Hope, 157
Capital city construction, 26–27, 29
Cardiac functions, and Five Phases theory, 212
Caron, François (1600–1675), 232
Caspar school (*Kasuparuryū* カスパル流), 285–286
Castle construction, 168, 203, 207
Castle towns. *See* Towns
"Cathartic school" (*Kung-hsia p'ai* 攻下派; *kōgeha* in Japanese), 212
Catholicism. *See* Roman Catholicism
Cavalry, 172
Cayenne pepper, 282
Celestial globe, 122
Center for Western Learning (Kaiseisho 開成所), 301, 302, 337, 359, 396–397
Center for Western Medicine (Seiyō igakusho 西洋醫學所), 336, 378, 394, 398
Centrifugal force, 353
Centripetal force, 353
Ceremonies. *See* Rites and ceremonies
Cha-ma-lu-ting. *See* Jamal al-Din
Ch'an 禪 sect. *See* Zen Buddhism
Chang. *See* Nineteen-year period
Chang 張 school of medicine, 139, 209, 277, 278, 279–280, 374
Ch'ang-an 長安, 17, 26
Chang Ch'iu-chien suan-ching 張邱建算經 (Mathematical manual of Chang Ch'iu-chien), 76
Chang Chung-ching 張仲景 (Chinese, med., early 3rd century), 85, 281, 374
Chang Ts'ung-cheng 張從正; adult name: Tzu-ho 子和; literary name: Tai-jen 戴人 (Chinese, med., 1156–1228), 212, 374
Chart of the human body, 45
Chemical research, 391

Chemical substances, 90
Chemistry, xxviii, 183, 305, 394, 396, 397
  terminology, 306
Ch'eng Ta-wei 程大位 (Chinese, math., 1533–ca.1592), 204, 265
Ch'i 氣 (ki in Japanese; yang energy), 61, 95, 212, 240, 259, 260, 261, 281, 282, 304
Ch'i-ku suan-ching 緝古算經 (Continuation of ancient mathematics), 76, 83
Ch'i-yao jang-tsai chueh 七曜攘災訣 (Formulas for avoiding calamities according to the seven luminaries), 69, 125
Ch'ien-hsiang li 乾象曆 calendar, 48
Chia-i ching 甲乙經 (Classic on acupuncture and moxibustion), 85
Chiba Tanehide 千葉胤秀 (math., fl. 1830), 272
Chih Ts'ung 知聰 (Chinese, med., fl. 562), 44
Chijiwa 千々岩, 189
Chikamatsu Monzaemon 近松門左衛門 (playwright, 1653–1724), 223
Childbirth, 215
Chin 晉 dynasty (280–420), 4, 16
Chin 金 dynasty (1126–1234), 104–105, 108, 139, 209
  medicine, 139, 209, 211–214, 277, 278, 280–283
Ch'in Chiu-shao 秦九詔 (Chinese, math., 1202–1261), 129, 130
Chin Chü-ch'a 金俱吒 (Buddhist astronomer, early 9th century), 69
Chin-kuei yao-lueh 金匱要略 (Essential [prescriptions] of the Golden Casket), 374
Chin shu 晉書 (History of the Chin dynasty), 54, 55, 79
China
  Buddhism in. *See* Buddhism
  cosmology. *See* Chinese cosmology
  belief in cultural superiority of, 149
  first emergence as a major seafaring power, 104
  gunpowder, 170
  hostility to things foreign, 149
  printing, 184
  relations with Japan, 73–74, 248
  smallpox vaccine discovered, 377
Chinese aristocrats, and political power, 106–107
Chinese astrology, 46, 47, 50, 55, 71, 73, 77, 198, 215
Chinese astronomical bureau
  Muslims in, 123
  in T'ang period. *See* Tai-shih chu
Chinese astronomy, xvii, 43, 46, 47, 50, 53, 71, 73, 77, 121, 125, 198, 215
Chinese books
  imported by Japan, 17–18, 156, 226, 236
  restrictions on import of, 261
Chinese calendars, 48, 51, 69, 70, 73, 123, 128
  adopted by the Japanese, 51–52, 74
  and dynastic legitimacy, 72, 121–122
Chinese characters, 29, 32, 113, 146
Chinese classical texts, reprinting of, 107–108
  in Japan, 185
Chinese cosmology, xiii, 53–54, 199, 251
Chinese cultural influences on ancient Japan, xviii, xxi–xxii, xxiv
  in ancient times, 1–6, 10–11, 14, 26, 29
  decline of, 28
Chinese emperor
  and portent astrology, 54
  and the production of the calendar, 52
  supremacy of, 6, 106–107
Chinese language, 21

448 Index

Chinese language (*continued*)
　pronunciation of, in ancient Japan, 32
Chinese learning, 114, 300–301
　challenge of Western learning to, 297
　decline of Japanese interest in, 114, 119
　in Ming dynasty, 109
　and the *ritsuryō* system, 29–30
　in the transitional period, 186–187
Chinese literature course. *See* Literature course
Chinese mathematics, xiii, 7, 74–84, 129–132, 135, 204, 265, 271, 274–275
　books on, brought to Japan, 133, 226, 231n.16
　essentially utilitarian, 78, 79, 82, 132, 204, 274–275
　government administration and, 74, 77
　Islamic influence on, 131n.36
　Japanese adoption of, 82–83, 132, 204, 215
Chinese medicine, xxviii, 7, 8, 36, 44, 77, 84–100, 138–140, 143, 209–213, 220, 223, 277, 279–281, 288, 323, 378, 381, 391, 394
　compared with early Japanese, 88, 100
　functional approach, 85, 379
　pain-killing drugs in, 388
　Taoist influence on, 87
　transmissions from India, 91
　transmission to Japan, 87, 88, 90, 99–100, 139–140, 141, 143, 146, 213–215, 217
　and *Wahō*, 309
　weaknesses of, 98–99
Chinese science, xiv–xvi, xxi, xxiv, xxvii, xxx, 9, 30, 45–46, 109, 187
　in Japan, xxvii, xxix, 27, 45, 121, 347–348
　decline of Japanese interest in, 28, 29, 114
　revival of Japanese study of, 103, 119, 121, 148, 187, 196–197, 204, 215, 223–224, 234, 249, 251, 277–278
Chinese ships, visiting Japan, 19, 28
Chinese technology, 169, 170, 231
Chinese traders, in Nagasaki, 224–225
Ch'ing 清 dynasty (1644–1912), 69, 236–237, 241
Ching-ch'u li 景初曆 calendar, 48
Ching mo 經脈 (main channels in human body), 95
Chinjufu 鎮守府 (military garrison), 124
Chirurgie (Heister), 329
Chiu-chang suan-shu 九章算術 (Nine chapters on mathematical art), 51, 75, 76, 77, 79, 82, 129, 130, 135, 207
Chiu-chih li 九執曆 (Calendrical system of the nine upholders), 49, 76
Chō 町 (area unit: approx. 2.45 acres), 112
Chōkei Senmyōreki sanpō 長慶宣明曆算法, 253
Chōshū 長州 domain, 382, 397, 399
Chōtei (or *Jūtei*) *kaitai shinsho* 重訂解體新書 revised edition of *Kaitai shinsho*), 330
Chou 州 (prefecture), 101
Chou 周 period (11th century B.C.–722 B.C.), 52, 127
Chou i 周易. *See I-ching*
Chou li 周禮 (Rituals of Chou), 8, 78
Chou-pi suan-ching 周髀算經 (Arithmetical classical of the Chou gnomon), 47, 50, 53, 54, 75
Chōzendō 超然堂 (Dutch Learning academy), 335
Christian Japanese priests, 188, 189
Christian realms, Chinese intercourse with, 109
Christianity, 158–160, 308, 312

Index   449

Christianity (*continued*)
  converts to, 158, 160, 195, 200, 201, 215
  edicts against, 161, 162, 167
  intolerance toward, 189, 258
  prohibited, 161, 163
Chronometer, 176, 341
Chu 朱 school of medicine, 139, 209, 213, 214, 215, 216, 250, 277, 278, 279
Chu Cheng-heng 朱震亨; adult name: Yüan-hsiu 元修; literary name: Tan-chi 丹溪 (Chinese, med., 1281–1358), 213
Chü-fang fa-hui 局方發揮 (Elucidation of the Official Prescriptions), 213
Chü-fang-p'ai 局方派 (Official Prescriptions school), 139
Chu Hsi 朱熹 (Chinese philosopher, 1130–1200), 108, 150, 167, 235, 240–241, 242, 243, 244, 247, 303
  cosmology of, 298
Chu Hsi synthesis, 296–299, 303
Chu Shih-chieh 朱世傑 (Chinese, math., Yüan era), 129, 131, 204, 264, 266, 271
Chu Shun-shui 朱舜水 (Chinese, Conf., 1600–1682), 237n.20
Chu-t'an Hsi-ta 瞿曇悉達 (Indian, cal., fl. 718), 49
Chu-t'an Lo 瞿曇羅 (Indian, cal.), 49
Chüan-t'i hsin-lun 全體新論 (New treatise on the human body), 385n.108
Chui-shu 綴術 (Coupling Method), 76, 83, 129
Ch'ung-chen li-shu 崇禎曆書 (Astronomical treatises of the Ch'ung-chen era), 368
Cinnabar, 90
Circle, 271
  area of, 362
  circumference, 271, 274
"Circular principles" (*enri* 圓理),

271, 274, 362, 366
Civil engineering. *See* Engineering
Civil service examination system, xiv, 6, 11, 33, 107, 149
Civil wars
  in China, 19
  in Japan, 148, 152, 159, 161
  wounds received in, 214
Clan leaders, 5, 6, 12, 13
  and irrigation projects, 25
Clavius, Christopher (fl. 1607), 259
Clocks
  astronomical, 350
  pendulum 356
  surveyor's, 357
  water, 50
Cloud formations, 53
Coefficients
  for solving algebraic unknowns, 129
  "heaven-origin," 131, 264
  numerical, 81, 82
Colonies and colonization, 156, 160, 292, 293, 395
"Combat medicine," 147, 220–221
Combat surgeons (*kinsōi* 金創醫), 214–215, 287
Comets, 53
Commerce, xiv, 7, 103, 148
  Chinese, 107
  control of, 164
Commercial households, overseas trading of, 155
Commercial printing, 184, 185, 186
Commodity-exchange economy, 112, 152, 163, 164
Commodity production, 163
Commoners
  education of, 275–276, 295
  medical care of, 101–102
Communications systems, 398
Compass. *See* Magnetic compass
Compass bearing, 175n.22
Compass card, 175n.22
Computational techniques, 134, 252, 298, 357, 361
  Western, 313, 315, 357
Confucian appendices to the

I-ching, 127
Confucian learning, 311
Confucian scholars and teachers, 223, 229
　secular, 242, 289
　and *wasan*, 364
Confucianism xxi, xxiv, 1, 5, 6, 7, 8, 9, 10, 11, 28, 30, 46, 47, 108, 109, 114, 135, 150, 158, 167, 185, 223–224, 235, 237, 288, 295, 296–297, 298, 395
　Buddhism and, 241, 242
　dissection of the human body and, 379
　government officials and, 39, 109, 110
　heliocentric theory and, 354
　and mathematics, 263, 274, 301–302
　and medicine, 84, 90, 213, 215, 251, 277–278, 283, 302–304, 374
　National Learning and, 307, 308
　samurai and, 229
　in the seventeenth century, 223, 239–240, 243, 245, 246, 247, 248, 249–251,
　shogunate's support of, 226
　study of, 32, 33, 36, 39, 41, 44
　unorthodox, 244
Construction projects, 33, 42, 44, 79, 83, 133
"Cooling medication school" (*Han-liang p'ai* 寒涼派; *kanryōha* in Japanese), 212
Copernicus (d. 1543) and the Copernican system, xii, xv, 171, 201, 221, 259, 313, 351–352
Copper, 152, 181, 183, 225, 226, 233
Copyright, 186
Corpses
　dissections of. *See* Dissections
　human bile collected from, 380
　observations of, 379, 380
Correspondence between celestial phenomena and human affairs, 298
Correspondences, system of, 85, 86
　yin-yang, 92
　Five phases, 93–94
Cosmological models, 54
Cosmology, 43, 47, 53, 202, 224, 240, 251, 258, 264, 299, 300, 314, 351–352
　Buddhist. *See* Buddhist cosmology
　Chinese. *See* Chinese cosmology
　Copernican. *See* Copernicus
　Greek, 54
　Indian. *See* Indian cosmology
　Ptolemaic. *See* Ptolemy
　Western. *See* Western cosmology
Council of Trent, 157
Counter-Reformation (European), 156, 157
Counting board (*sanban* 算盤), 72, 79, 80, 132, 205, 264
　arrangement of four unknowns on, 131, 271
　defects of, 266–267
　as inhibiting factor in development of abstract mathematics, 275
Counting rods, 79, 267
Cowpox scab, 378
Crafts and craftsmen, 26–27, 163, 168, 207
　low-ranking samurai as, 294
Crane, 232
Criminals
　executions of, 380
　post-mortem dissections of, 317, 322, 380, 382
Critical linguistic and textual study (*kobunjigaku* 古文辭學), 248
Crusades, 175
Cube roots, 74, 75, 81, 132, 206, 208
Cultural contacts
　Chinese-Japanese, xviii, xxi–xxii, xxiv, 1–6, 11, 50, 73, 106, 121, 122, 135, 168–169, 187, 224, 236–237
　Korean-Japanese, xviii, 5, 25
　Indian-Chinese, 49–50, 69, 76–

77, 91
  Indian-Japanese, 69
  Japanese-Western 179–180
Cupellation, 183
Curricula
  in Jesuit college, 194
  in Jesuit primary schools, 188
  in Jesuit seminaries, 189, 194
  in *ritsuryō* academic program, 32–34, 35, 36, 38, 39, 41, 83
  of the Shugei shuchiin (academy), 41
  in *terakoya*, 276
Curriculum reform (University)
  A.D. 728, 34, 39
  A.D. 808, 34, 83
  A.D. 834, 34
Curses, 114
Cyclic variation. *See* Law of variation
Cylinders problem, 272

Da Nang, Vietnam, 155
Dai Nihonshi 大日本史 (History of Great Japan), 238
Daidō ruijūhō 大同類聚方 (Classified prescriptions [collected] in the Daidō era), 88, 100, 309
Daigaku nankō 大學南校, 359
Daigakuryō 大學寮 (University), 31, 32–34, 39–40, 41, 74, 84, 118, 123
  alternatives to, 115–117
  curriculum, 32–34
  decline of, 115, 116, 117
  destroyed by fire, 115, 118, 124
Daigakuryō bessō 大學寮別曹 (University affiliate schools), 40, 115, 118
Daisokuhyō 大測表 (Tables of large measurements), 371
Dajōkan 太政官 (Grand Council of State), 12
Dangerous directions, 56, 66
Darani 陀羅尼 (printed charms), 22, 184n.36
Date Masamune 伊達政宗 (feudal lord, 1567–1636), 177

Datura alba (white datura), 388
Day and night, lengths of, 262
Day-length, seasonal differences in, 259
Dazaifu 大宰府, 124
  regional college at, 38
De fabrica corporis humani (Vesalius), 221, 381
De motu cordis (Harvey), 221
De sphaera (Gomez), 260
Decimals, 80
Declination, 107
  tables of, 176
Deerskins, 226
Defense. *See* National defense
Definite integrals, 362
  tables of, 362
Dejima 出島, 225, 226, 227, 237, 239, 262, 292, 338
  Dutch head of, 311
  medical officers on, 228, 238, 285, 286–287, 288, 315, 316, 340
  Siebold confined on, 343
D'Elia, Fr., xvii
Demons and demonic possession, 97, 99, 114
Dental hygiene, 214, 217
Descartes, René (1596–1650), 364, 370
Determinants, to solve equations with two or more unknowns, 271
Diagnosis, 85, 97, 281–282, 332, 394
Dias, Manoel (Portuguese Jesuit, 1574–1648), 261
Dictionaries, 331, 383
  Dutch-French, 332
  Dutch-Japanese, 332, 333
  Dutch-Latin, 333n.38
  English-Japanese, 333
  Japanese-English, first printed, 333n.40
  Japanese-Portuguese, 194
  Latin-Portuguese-Japanese, 194
Differential calculus, 361
Digestion and the digestive system, 95, 391

452   Index

Dioscorides (1st century A.D.), 87
Diplomatic missions and envoys, 4,
    11, 16, 44, 106, 119, 226n.5
  from Korea, 242
  to Ming court. *See* Tally Trade
  to Sui court. *See* Kenzuishi
  to T'ang court. *See* Kentōshi
Diseases, 221
  causes of, 43–44, 97, 142–143,
    211–212, 374, 394, 395
  classification of, 146, 281
  cure of, 97, 212, 220, 381
  treatment of. *See* Medical
    treatments
Dislocations, 221
Disputations between Jesuit and
    Japanese scholars, 208
"Dissection booms," 317, 323, 383,
    391
Dissections. *See also* Vivisection
  animal, 378, 379, 381
  of condemned criminals, 317,
    322, 380, 382
  human, 96, 317, 322, 323, 376,
    378–381
  official permission for, 380, 382
  performed by outcasts, 380, 382
  research based on, 381
Distance, measurement of, 75–76
Divination, 7, 10, 31, 35, 44, 49, 53,
    56, 61, 77, 144. *See also*
    Onmyōdō
  Buddhist, 125, 126
  calendrical, 56, 74, 124
  and the *I-ching*, 127
  on matters of national interest,
    124
"Divine Cultivator's classic on
    materia medica." *See* Shen-
    nung pen-ts'ao ching
Division, 74, 75, 76, 79, 206
  on the abacus, 205
Dōbutsugaku 動物學 (zoology), 306
Dodoens, Rembert (Flemish
    botanist, 1517–1585), 315
Doeff, Hendrik (head of Dejima
    company, 1803–1817), 333,
    334
Dōkaisho 童介抄, 265

Domain schools. *See* Schools (fief)
Dominican missionaries, 157, 219
Drainage, 183, 234
Drugs, xxviii, 44, 87, 91, 97–98, 99,
    100, 146–147, 156, 213, 220,
    280, 283, 374, 387, 393
  animal substances used in, 85
  body's resistance to, 212
  cathartic, 212
  Chinese, 143, 218, 226, 395
  dangerous use of, 212
  dosage, 220
  drying, 213
  used in Dutch medical practice,
    385
  emetic, 212
  for general anesthesia, 388
  for long life and immortality. *See*
    Elixirs of longevity and
    immortality
  ophthalmological, 343
  pain-killing, 388
  primitive, 146
  standardized, 138
  sudorific, 212
  Taoist, 140
Dutch, 155, 157, 160–161, 174, 178,
    237, 239, 262, 292, 349
  social restrictions on, 226–227,
    237
Dutch astronomy, 300, 314, 349,
    351, 357, 359
Dutch-Chinese eclectic medicine,
    383, 387–391, 393
Dutch colonies, 338
Dutch East India Company, 338
Dutch East Indies, 343
Dutch interpreters, 284. *See also*
    Nagasaki, interpreters
Dutch-Japanese relations, 226–228,
    284
Dutch-Japanese trade. *See* Overseas
    trade
Dutch language, 237, 284, 291, 315,
    316, 317, 322, 357, 371, 398
  study of, 350, 372, 383, 393
  teaching of, 316, 334
"Dutch Learning" (*Rangaku* 蘭學),
    228, 278, 284, 291, 295, 299,

Index   453

301, 303, 314, 315–316, 333, 334, 340, 344–345, 347, 360, 372, 386
  political implications of, 344
  repressive measures against, 337, 347
  scholars, mobilized for defense, 396
  shogunate policy toward, 336–337, 338, 394
Dutch Learning academies (*Rangakujuku* 蘭學塾), 334–336
Dutch medical studies, 384, 393
"Dutch Medicine" (*Ranpō* 蘭方), 219, 298, 303, 316, 323, 336, 347, 372–373, 376, 388, 391
  instruction in, 373
Dutch navigational table, 371
Dutch surgery. See *Kōmō geka*
"Dutch" pharmacopeia, 385
Dutch traders, 178, 219, 224, 356
Dutch word lists, 316
Dynamics, xxix, 173–174, 201, 232

E 兄 (*yang* aspect), 57
the Earth
  as a member of the solar system, 352
  position of, 54, 201
  revolution of, round the sun, 70
  rotation of, 69
  shape of, 53, 54, 200
  sphericity of, 178, 200, 201, 260, 298, 299, 300
"Earth-supplementing school" (*Pu-t'u p'ai* 補土派; *hodoha* in Japanese), 213
Earthquakes, 53, 200
East Asia during Chinese Culture Wave I, map of, xxxv
East Asian Scientific Revolution, xvi
East China Sea, 19, 175
East India Companies, 157
Eclectic School of Confucianism, 375
Eclectic School of Medicine (*Setchūha* 折衷派), 278, 302, 375, 376
Eclecticism, 295, 296, 298, 299, 302, 372, 373
Eclipse meter, 356
Eclipses, 7, 38, 48, 53, 59, 71, 200, 252, 260
  lunar, 55, 255, 256, 300, 351
  solar, 49, 55, 72, 126, 254, 255, 260
  prediction of, 72, 73, 74, 126, 128, 133, 252, 255, 256, 257, 258, 260, 351
Ecliptic, determination of, 262
Edo 江戸 period (1603–1868), 64, 99, 136, 138, 165, 166, 225, 227, 228, 229, 230, 235, 237, 280, 299, 303, 334. *See also* Tokugawa era
  mathematics, 266, 274, 302, 364, 366–367
Edo 江戸 (national capital), xxiii, 164, 230, 255, 256, 258, 293, 346, 358, 378
  annual visits of Dejima doctors and factory heads to, 238–239, 287, 315, 340, 353, 375
  astronomical observatory, 312
  detention house for Christians, 310
  Dutch study boom at, 383
  Hayashi academy, 243–244
  Lecture Hall for Japanese Learning, 308
  private academies for Dutch Learning, 334, 335, 336
  private academy for National Learning, 307
  Seijukan private academy (medicine), 302, 323, 375
  Siebold's trip to, 341
Edo Bay, 292, 342, 345, 395
Edo haruma 江戸波留麻 (The Edo "Halma"), 332, 333, 336, 383
Education
  Buddhism and, 41, 119
  of children of the Japanese court

nobility, 134, 135
of commoners, 275–276, 295
of girls, 185
Jesuits and. *See* Jesuit educational institutions
in medicine. *See* Medical training
under the *ritsuryō* government, 18–19, 31–42
of samurai, 166, 188, 224, 295
of sons of samurai, 134, 135
in the Tokugawa era, 295
Egawa Hidetatsu 江川英龍, or Tarōzaemon 太郎左衛門 (shogunate official, 1801–1855), 344–345, 346
Egg-yolk model. *See* "Hen-egg" theory
Egyptians, 175
Eisai 榮西 (Buddhist priest, 1141–1215), 143
Eki 易 divination, 127–128
Electric telegraphy, 398
Elements (Euclid), 202, 208, 361, 370
Elixirs of longevity and immortality (tan 丹), 87, 90, 98
techniques for refining, 90
Ellipses, 352, 353, 362
Elliptic, inclination of, 355
Empirical investigation, 99, 261–262, 303, 373, 378, 379, 394
Enami Kazusumi 榎並和澄 (math., fl. 1650s), 253, 265
Enchiridon medicum (Hufeland), 386
Energetic circulation system, 96
Engetsu 圓月 (Buddhist priest, 1300–1375), 135
Engineering, xiv
civil, 24, 133, 168, 206
Engishiki 延喜式 (Procedures of the Engi era; an elaboration of the Yōrō Code of 718), 110, 147
England, 347
Treaty of amity (1854), 395
English language, 398
grammar, 333
English pilots, 178

English traders, 157, 160–161, 219
Enichi 惠日 (Buddhist priest, fl. 608), 87
Ennin 圓仁 (Buddhist abbot, 794–864), 42
Enpō shikanki 圓方四卷記, 265
Enri. See "Circular principles"
Enri kenkon no maki 圓理乾坤之卷 (Monograph on circular principles), 362
Enri shohyō 圓理諸表 (Tables of circular principles), 362
Enryakuji 延暦寺 (monastic center), 42
Entsū 圓通 (Buddhist monk, 1754–1834), 301, 354
Environment
and the human organism, 394
and causes of disease, 394, 395
Envoys. *See* Diplomatic missions and envoys
Ephemerides, 49, 72, 122, 252, 351
Equations
algebraic, 81, 82, 264
biquadratic, 131
cubic, 74, 76
of degrees higher than the second, 129–130, 266
determinate, 130
linear, 131
numerical, 130
quadratic, 75, 76
simple, 266
of the tenth degree, 129
with several unknowns, 131, 267, 271
Equator, determination of, 262
Equinoxes, 47, 59, 122
Era names (*nengō* 年號), 63
changes of, 63–64
Eta 穢多 (outcast group), 380
Ethics, 108, 159
Etiology, 97, 281
Euclid, 78, 82, 202, 208, 361, 370
Europe, 149. *See also* topics under Western
use of gunpowder in, 170–171

Europe (*continued*)
  modernization in, 156
  and the Mongol threat, 105
  printing, 185
  trade, 226
  transmission of Japanese medical techniques to, 287
European doctors, 285, 286–287, 288
European traders and mariners, 197, 221
  medical care of, 219
European trading companies, 155
Evil spirits, 114
Examinations
  civil service, xiv, 6, 11, 33
  and social class, 40
  University entrance, 39
Excavation techniques, 168
Excretion, 95, 212
Exorcism, 90, 91, 94, 98, 99, 100, 114, 120, 125, 141
Exorcistic incantations, 30n.36, 36, 44, 87
Experimentation, in medicine, 221, 391
Exploration, 157, 176
Explosives, 107, 232
Exports, 226
  to China, 104n.2, 225
  to Southeast Asia and Europe, 155n.10
Eye diseases. *See* Ophthalmology

Fabian, Fucan 吧毗庵不干 (Japanese Jesuit, fl. 1605), 201
Family system, 41
Fang-shih 方士 (magicians and medical practitioners), 90
Farming and farmers, xxiii, 7, 13, 20, 43, 112, 163, 164, 168, 229
  Tenant, 294
Fate calculation. *See* Prognostication
Fernandes, Juan (lay brother, 1525/6–1587), 194
Ferreira, Christavão. *See* Sawano Chūan
Fetus, position of, in the womb, 388, 391, 392
Feudal system, xxiii, 103, 109, 110
  abolition of, 399
  and Chu Hsi, 243
  erosion of, 294, 299, 308
  learning and, 236
Fief schools (*hankō*). *See* Schools
Filial piety, 41, 379
Finance, and mathematics, 79, 83, 133
Financial crisis, 293
Firearms, 168, 169, 170–174
Fireboats, 232
Fireworks, 107
Fishing, 43, 168
Five Circuit Phases (*wu-yün* 五運), 86, 211
Five Classics, 8, 149
Five Dynasties period (907–960), 19, 107, 143
Five Phases (*wu-hsing* 五行), 7, 8, 9, 46, 55–57, 61–62, 77, 240, 259, 260
  combined with planets and colors, 65
  Japanese names of, *ki* 木 (Wood), *hi* 火 (Fire), *tsuchi* 土 (Earth), *ka(ne)* 金 (Metal), *mizu* 水 (Water), 57
  correspondences, 93–94
  lunar eclipses and, 300
  medicine and, 84, 85, 91–94, 95, 97, 210–212, 280, 281, 283, 373
"Five Viscera" (*wu-tsang* 五臟), 95, 144, 379
Floods, 53
Flower-arranging, 224, 231, 364
Folk medicine, 98
  folkloric pathology, 43
Folklore, 43, 127
Food, fermenting of, 95
Foot soldiers, 172
Force, concept of, 173, 253
Foreign policy
  criticism of, 345, 346

Foreign policy (*continued*)
  of Hideyoshi, 164
Foreign trade. *See* Overseas trade
Foreigners, expulsion of, 161, 174, 228
Forging, 169, 172
Former Han dynasty, 2, 55
Fortnightly periods. *See* Twenty-four Fortnightly Periods
Fortune-telling. *See* Prognostication
Foundling home, 218
Four Books, 149
Four elements, xvi, 87, 260
  and Buddhist thought, 142, 143
"Four unknowns," 131
Fractions, 74, 80
Fractures, compound, 388
Franciscan missionaries, 157, 219
French language, 398
Fu-t'ien li 符天曆 calendar, 125
Fugaku 府學 (regional college), 38, 101
Fujibayashi Fuzan 藤林普山 (med., 1781–1836), 333
Fujita Sadatsugu 藤田貞資 (math., 1734–1804), 363, 364–365, 366, 369
Fujiwara 藤原 clan, 39, 102, 115, 116
Fujiwara Seika 藤原惺窩 (Conf., 1561–1619), 241, 242, 243
Fujiwara Shunzei 藤原俊成 (poet, 1114–1204), 248–249
Fujiwara Teika 藤原定家 (poet, 1162–1241), 249
Fujiwarakyō 藤原京, 26
Fukane Sukehito 深根輔仁 (med., fl. 920), 147
Fukudai 伏題 ("covered problem"), 271
Fukudenhō 福田方 (Prescriptions gathered for [Buddha's] blessing), 146
Fukui 福井, 378
Fukuin 福因 (Buddhist priest, fl. 608), 87
Fukyū sanpō 不朽算法 (Immortal mathematical method), 362

Funai 府內, Bungo 豊後 district, 171, 172
  foundling home and hospital at, 218
  invasion of, 189, 219
  Jesuit school at, 188
"Functional orbs," 95, 96–97, 379, 391
Funeral services, 102
Furubayashi Kengi 古林見宜 (med., 1597–1657), 279
Furukawa Jirōemon 古河治郎衛門 (compiler, gunnery), 232
Fuseya Soteki 伏屋素狄 (med., 1747–1811), 391
Fushi keiken ikun 扶氏經驗遺訓 (Lessons from Mr. Fu's experience), 386
Fuyukage 冬景 (son of Kajiwara Shōzen), 146

Gakkan'in 學館院 (University affiliate school), 116
Galen and Galenic doctrines, xv, xvi, 87, 218, 221, 381
Galileo, G., xv, 173, 201, 304
Galileo in China (D'Elia), xvii
Gall bladder, 95, 97
Gama 蒲 (bulrush), 43
Games and puzzles, 77, 78, 135, 136, 206
Ganjin 鑑眞 (Buddhist priest), 17
Genbyō yakusetsu 原病約說 (Summary of pathology), 385
Genji 源氏 (Minamoto 源) clan, 111
Genji monogatari 源氏物語 (The Tale of Genji), 113
Genna 元和 era (1615–1624), 220
Genna kōkaisho 元和航海書 (Navigation treatise of the Genna era), 178–179, 180, 203
Genshō 元正, Empress (r. 715–724), 55
Geocentric theory, 201, 353, 358
Geographical measurements, for

determination of the meridian, the equator, etc., 262
Geography, 45, 178, 310, 311, 396
Geomancy, 9
Geometric models, 54, 200, 258
Geometrical figures, 82
Geometry, 74, 78, 82, 305, 370
 analytical, 173
 spherical, 176
Geophysical abnormalities, 36
German language, 398
Gimbals, 175
Girls, education of, 188
Glass, 181
Gnomon, 47, 50, 258, 356
 shadow lengths, 250
Go 碁 (game), 255
God of the Year's Virtue (Saitokujin 歳德神), 57
Gokurakuji 極樂寺 (Paradise Temple), Kamakura, 141
Gold, 226, 233
 artificial, 90
 mining, 181, 182
 refining of, 183
 trade, 181
Gomez, Pedro (Spanish missionary, 1535–1600), 260
Gonda Naosuke 權田直助 (med., 1809–1887), 309
Gonzalvez, Manuel (Spanish ship captain), 179
Gorter, Johannes de (fl. 1731), 329, 385n.109
Goseihōha 後世方派 (name for the Li-Chu 李·朱 medical tradition), 279, 280, 281, 282, 285, 297–298, 302, 373, 374, 375, 378, 391
Goseihō beppa 後世方別派 (name for the Liu-Chang 劉·張 medical tradition), 280, 282, 373, 374, 375, 378
Gotō Boan 後藤慕庵 (med., 18th century), 374
Gotō Gonzan 後藤艮山 (med., 1659–1733), 282–283, 374, 378
Gotō Nashiharu 後藤梨春 (writer, 1696–1771), 323
Government administration
 and advanced mathematics, 83
 centralized, 164
Government office and officials
 and Confucianism, 39
 and portent astrology, 55
 ranks and responsibilities of scholars and scientists, 8
 student-monks and, 42
 training of, 31, 32, 39
Government posts
 monopolization of, by Japanese nobility, 28, 40, 110, 115
Government promotion of learning, 229, 296
Government policy, criticism of, 345
Government regulations
 and the publication of translated works, 331
 and Siebold's information-gathering, 341–343
Gozan 五山 (temple complexes), 120, 121, 127, 184, 239, 242
Grammars, 331, 332
 English, 333
 Japanese, 194–195
Grand Polarity (*T'ai-chi* 太極), 240
Grand Shrine, Ise 伊勢, 257
Gratama, W. K. (fl. 1866), 397
Gratianus, Justus, 382
Gravitation, 353
 center of, 371
"Greater Heat" (*Ta-shu* 大暑), 62
Greek astronomy, 69
Greek calendars, 70
Greek cosmology, 54
Greek mathematics, 49, 275
Greek mechanics, 173
Greek medicine, xv, xvi, 86
Greeks, 175
Gregorian calendar, 69, 123, 360
Grenades, 107
Gronbeginsels der Steerekunde (Steenstra), 359

458　Index

Guchūreki. *See* Almanac
Gun casting, 396
Gunkan kyōjusho. *See* Warship training center
Gunnery, xxix, 173, 183, 232, 311, 397
Gunpowder, 107, 109n.7, 160, 170–171, 175
Gunshot wounds, 221, 377
Gynecology, 147, 217, 386, 393

Habu Genseki 土生玄碩 (med., 1762–1848), 341, 343
Hagi 萩, post-mortem dissection at, 382
Hai-chin 海禁 (maritime prohibitions), 106
Hai-tao suan-ching 海島算經 (Sea island mathematical manual), 75
Hai yaso 排耶蘇 (Refutation of Jesus), 244
Hakase 博士 (doctor [academic]), 35
Hakata 博多, 163
Halma, François (1652–1722), 332, 333
Han 漢 dynasty, 2, 6, 8, 9, 12, 20, 45, 77, 127, 239, 241
　astrology and astronomy, 47, 55, 56
　collapse of, 24
　mathematics, 74, 75, 79
　medicine, 84, 90
Han shu 漢書 (History of the Han dynasty), 55, 379
Hanaoka school of surgery, 388
Hanaoka Seishū 華岡青洲 (med., 1760–1835), 387–388, 389, 390
Hanawa Hokiichi 塙保己一 (National Learning, 1746–1821), 308
Handa Jun'an 半田順庵 (med., fl. early 17th century), 220
Hankō. *See* Schools (fief)
Hara Shōan 原松庵 (med., fl. 1754), 380

Hari 針 (acupuncture) course, 37
Harmonics, 50
Haruma wage 波留麻和解 (Japanese version of Halma), 332. *See also Edo haruma*
Harvey, William (1578–1657), 221
Hasegawa Hiroshi 長谷川寛 (math., 1782–1838), 272, 366, 369
Hasekura Tsunenaga 支倉常長 (1571–1622), 177
Hashimoto Sōkichi 橋本宗吉 (Dutch Learning, fl. 1801), 335
Hata Kōzan 畑黄山 (med., 1721–1804), 374
Hatsubi sanpō 發微算法 (Detailed explanation of mathematical method), 265, 271
Hatsubi sanpō endan genkai 發微算法演段諺解 (Detailed explanation of calculations in the *Hatsubi sanpō*), 268, 271
Hatsusaka Shigeharu 初坂重春 (math., fl. 1657), 265
Hattori Nankaku 服部南郭 (1684–1759), 248n.29
Hayashi academy. *See* Kōbun'in
Hayashi 林 family, 295
Hayashi Dōkai 林洞海 (med., 1813–1895), 385
Hayashi Jussai 林述齋 (head of the Shōheikō, 1776–1841), 345
Hayashi Kichizaemon 林吉左衛門 (?–1616), 260
Hayashi Nobutaka 林信敬 (head of the Shōheikō, fl. 1790), 298
Hayashi Razan 林羅山 (Conf., 1583–1657), 201, 234–235, 238, 241, 242–243, 244, 245, 246, 247, 282, 289, 299, 306
　academy established by, 243
Hayashi Tsuruichi 林鶴一 (math., 1873–1935), 369
Hazama Shigetomi 間重富 (astron., 1756–1816), 314, 315, 337, 355, 356, 357, 359

Head, 381
Health care. *See* Hygiene
Health services, 91
Heart, 61, 93, 95
  function of the, 96, 221
Heaven, concept of, 7
"Heaven-origin algebraic method" (*t'ien-yüan shu* 天元術; *tengenjutsu* in Japanese), 129, 130, 131, 204, 264, 266
  defects of, 266
  the Heavens, shape of, 53, 54, 200
Heaven's mandate, and imperial rule, 47, 52, 72
Heelkundige anderwyzingen, 329
Heian 平安 era (794–1192), xxvi, 27, 28, 32, 67, 112, 136, 141
  reforms, 1
Heiankyō. *See* Kyoto
Heights, measurement of, 75–76
Heijōkyō. *See* Nara
Heike 平家 (Taira 平) clan, 111
Heike monogatari 平家物語 (Tale of the House of Taira), 185
Heister, Laurens (fl. 1718), 329
Heizei 平城, Emperor (r. 806–809), 88
Heliocentric theory, 201, 266, 298, 313, 352, 353, 354, 355
Hellenistic astrology, 125
Hemerological annotations, 257, 359, 360
"Hen cgg" theory, 54, 260
Henry the Navigator, 157
Herbs, used in medicine, 36, 85
Hereditary professorships, 116–117, 119, 132
Hereditary right to rule, and education, 237
Hereditary succession to socio-economic roles, 167
Heterodoxy, measures to eliminate, 298, 302
Hi "*tenkyō wakumon*" 非天經或問 (Against the *T'ien-ching huo-wen*), 354
Hi "*zōshi*" 非臟志 (Against "Zōshi"), 381

Hiden 秘傳 (guarded transmission [of knowledge, skills]), 166
Hiden'in 悲田院 (medical facility), 101, 102
Hideyoshi. *See* Toyotomi Hideyoshi
Hiei 比叡, Mt., monastic center on, 42, 120
Higuchi Kanetsugu 樋口兼次 (math., fl. 1670), 265
Himiko 卑彌呼, Queen of Yamatai, 3n.5
Hirado 平戶, 225, 232, 286
Hiraga Gennai 平賀源内 (Western Learning, 1728?–1779), 297
Hirata Atsutane 平田篤胤 (National Learning, 1776–1843), 308, 309
Hirata Kanetane 平田鐵胤 (adopted son of Atsutane), 308
Hiroshima 廣島, 230n.11
Hisada Gentetsu 久田玄哲 (fl. 1658), 204
History course (*kidendō* 紀傳道), 32
History of Japan (Kaempfer), 287
Hizen 肥前 domain, 399
Ho-chi-chü fang 和劑局方 (Official prescriptions), 138, 139
Hoashi Banri 帆足萬里 (Western Learning, 1778–1852), 305, 314
Hobson, Benjamin (British missionary, 1816–1873), 385n.108
Hodoha. *See* "Earth-supplementing school"
Hōen hikenshū 方圓祕見集, 265
Hoffman, Willem (Dejima doctor, 1671–1675), 284
Hōjō 北條 household, 119
Hōjō Ujinaga 北條氏長 (military instructor, 1609–1670), 232
Hoki genkai taizen 簠簋諺解大全, 253
Hoki naiden 簠簋内傳 (Ritual implement tradition), 124, 199, 253

460 Index

Hokkaido 北海道, map of, 343
Hokke. See "Lotus Sutra" sect
Holland, 329, 338, 349
  Treaty of amity (1855), 395
  warships and arms ordered from, 395
Holland, King of (1844), 292
Hollyhock (aoi 葵) design of Tokugawa household, 343
Honda Toshiaki 本多利明 (Dutch Learning, math., 1744–1821), 297, 301, 369, 371
Hondō 本道 (internal medicine), 289
Honma Sōken 本間棗軒 (med., 1804–1872), 388
Honorary titles, 110
Honshu 本州, 3
  Jesuit schools in, 188
Honzō. See Materia medica
Honzō wamyō 本草和名 (Japanese names of drugs), 147
Horiuchi Sodō 堀内素堂 (med., 1801–1854), 386, 393
Horner's Method, 131
Hōrōki 放浪記 (Japanese translation of Fernão Mendes Pinto's journal), 172n.19
Horoscopes, 56, 69, 125–126
Hōryakureki 寶曆曆 (calendrical system), 254, 313, 350, 351
Hōryūji 法隆寺 temple, 21, 25, 184n.36
Hoshina Masayuki 保科正之 (daimyo, 1611–1672), 247, 256
Hospitals, 141, 159, 218–219
Hot spring baths, 282
Hou Han shu 後漢書 (History of the Later Han; 5th century), 3n.4
Hourglass, 176
House Learning (Kagaku 家學), 117–119, 121, 123, 138, 140, 141, 166, 200, 203, 239, 242
Hsi-fang li-suan-hsüeh chih shu-ju 西方曆算學之輸入 (The introduction of Western astronomical and mathematical sciences into China), xvii
Hsia 夏 dynasty, 64
Hsiang-chieh chiu-chang suan-fa tsuan-lei 詳解九章算法纂類 (Detailed analysis of the mathematical rules in the Nine Chapters and their reclassification), 131
Hsiao ching 孝經 (Classic of Filial Piety), 135
Hsin-hsiu pen-ts'ao 新修本草 (Newly revised materia medica), 86
Hsin-hsüeh. See "Learning of the mind"
Hsin-kan ming-fang lei-cheng i-shu ta-chüan 新刊名方類証醫書大全, 214n.70
Hsin-ku p'ai 信古派 (Conservative School), 281
Hsin pien Ching-i-hsüeh kai yao 新編中醫學概要 (traditional Chinese medicine textbook), 95
Hsin-yu 辛酉 years, 63–64
Hsiu-yao ching 宿曜經 (Canon of lunar mansions and planets), 50, 69, 125, 253
Hsiung Ming-yu 熊明遇 (Chinese scholar), 261
Hsu Ang 徐昂 (compiler of the Hsüan-ming li), 49, 253, 254
Hsüan-ming li. See Senmyōreki
Hsüan-ming li-ching 宣明曆經 (calendrical system), 253, 254
Hsueh 血 (yin energy or blood), 96
Hua T'o 華陀 (lengendary doctor), 98, 387
Huang-ti nei-ching 黃帝內經 (Yellow emperor's inner classic), 84–85, 94, 138, 209–210, 281–282, 283
Hufeland, C. W., 386
Human body, 84–85
  chart of, 45
  cyclical functions of, 97
  dismemberment of, 317, 376. See

Index   461

also Dissections
and the four elements, 142
malfunctions of, 91–92
manipulations of, 394
as microcosm, 91
structure of, 378
surgery and, 394
Humors, Galenic, 87, 218
Hun'i 渾儀 (armillary sphere), 54
Hun-t'ien 渾天 ("spherical heaven") theory, 54
Huang-fu Mi 皇甫謐 (Chinese, med., 223–282), 85
Hyakushō ikki 百姓一揆 (peasant revolts), 293n.2
Hygiene, 305, 394

I 醫 (medicine) course, 37
I-ching 易經 (Book of changes), 8, 56, 77, 127, 240
I-feng li 儀鳳曆 calendar, 51
I-hsing 一行 (Buddhist monk, cal., 683–727), 49, 72, 81, 129
I-ku yen-tuan 益古演段 (New steps in computation), 130
I-shu ta-ch'üan 醫書大全 (Encyclopedia of medicine), 214
Idai. See "Unsolved problems"
Idai keishō. See "Problem succession"
Idan 醫斷 (New perspective on medicine), 374, 381
Idealist philosophy, 297, 299
Iemitsu. *See* Tokugawa Iemitsu
Iemoto seido 家元制度 (secretive schools), 166, 249
Ienari. *See* Tokugawa Ienari
Ienobu. *See* Tokugawa Ienobu
Ietsugu. *See* Tokugawa Ietsugu
Ietsuna. *See* Tokugawa Ietsuna
Ieyasu. *See* Tokugawa Ieyasu
Igakkan 醫學館 (Institute of Medicine [Tokugawa]), 302–303, 323, 347, 375, 376, 377, 378, 385
smallpox section established, 377
Igen sūyō 醫原樞要 (Principles of physiology), 384

Ihara Saikaku 井原西鶴 (novelist, 1642–1693), 223
Ikai Toyojirō 猪飼豐次郎 (cal., ?–1741), 348–349
Ikeda Kōun 池田好運 (writer on navigation, fl. 1618), 178
Ikeda Mitsumasa 池田光政 (daimyo, 1609–1682), 245–246
Ikeda Shōi 池田昌意 (math., fl. 1670s), 255
Ikuno 生野 silver mine (Tajima 但馬 district), 183
Illustrations. *See* Anatomical illustrations
Imamura Tomoaki 今村知商 (math., fl. 1639), 205, 207, 253, 255, 263
Imitation of Christ (Thomas à Kempis), 185
Immigrants, 24, 29
from Korea to Japan, 2, 19–20, 21, 25
from mainland to Japan, 87, 237n.20
Immortality concept, 30n.36
Immortality drugs. *See* Elixirs of longevity and immortality
Imperial authority
and astrology, 121–122
Buddhism and, 41
and calendars, 69
and compilation of standard texts, 90
desacralization of, 167
under heaven's mandate, 47, 52
Imperial polity, 308
Import restrictions, 104, 225. *See also* Book import ban
Imports, 226
from China, 104n.2, 110, 225–226
from Southeast Asia and Europe, 155n.10
In sphaeram Ioannis de Sacro Bosco, commentarius (1607), 259
Inamura Sanpaku 稻村三伯 (Dutch Learning, 331, 333,

336
Inamurajuku (?) 稲村塾 (Dutch Learning academy), 335
Incendiary flares, 107
Indeterminate analysis, 74, 76, 129, 130
India and the Indians, 6, 142, 157, 388
Indian astrology, 46, 49–50, 56, 69
Indian astronomy, 354
Indian Buddhism, 10, 49, 91
Indian calendars, 70
Indian cosmology, 54
Indian mathematics, 76
Indian medicine, 86, 91, 140, 143
Indian numerals. *See* Numerals
Indian Ocean, 104
Indigo dye, 226
Individual fate calculation. *See* Prognostication
Indo-China, 6
Indonesia. *See* Batavia
Industrial development, 395
Industrial Revolution, xiv
Industry, xiv, 148
Ingyō 允恭, Emperor (fl. 414), 44
Inki sanka 因帰算歌 (Mathematical verse for memorization), 205, 207, 255
Inō Tadataka 伊能忠敬 (or Chūkei; cartographer, 1745–1818), 342, 343
Inoue Gentetsu 井上玄哲 (med., 1602–1686), 279
Inoue Kinga 井上金峨 (Conf., 1739–1784), 375
Inscriptions, on oracle bones, 7
Institute of Astrology. *See* T'ai-shih chu
Institute of Divination. *See* Onmyōryō
Institute of Medicine (*ritsuryō*). *See* Ten'yakuryō
Institute of Medicine (Tokugawa). *See* Igakkan
Institutions of science and learning, state support of, 30
Integral calculus, 361

Integrals, definite, 362
Intellectual renaissance, or outburst (17th century), 100, 138, 148, 156, 186, 187, 223–224, 226, 230, 231, 234–236, 239, 249, 291, 294, 299, 304, 306
Intercalary cycles, 48
Intercalary (leap) months, 47, 57, 70, 71, 199, 257, 360
Internal medicine. *See* Medicine
Interpolation formulas
for equal intervals, 48, 72, 81
for unequal intervals, 49, 72, 81
Introductiones ad veram Physicam et veram Astronomiam . . . (Keill), 353
Ippondō 一本堂 (Hall of the Single Basis: literary name of Kagawa Shūan, *q.v.*)
Ippondō yakusen 一本堂藥選 (Selected drugs of Ippondō), 283
Irie Yoriaki 入江頼明 (med., fl. 1590s), 217
Iron, 3, 7, 152
production, 396
Irrigation works, 24, 25, 168
Iryakushō 醫略抄 (Selected therapies), 141
Ise 伊勢 district
calendar printed in, 199
Grand Shrine, 257
Isei teikin ōrai 異制庭訓往來 (Extraordinary exchange of letters), 136
Ishiguro Nobuyoshi 石黑信由 (math., fl. 1813), 264n.41
Ishii Sōken 石井宗謙 (Dutch Learning, early 19th century), 341n.45
Ishii Tsuneemon 石井恒右衛門 (interpreter; late 18th century), 332
Ishikawa Genjō 石川玄常 (med., 1754–1816), 322
Ishinpō 醫心方 (sometimes *Ishinhō*; Prescriptions at the heart of

Index   463

medicine), 88, 89, 90, 140, 141, 143, 375
  printing of, 376
Ishiyama Honganji 石山本願寺 temple, Osaka, 173
Islam, 105, 109
Islamic astronomy, 122–123
Islamic calendar, 70
Islamic scholars, 122
Isoho monogatari 伊曾保物語 (Aesop's Fables), 185
Isolation decrees, 155, 219, 232, 233
Isolation period (1639–1854), 148
Isolation policy, xxv, 149, 156, 162, 165, 196, 347
  and ban on import of mercury, 183
  and calendrical astronomy, 198, 203, 251
  challenge to, 291–294
  and Chinese Cultural Wave II, 223
  and the curtailment of Western influence, 224, 237, 284
  ending of, 294, 296, 361, 372, 373, 395, 399
  and innovations in techniques and the sciences, 163
  and intellectual revival, 187
  and mathematics, 206, 208, 262–263
  and medicine, 209, 220, 221
  and navigational techniques, 176, 179, 233
  and restrictions on the import of Chinese texts, 261
  and sociopolitical affairs, 228ff.
  and trade relations, 224–225
  and warfare techniques, 232
  and Western sciences, xxii, 197, 208, 258, 329, 347
Isomura Yoshinori 磯村吉徳 (math., fl. 1661), 264, 265, 266
Italian Renaissance, 173
Italy, 175
Itō 伊東, ships built at, 177
Itō Genboku 伊東玄朴 (med., 1800–1871), 336, 377, 378
Itō Jinsai 伊藤仁齋 (Conf., 1627–1705), 238, 241, 248, 249, 277, 282, 283, 297, 304
Itō Keisuke 伊藤圭介 (med., fl. 1829), 330
Itō Tomonobu 伊藤友信 (med., fl. 1754), 380
Izu 伊豆 peninsula, 177, 342
Izumo no Hirosada 出雲廣貞 (med., fl. 808), 88, 309

Jacket (*kosode* 小袖), gift to Siebold, 343
Jamal al-Din (Cha-ma-lu-ting 札馬魯丁; Muslim, astron., fl. 1267), 122
Japan
  coastal map of, 342
  first centralized state of, 29
  map of, during Chinese Cultural Wave II, xxxvi–xxxvii
  unification of. *See* National unity
Japanese court, 111, 112
  medical services and, 100–101
  tax-free estates and, 110–111
Japanese court nobility, 104, 110, 111, 170, 239
  aesthetic refinement of, 114
  government and, 40, 110, 114, 115
  interest in divination, 124–126, 128
  and the intellectual renaissance, 230–231
  and *kagaku* ("poetics"), 248–249
  and mathematics, 135
  and medicine, 140
  and political power, 133
  and social change, 103
  tax-free estates and, 110, 111
Japanese culture
  indigenous, 22–23
  samurai, 166
  secularization of, 167
  specialization of, 167
Japanese interpreters, 227, 237, 285, 287

Japanese language, 194, 195, 287, 331
"Japanese Learning" (*Wagaku* 和學), 238, 249, 306
Japanese living abroad, prohibited from returning to homeland, 225
Japanese-Portuguese dictionaries, 194
"Japanese towns" (*Nihonmachi* 日本町), 154–155, 161, 165
Java, 338
Jenner, Edward (1749–1823), 220, 377
Jesuit college, 186, 188, 189, 194, 260
Jesuit educational institutions, 159, 188, 189, 194
   for training Japanese priests, 188
   frequent relocations of, 197
   curricula of. *See* Curricula
Jesuit translations of scientific texts into Chinese, 260–261, 262
Jesuits, xvi, xxv, xxvii, 157, 158, 181, 185, 188, 197–198, 200–203, 208, 217–218, 299, 331, 361
   in China, 196, 202, 349, 351, 356, 357, 370, 371
   forbidden by Rome to practice medicine, 218
   forbidden works of, 312
   impact on Japanese intelligentsia, 196
   persecution of, 195, 196, 197, 201, 202, 219
Jimsonweed, 388
Jinkōki 塵劫記 (Treatise on [numbers] great and small; sometimes *Jingōki*), 205, 206, 207, 255, 264, 271, 361, 366
   Eitai 永代 (eternal), plagiarized edition of, 206
   Fūki 富貴 (rich and noble), plagiarized edition of, 206
Jinsei kyūri 盡性窮理 ("exhaustively studying the characteristics [of entities] to grasp their ri"), 303
Jinshin kyūrigaku 人身窮理學 (term for "physiology"), 385
Jisha bugyō. *See* Office of Temples and Shrines
Jitsugaku. *See* "Practical learning"
Jōdo. *See* "Pure Land" sect
Jōkyōreki 貞享暦 (calendrical system), 238, 254, 255, 256, 258, 311, 348, 349, 350
Jōkyōrekisho 貞享暦書, 254
Jōmon 繩文 period (ca. 8000 B.C. to third century B.C.), 2
Jugairoku 豎亥錄 (Record of a young boar), 205, 207, 255
Jugon 呪禁 (ritual healing), 30n.36
   course of study, 37
Juji hatsumei 授時發明, 253
Jujireki 授時暦 (*Shou-shih li* in Chinese), 122, 252, 255, 256, 258, 312, 355
Jujireki kaigi 授時暦解義, 253
Julian calendar, 47
Junsei Shoin 順正書院 (Dutch Learning academy), 335
Jupiter (planet), 356
Jürched clans, 104, 107
Jürched rule, 209
Jūsō sagen 銃創瑣言 (Trivial words on gun wounds), 377

Kada no Azumamaro 荷田春滿 (National Learning, 1669–1736), 249, 307
Kaempfer, Engelbert, (Dejima doctor, 1651–1716), 228, 287
Kagaku. *See* House Learning
Kagaku 化學 (chemistry), 306
Kagaku 歌學 ("poetics"), 248–249
Kagawa Gen'etsu 賀川玄悦 (med., 1700–1777), 386, 388, 391
Kagawa Genteki 賀川玄廸 (med.), 392
Kagawa Shūan 香川修庵 (med., 1683–1755), 283, 373, 378
   school of, 341n.45
Kageyukōji Aritomi 勘解由小路在

Index 465

富 (astrol., fl. 1560), 199
Kagoshima 鹿兒島, 157, 184, 346
Kai fukudai no hō 解伏題之法 (Method for solving *fukudai*), 271
Kai-t'ien 蓋天 ("sky as a cover") theory, 53, 54
Kaibara Ekken 貝原益軒 (Conf., 1630–1714), 238, 251, 304
Kaigunsho. *See* Naval Academy
Kaisei sanpō 改精算法 (Reformed mathematical methods), 363
Kaiseisho. *See* Center for Western Learning
Kaishihen 解屍篇 (On dissection of corpses), 382
Kaitai shinsho 解體新書 (New text on anatomy), 286, 323, 324, 326, 328, 332, 340, 351, 372, 373, 376, 382–383, 386, 387
  permission to publish, 337
Kaitai yakuzu 解體約圖 (Concise chart of anatomy), 323, 383
Kaitokudō 懷德堂 academy, Osaka, 305
Kaizanki 改算記, 265
Kajiwara Shōzen 梶原性全 (med., 1266–1337), 143, 144, 146, 379
Kakubutsu kyūri 格物窮理 ("investigating things to penetrate the *ri*"), 303, 304
Kakuchi sanpō 格致算法, 265
Kamakura 鎌倉, 111, 119, 141, 214
  Gokurakuji (Paradise Temple), 141
  temple complex at, 120, 121
Kamakura 鎌倉 period (1192–1336), 103, 109, 110, 111–112, 116, 136, 138
Kamakura 鎌倉 shogunate, 103, 104, 105, 119, 120, 141
Kami 神 (gods), 44
Kamigamo 上賀茂 Shrine, Kyoto, 23
Kamo 賀茂 household, 123, 199
Kamo no Arikata 賀茂在方 (astrol., fl. 1414), 54n.72, 198
Kamo no Mabuchi 賀茂眞淵 (National Learning, 1697–1769), 307
Kamo no Mitsuyoshi 賀茂光榮 (astrol.), 123
Kamo no Yasunori 賀茂保憲 (astrol., d. 987), 123
Kana 假名 (Japanese syllabic script), 112–113, 146, 316, 332
Kanagawa 神奈川 (Yokohama 橫濱) Treaty (1854), 293
Kanagoyomi 假名曆 (*kana* version of the almanac), 128
Kanazawa 金澤, 230n.11
Kanazawa Library (Kanazawa Bunko 金澤文庫), 119
Kanbun 漢文 script, 322
Kanbun Nagasaki zu byōbu 寬文長崎圖屏風 (A screen painting of Nagasaki in the Kanbun era [1661–1673]), 227
Kanda 神田, Edo, 312
Kan'ei 寬永 era (1624–1644), 206
Kang Hang 姜沆 (Korean, Conf., fl. 1590s), 242
Kangakuin 勸學院 (private academy), 39–40, 115, 116
Kanmu 桓武, Emperor (r. 781–806), 88
Kanpaku 關白 (ruler "on behalf of" the emperor), 162
Kanpōyaku 漢方藥 (Chinese-style medicines), 395
Kanryōha 寒涼派. *See* "Cooling medication school"
Kansei 寬政 calendar, 337, 357
Kanseireki 寬政曆 (calendrical system), 254, 313, 314
Kanseirekisho 寬政曆書, 254
Kantō 關東 plain, 120, 164, 214, 215
Karak. *See* Mimana
Karin sankasho 訶倫產科書 (obstetrics text), 330
Karma, 97
Karō 家老 (elder), 322

Kasama 笠間 domain, Hitachi 常陸 province, 375n.94
Kasuga 春日 Shrine, Nara, 23
Kasuparuryū. *See* Caspar school
Katayama Kenzan 片山兼山 (Conf., 1730–1782), 298
Katsuragawa 桂川 household, 286
Katsuragawa Hochiku 桂川甫筑 (med., 1661–1747), 286
Katsuragawa Hoken 桂川甫賢 (med.), 286
Katsuragawa Hoshū 桂川甫周 (or Kuniakira 國瑞, med., 1751–1809), 286, 322, 323, 385
Katsuragawa Hoshū 桂川甫周 (or Kunioki 國興, med., 1826–1881), 333
Katsusa 加津佐, 189
 Jesuit college at, 185–186
Katsuyō sanpō 括要算法 (Essentials of mathematical method), 368
Kawaguchi Nobutō 河口信任 (med., 1736–1811), 382
Kawahara Keiga 川原慶賀 (painter, 1786–?), 339
Kawakita Chōrin 川北朝鄰 (math., 1841–1919), 369
Kawamoto Kōmin 川本幸民 (Dutch Learning, 1809–1871), 305
Kegare 穢 (personal impurity), 44
Keichō 慶長 era (1596–1615), 163, 183, 220
Keiden'in 敬田院 (medical facility), 101
Keii 敬尹, 349n.54
Keiki 形氣, heaven of, 261–262
Keill, John (1671–1721), 353, 361
Keisei hisaku 經世祕策 (Secret plan for governing the country), 371
Keitekiin 啓迪院 (private medical academy), 216
Keitekishū 啓迪集 (Collected orientations), 216
Kempis, Thomas à, 185

Kenchōji 建長寺 temple, Kamakura, 105
Kenkon bensetsu 乾坤辯說 (Critical commentary on cosmography), 244, 258–259, 260, 261
Kenkyusha, xxx
Kenninji 建仁寺 temple, Kyoto, 135, 242
Kentōshi 遣唐使 missions, xvii, 13, 14–19, 20, 29, 41, 42, 103
 discontinued, 19, 28, 104
 personnel, 16, 17, 18
Kenzuishi 遣隋使 missions, 14, 16, 18
Kepler, Johannes, 201, 304, 352, 355
Ketsugishō ippyakumon tōjutsu 闕疑抄一百問答術, 265
Ketu ("invisible" planet), 49
Ki. *See* Ch'i
Kiba minzoku 騎馬民族 ("Horse-riding people"), 4n.6
Kidendō. *See* History course
Kidney, 61, 93, 95
 filtering function of, 391
Kii 紀伊 province, 341n.45
Kikai kanran 氣海觀瀾, 330
Kikai kanran kōgi 氣海觀瀾廣義 (Extended explanation of *Kikai kanran*), 305
Kikanzu 奇患圖 (Illustrated report of a rare disease), 389
Kimura 木村, Sebastio セバスチオ, 189n.46
"Kinchū narabi ni Kuge Shohatto" 禁中並公家諸法度 (Ordinances for the Imperial Household and Court Nobles), 231n.13
Kinoshita Jun'an 木下順庵 (Conf., 1621–1698), 244
Kinranpō 金蘭方 (ancient medical text), 309
Kishiwada 岸和田 domain, 344
Kissa yōjōki 喫茶養生記 (Guide to good health through tea-drinking), 143

Kizuki 杵築 fief, 354
Kō Ryōsai 高良齋 (Dutch Learning, fl. 1836), 335, 341n.45
Kobayashi Yoshinobu 小林義信 (astron., 1601–1684), 260, 261
Kōbun'in 弘文院 (Hayashi academy), 238, 243, 295
Kōbun'in 弘文院 (Wake household's library), 116n.15
Kobunjigaku. See Critical linguistic and textual study
Kōbusho. See Military training center
Koch, Robert (1843–1910), 220
Kōfukuji 興福寺 temple, 21, 102
Kofun 古墳 period (Ancient Tomb period; ca. 250–552), 22
 See also Burial mounds
Koga 古河, 214, 215
Kogaku(ha). See Ancient Learning
Kōgeha. See "Cathartic school"
Kogidō 古義堂 (Hall of Ancient Principles), 247
Kögler, Ignatius (German missionary), 355
Koguryŏ 高句麗 state, 4, 12, 20, 21, 44
Kohōha. See Ancient Practice School
Kōi geka sōden 紅夷外科宗傳 (Orthodox tradition of red [-hair] surgery), 284–285
Kojiki 古事記 (Record of ancient matters; compiled in A.D. 712), 44, 249, 307, 309
Kojikiden 古事記傳 (commentary on the *Kojiki*), 307
Koishi Genshun 小石元俊 (Dutch Learning, fl. 1786), 335
Kokan Shiren 虎關師錬 (Zen priest, 1278–1346), 136
Kokon sanpōki 古今算法記 (Old and new methods of mathematics), 265, 266
Koku 石 (unit of measure), 232
 for grain, 283
 for ships, 232, 233

Kokubunji 國分寺 (provincial temples), 21
Kokugaku. See "National Learning"
Kokugaku 國學 (provincial colleges), 32, 36, 38, 41
 medical courses in, 88, 101, 140
Kokura 小倉, map of, 343
Kōkyō. See *Hsiao ching*
Kōmō geka 紅毛外科 ("Red-hairs" surgery), 220, 284–288, 387, 393
 low standards of, 285, 288
Kōmō kajutsuroku 紅毛火術錄 (Notes on Dutch gunnery), 232
Kōmyō 光明, Empress (701–760), 102
Korea and the Koreans, 2, 3, 6, 14, 110, 183, 309, 329
 books brought to Japan from, 264
 Buddhism in, 21, 25, 26, 110
 Chinese settlements in, 2
 cultural influences on Japan, xviii, 5, 12, 45, 187
 decline of Chinese control over, 4
 immigrants from, 2, 19–20, 21, 25
 invasions of, 26, 105, 153, 156, 164, 169, 173, 177, 185, 187, 204, 242, 264
 Japanese troops in, 4, 5, 14
 medical practitioners from, 87
 overseas trade, 104
 printing, 184–185, 186
 unification of, 20
Korea Strait, 2, 177
Korean astronomy, 50
Korean diplomatic mission, 242
Korean technicians, in Japan, 169
Koreki benran 古曆便覽, 253
Koryŏ 高麗 dynasty (935–1392), 110
Kōsei shinpen 厚生新編 (household encyclopedia), 330
Kosugi Genteki 小杉玄適 (med., fl. 1754), 380
Kōtokui 幸徳井 household, 123

Kōya 高野, Mt., Buddhist study center on, 42, 120
Kozukahara 小塚原 execution grounds, 322
Krusenstern, A. J. von (Russian admiral), 343
Ku Yen-wu 顧炎武 (Chinese, Conf., 1613–1682), 241
Kuchizusami 口遊 (lit., "singing to oneself," i.e., impromptu), 134
Kūkai 空海 (Buddhist priest, 774–835), 41
Kulmus, Johan Adam, 317n.32
Kumazawa Banzan 熊澤蕃山 (Conf., 1619–1691), 238, 244, 245–246
Kunaishō 宮內省 (imperial household ministry), 36
Kung-ho 光和 era (178–183), 48n.61
Kuni 國 (province), 36
Kunitomo 國友 (Ōmi 近江 district), 171
Kuo Shou-ching 郭守敬 (Chinese, cal., fl. 1280), 122, 254, 258
Kuo-tzu-chien. *See* T'ang university
Kurisaki Dōki 栗崎道喜 (med., 1566–1651), 220
Kuriyama Kōan 栗山孝庵 (med., 1728–1791), 382
Kurume 久留米 domain, 365
Kusaka Makoto 日下誠 (math., 1764–1839), 366, 369
Kwal-luk 觀勒 (Korean, Buddhist priest, fl. 602), 45
Kyōhō 享保 era (1716–1736), 293n.1
Kyoto 京都 (Heiankyō 平安京), 16, 20, 22, 26, 111, 119, 120, 126, 133, 141, 165, 167, 184, 188, 197, 202, 215, 230, 242, 255, 256, 257, 258, 282, 307, 350, 378
  academy founded by Yamazaki Ansai, 245
  calendars, 198, 199
  execution of thieves at, 380
  gozan temples in, 121, 151
  hospital and leprosarium, 219
  Institute of Divination, 123–124
  Kamigamo Shrine, 23
  Kenninji temple, 135, 242
  medical academy of Manase Dōsan, 279
  monastic center, 42
  post-mortem dissection at, 382
  Ryōrenji temple, 354
  school founded by Kagawa Gen'etsu, 388
  Shimogamo Shrine, 23
  Shōkokuji temple, 242
  Tenryūji temple, 105
  Tōfukuji temple, 136, 204
Kyū 球 (technical term for sphere), 206
Kyūji 舊辭 (Collected tales of the Yamato court), 42n.51
Kyūri 窮理 (process of perceiving and comprehending the *ri*; *ch'ung-li* in Chinese), 303–304
  equated with Western science, 305, 306
  as equivalent for Western terms, 385
Kyūridō 窮理堂 (Dutch Learning academy), 335
Kyūrigaku 窮理學 ("the study of *kyūri*, i.e., physics"), 305, 306
Kyūritsū 窮理通 (General investigation [of science]), 305
Kyushu 九州, 2, 3, 38, 101, 104, 157, 160, 171, 189, 242
  Japanese ports in, 155
  Jesuit schools in, 188
  map of, 343

Laboratory for physics and chemistry (Bunseki kyūrisho 分析窮理所), 397
Lacerations, 87, 214
Lalande, J. J. L. de, 314, 337, 357, 359, 371

Land
　area under cultivation, 112, 149, 163
　measurement, 33
　owned by the Institute of Divination, 123
　owned by temples, 21, 22, 27
　redistribution of, 6, 13
　reform, 13
　sales, prohibition of, 165
　tax-exempt, 19, 27, 110–111
Land-surveying, xxviii, 79, 83, 183, 204, 205, 274, 361, 370, 371
　tax system and, 164
　of Uraga coast, 345
　Western, 346
Landowners, 293, 307
Language barrier. *See* Linguistic problems
Languages. *See also* Linguistic problems
　competence in, 334, 337
　study of, 398
Lao Tzu 老子 (Chinese philosopher), 9
Later Han dynasty (A.D. 25–220), 2, 6, 9
　calendars, 47, 48
　mathematics, 77
　medicine, 85, 95
Latitude
　determination of, 176
　differences of, between Peking and Kyoto, 258
Law course (*myōbōdō* 明法道), 32, 38, 39
Law of variation (*shōchōhō* 消長法), 355–356, 357
Laws of nature, mechanistic, 353
Learning
　Chinese and Japanese institutions of, contrasted, 237
　Chinese-style, 299
　and government, 6–7, 8, 229, 233
　and the *kana* script, 113
　revival of, 165–167
　samurai and, 119, 120

"Learning of the mind" (*Hsin-hsüeh* 心學), 241
Lecture Hall for Japanese Learning (Wagaku kōdansho 和學講談所), 308
Legalists, 1, 7
Leibniz, 271, 364
Length, calculations of, 82, 361
Lens-grinding, 356
Leprosaria, 141, 159, 218, 219
Leprosy, 97, 219
Li 理 (*ri* in Japanese; the pattern which underlies all phenomena), 240, 259, 260, 261, 303, 304, 306
Li 李 school of medicine, 139, 209, 213, 214, 215, 216, 251, 277, 278, 279
Li Ai 李杲; adult name: Ming-chih 明之; literary name: Tung-yüan 東垣 (Chinese, med., 1180–1252), 212, 213
Li chih 曆志 (Calendrical astronomy), 48
Li-Chu 李·朱 medical traditions. *See* Goseihōha
Li-hsiang k'ao-ch'eng 曆象考成 (Compendium of calendrical phenomena), 355
Li-hsiang k'ao-ch'eng hou-pien 曆象考成後編 (Sequal to the *Li-hsiang k'ao-ch'eng*). 355, 358
Li suan ch'üan-shu 曆算全書 (Comprehensive collection of works on astronomic computation), 349, 350, 355, 368
Li Yeh 李冶 (Chinese, math.), 129, 130
Liefde (Dutch ship), 177
Lightning, 200
Ling-shu 靈樞 (part of the "Yellow Emperor's inner classic"), 85
Linguistic problems, 170, 197, 237, 329
Lister, Joseph (1827–1912), 221
Literacy, xxiii, 113, 295
Literary competence, to deal with

470 Index

scientific materials, 329, 330
Literati, 6, 149
Literature course (*monjōdō* 文章道), 32–34, 39, 115
Liu 劉 school of medicine, 139, 209, 211, 277, 278, 279–280
Liu-Chang 劉·張 medical tradition. *See Goseihō beppa*
Liu Ch'o 劉焯 (Chinese, cal., fl. 600), 48, 72, 81
Liu Hui 劉徽 (Chinese, math., fl. 263), 75
Liu Wan-su 劉完素; adult name: Shou-chen 守眞; literary name: Ho-chien 河間 (Chinese, med., 1110–1200), 211, 212
Liu Wen-shu 劉溫舒 (Chinese, med., fl. 1099), 210
Liver, 61, 93, 95
  functional orb of, 96
Lo 洛, River, 64
Lo mo 絡脈 (branch channels in the human body), 96
Lo Shih-lin 羅士林, 131
Location at sea, determination of, 178, 179, 203
Logarithmic tables, 176, 370, 371
Logarithms, 180, 361, 370
Logic, 78
Longitude
  determination of, 176
  differences of, between Peking and Kyoto, 258
  and latitude, surveyed by Siebold, 341
"Lotus Sutra" (*Hokke* 法華) sect, 113, 184
Lo-yang 洛陽, 3n.5
Loyola, Ignatius, 200
Lu Chiu-yüan 陸九淵 (also, Lu Hsiang-shan 陸象山; Chinese, Conf., 1139–1193), 240
Lu-li chih 律曆志 (Treatise of harmonics and calendrical astronomy), 48, 50

Lucky and unlucky days or periods, 56, 57, 63, 67, 126
Lulof, Johan, 352
Lun yü 論語 (Analects), 135
Lunar and solar anomalies, 49, 72
Lunar cycle, 70
Lunar mansions, 50, 69
Lungs, 61, 93, 95, 380

Ma-fei-t'ang 麻沸湯 (*mafutsutō* in Japanese; a drug for general anesthesia), 388
Macao 澳門 or 媽港, 162, 186, 220
Machi bugyō 町奉行 (town commissioners), 257
McCune-Reischauer system of romanization, xxi
Macrocosmic-microcosmic correspondences, 86, 91, 210, 280
Maeda Noriyuki 前田憲舒 (math., fl. 1673), 264
Maeno Ryōtaku 前野良澤 (med., 1723–1803), 316, 317, 322, 332
Magic, 45, 79, 94, 135
Magic squares, 64–66, 77
Magical cures and rituals, 10, 42, 87, 91
Magnetic compass, 107, 109n.7, 169, 175, 181, 341
  floating needle, 175n.22
Magnetic polarity, 107
"Main way" (*hondō* 本道) to health, 289
Majima Seigan 馬島清眼 (med., ?–1379), 147
Mamakodate 繼子立 problem, 136–137
Man'anpō 萬安方 (Prescriptions for all occasions[?]), 146
Manase Dōsan 曲直瀨道三 (med., 1507–1594), 139, 214, 215–216, 277, 279, 282
Manase Gensaku 曲直瀨元朔 (med., 1548–1631), 216, 279
Manbyō ichidoku setsu 萬病一毒說

("single poison" theory of
    disease), 374
Manchu 滿洲 dynasty, 149, 236,
    241
Manchuria, 6
Mandarage 曼陀羅華, 388
Mandate of heaven. *See* Heaven's
    mandate and imperial rule
Man'en 萬延 era (1860–1861), 64
Manila, 154, 161, 177
Man'yōshū 萬葉集 (Collection of
    Myriad Leaves), 307, 309
Mao 茅, Mt., Taoist sect, 91
Maps and map-making, 176, 342,
    343, 357
Marin, P., 331
Maritime trade. *See* Overseas trade
Mass, concept of, 173, 353
Massage (*anma* 按摩), 36, 37, 85,
    87, 96, 98, 99, 100
Materia medica (*pen-ts'ao* 本草;
    honzō in Japanese), xxviii,
    43, 45, 85, 86, 88, 98, 138,
    146–147, 223, 304, 311
Mathematical games, 206. *See also*
    Games and puzzles
Mathematics, xvii, xxvii, xxix, 32,
    33, 43, 47, 74–79, 88, 112,
    129–138, 173–174, 183, 202,
    203–207, 223, 224, 231, 239,
    255, 262–277, 298, 348, 397.
    See also Wasan
  and astronomical data, 72, 74
  and bureaucracy, 275
  and calendrical research, 251,
    252, 275
  Chinese. *See* Chinese mathematics
  for commercial, military, and
    technical purposes, xxviii,
    78, 79, 137, 207, 263, 276,
    301, 361, 396
  Confucianism and. *See*
    Confucianism
  in curriculum of Jesuit college,
    194
  decline of, 84, 117, 132, 133
  first known Japanese monograph
    on, 88

as a government monopoly, 83
  Greek. *See* Greek mathematics
  hereditary professorships in, 117,
    132
  Indian. *See* Indian mathematics
  logical aspect of, xxvii–xxviii, 78
  low value accorded to, in ancient
    Japan, 78–79
  and the merchant class, 84, 274,
    302
  practical, 263, 274, 364, 365, 366,
    371
  printed books on, 138
  "pure," 78, 364
  recreational, 77, 78
  revival of, 137–138
  in the *ritsuryō* academic system,
    79
  samurai and, 207, 274
  technical terms, 205, 206
  traditional Chinese-style, 208,
    262–263
  verses for memorizing rule of, 207
  Western. *See* Western
    mathematics
Mathematics course (*sandō* 算道),
    32–34, 77, 79, 83, 126, 133
Matsukawa Tsurumaro 松川鶴麿
    (med., 1791–1831), 309
Matsumoto Ryōjun 松本良順
    (med., 1832–1907), 397
Matsunaga Yoshisuke 松永良弼
    (math., ?–1744), 362, 364,
    369
Matsuo Bashō 松尾芭蕉 (poet,
    1644–1694), 223
Matsuoka Joan 松岡恕庵 (med.,
    1669–1747), 304
Matsusaka 松坂 or 松阪, Ise
    district, 307
Mechanical watch. *See* Watches
Mechanics
  Galilean, xv
  Greek, 173
  Newtonian, 314, 352, 353, 371
Medical books, 141, 214
  earliest Japanese compilations,
    88, 100

472   Index

Medical books (*continued*)
  from Chinese sources, 140, 143
  control of printing of, 347, 377
  first medical book printed in
    Japan, 214
  by priest-practitioners, 142
  Western, translations of, 330–
    331, 384–387, 393
Medical care, 100
  and Buddhist compassion, 142
  for commoners, 101–102, 138
  of European traders and
    mariners, 219
  for government officials, 101
  for the imperial family and
    nobility, 140
  missionaries and, 159
Medical classics, 280, 281, 375, 376
Medical instruments, 388. *See also*
    Surgical instruments
Medical research, 121, 140, 221
Medical schools. *See* Schools
Medical specialists, and
    specialization, 147, 213,
    214–215, 217, 289
  legal ranking of specialists
    serving the shogunate, 289
Medical training, 101, 216, 218,
    288, 295, 397
Medical treatments, 43, 88, 97–98,
    140, 209, 210, 280, 282, 309,
    332, 379
  Buddhist, 143
  magical and religious, 87, 91
  of the whole person, 394, 395
Medicinal compendium (de Gorter),
    322, 385n.109
Medicinal prescriptions, 85, 88, 91,
    97, 99, 138, 147, 210, 211,
    280, 281, 309, 374, 376, 378
  animal, herbal, and mineral
    ingredients of, 85
  Buddhist, 143, 146
Medicine, xxvii, 19, 31–32, 36, 43,
    46, 117, 133, 209–218, 224,
    240, 277–290, 307, 315, 348,
    360
  apotropaic, 36

Buddhism and, 94, 100, 102, 121,
    139, 141, 282–283
Chinese. *See* Chinese medicine
Confucianism and. *See*
    Confucianism
Dutch. *See* "Dutch medicine"
Dutch Learning and, 315, 316ff.
earliest known medical work
    produced in Japan, 88
eclecticism and, 375, 383
folk, 98
government policy on, 375
Greek. *See* Greek medicine
internal, 82, 218, 221, 285, 289,
    329, 332, 375, 385, 393
modern, 323
preventive, 216n.72
secularization of, 147, 213, 215,
    217, 283
and social status, 88
Taoism and. *See* Taoism
technical terms, 384
traditional, 317, 323, 372, 373,
    378, 380, 383, 386, 387
  attack on, 298
Wahō school of medicine. *See*
    Wahō
Western. *See* Western medicine
Medicines. *See* Drugs
Mediterranean Sea, 175
Meerdervoort, Pompe van (Dejima
    doctor, 1829–1908), 397
Mei Wen-ting 梅文鼎 (Chinese,
    math., 1633–1721), 349, 368
Meiji bureaucracy, xxiii, xxvi
Meiji 明治 era, 1, 111, 123, 156,
    174, 186, 206, 207, 247, 278,
    291, 294, 306, 308, 309, 360,
    361, 377, 384, 388, 394, 397,
    399
Meiji institutions of higher learning,
    397, 398, 399
Meiji Reforms, 196, 299, 303, 306,
    398
Meiri 命理, heaven of, 261–262
Mencius, 247
Mensuration, xxix
Merchant ships. *See also* "Red Seal

Index   473

Ships"
  Chinese, 19, 28, 104, 105, 152
  European, 168
  Japanese, 104, 105, 119, 135, 151, 178, 179
  Tally Trade, 151, 152
Merchants, xxiii, 79, 83, 84, 104, 105, 108, 112, 138, 163, 164, 165, 166, 180, 187, 195, 207, 230
  Arab, 170, 175
  Chinese, 225, 236
  and cultural influx, 148–149, 156, 196
  and mathematics, 204, 274, 276, 302
  restrictions on, 230
  scholarship and, 235–236
  and the status system, 294
  technology and, 169, 170
  and trade relations with China, 148, 151
Mercuric oxide, 90
Mercuric sulfide. *See* Cinnabar
Mercury, 90, 183
Mercury amalgam, 183
Meridians, 260
  determination of, 262
Merit system, and official rankings, 12–13, 28, 40, 237
Metabolism, 95
Metal type. *See* Printing
Metallurgy, xxiv, 169, 181–183
Metalworking, 2, 24
Metaphysics, 197, 303, 306
Meteors, 53
Meton of Athens (fl. 432 B.C.), 47
Mexico, 177
Midwifery, 215
Military Academy (Rikugunsho 陸軍所), 398
Military class, 103, 164, 229
  and political power, 133
Military cliques, 148
Military code, 165
Military government, 164, 166, 167
Military maneuvers, and mathematics, 207

Military regimentation, 112
Military technology, xxii, xxvi, xxix, 293
  Western. *See* Western military technology
Military training center (Kōbusho 講武所), 398
Mima Junzō 美馬順三 (med., 1795–1825), 341n.45
Mimana 任那 (Karak 加羅), 4, 5
Minagawa Kien 皆川淇園 (Conf., 1734–1807), 391
Minamoto clan. *See* Genji clan
Minamoto no Sanetomo 源實朝 (Kamakura shogun, 1192–1219), 143
Minamoto no Tamenori 源爲憲 (math., fl. 970), 134
Mineral substances, used in drugs, 85
Ming 明 court, 19, 156
Ming 明 dynasty (1368–1644), 106, 109, 110, 153, 167, 172, 175, 229, 235, 241, 242
  astronomy, 121–122
  and foreign trade, 150ff
  mathematics, 132, 204
  medicine, 139, 209, 210, 213
Mining, xxix, 163, 164, 168, 181–183, 204, 234, 236
Minting coins, 225n.3
Mishima 三島, calendars issued at, 199
Missionaries, xv, 149, 157, 159, 160, 162, 167, 221, 259
  Jesuit, xxv, 157, 161, 188, 189, 195, 200, 218
Missions. *See* Diplomatic missions and envoys
Mito 水戶 domain, 237n.20
Miura Baien 三浦梅園 (Western Learning, 1723–1789), 297
Miura Seiin 三浦靜陰 (Conf., fl. mid-18th century), 300
Miyagi Kiyoyuki 宮城清行 (math.), 265
Miyazaki Yasusada 宮崎安貞 (agriculture, fl. 1696), 238

Miyoshi 三善 household, 132, 203
Mo ching 脈經 (Classic on pulsation), 85
Mo-teng-ch'ieh ching 摩登伽經 (Canon of astrology based on lunar mansions), 49, 69
Mobility, restrictions on, 165
Mochizuki Rokumon 望月鹿門 (med., 1697–1769), 374, 375
"Modernist" medical traditions, 280–281, 373, 374
Modernization, xii–xiv, xvii–xviii, xxi, xxii–xxiii, 1, 155, 156, 157, 167, 168–170, 231, 399
  European, 156, 168–169
Mohnike, Otto (Dejima doctor, fl. 1849), 378
Moku 目, *ji* 耳, *kō* 口, *shi* 齒 (Eye, ear, mouth, teeth) course, 37
Momentum, 353
Momokawa Chihei 百川治兵衛 (math., fl. 1622), 205
Money economy, 163
Mongol invasions, 26, 105, 107, 111
Mongols, 109, 110, 122, 149, 209
Monjōdō. See Literature course
Monjōin 文章院 (University affiliate school), 39, 115, 116, 118
Monnō 文雄 (Buddhist monk, cosmology, fl. 1756), 301, 354
Mononobe no Hiroizumi 物部廣泉 (med., fl. 868), 309
Monopolistic privileges, 230n.12
Months
  long and short, 46, 56, 70, 257
  intercalary, 47, 48, 57, 70, 257
  synodic, 48, 69, 70
  in lunar calendar, 70
Moon, 49
  conjunction of sun and, 48
  motion of, 48, 52, 72
  orbit of, 54
  position of, 50, 121
  shape of, 70
  surface of, 356
  waxing and waning of, 69, 70, 200
Morality, 241, 247, 248
Mōri Shigeyoshi 毛利重能 (math., fl. 1622), 205, 206, 207, 254
Morikawa Sōen 森川宗圓 (med.), 309
Morrison (American merchant ship), 345
Mortars, 232
Motion
  laws of, 353
  relativity of, 353
Motoki Ryōei 本木良永 (astron., 1735–1794), 304, 313, 314–315, 331, 351–352, 353
Motoki Ryōi 本木良意 (med., 1628–1697), 285n.67, 287, 304, 331, 382–383
Motoki Shōzaemon 本木庄左衞門 (Dutch Learning, 1767–1822), 333
Motoori Norinaga 本居宣長 (National Learning, 1730–1801), 248, 307, 308
Mouth diseases, 147
Movable type. *See* Printing
Moxibustion, 36, 45, 85, 87, 96, 98, 99, 100, 217, 282, 287, 289
Mukai Genshō 向井元升 (Conf., med., astron., 1609–1677), 244, 259, 260
Multiplication, 74, 75, 76, 79, 206
  on the abacus, 205
Multiplication table, 134
Muramatsu Shigekiyo 村松茂清 (math., fl. late 17th century), 265, 274
Muro Kyūsō 室鳩巣 (Conf., 1658–1734), 244, 251
Muromachi 室町, 161
Muromachi legacy, 165, 166
Muromachi 室町 period (1336–1467), 103, 112
Muscles, 381
Music, teaching of, 44, 189, 194
Muslim. *See* Islamic
Muyō no muyō 無用の無用 (mathematics "not useful

for no use"), 365
Muyō no yō 無用の用 (mathematics "useful for no use"), 365
Mu-yung 慕容 clan of the Hsien-pi 鮮卑 people, 26
Myōbōdō. *See* Law course
Myōtei mondō 妙貞問答 (A dialogue between Myōshū 妙秀 and Yūtei 幽貞), 201

Nagasaki 長崎, 161, 189, 220, 224, 225, 226, 228, 232, 236, 260, 284, 285, 293, 317, 378
  Dutch medical officer at, 219, 220
  interpreters, 258, 278, 304, 315, 316, 317, 333, 372, 383
  trade, 334, 338
Nagasaki harbor, 292
Nagasaki haruma 長崎波留麻 (The Nagasaki "Halma"), 333, 383
Nagasaki Naval Training Center (Nagasaki kaigun denshūsho 長崎海軍傳習所), 302, 394, 397
Nagashino 長篠, battle of (1575), 172
Nagoya 名古屋, 230n.11
Nagoya Gen'i 名古屋玄醫 (med., 1629–1696), 282
Naibara 内原, Luis ルイス, 189n.46
Naishō dōbanzu 内象銅版圖 (Copperplate charts of anatomy), 384
Naitō Masaki 内藤政樹 (math., 1703–1766), 369
Naiyakushi 内藥司 bureau, 101
Naka Ten'yū 中天游 (Dutch Learning, fl. 1817), 335
Nakae Tōju 中江東樹 (Conf., 1608–1648), 238, 246, 297
Nakagawa Jun'an 中川淳庵 (med., 1739–1780), 322
Nakajuku 中塾 (Dutch Learning academy), 335
Nakane Genkei 中根元圭 (math., 1661?–1733), 312–313, 349, 350, 368, 369
Nakarai 半井 family, 288
Nakatsu 中津 domain, 317
Nakatsukasashō 中務省 (central administrative ministry), 33
Nakayama Shigeru 中山茂, xvii, 45, 125–126
Nanban byōbu 南蠻屏風 ("Southern barbarians" screens), 158
Nanban geka. *See* "Southern barbarians' surgery"
Naniwa 難波 (Osaka). *See* Osaka
Naniwa no kusushi 難波藥師 (medical practitioners descended from Te Lai), 44
Napoleonic wars, 292, 334, 338
Nara 奈良 (Heijōkyō 平城京), 20, 21, 26, 133
  calendar, 199
  Kasuga Shrine, 23
  Kōfukuji temple, 102
  Tōdaiji temple, 27
Nara 奈良 era (710–794), xxvi, 16, 25, 26, 27, 28, 42
  reforms, 1
Narabayashi Chinzan 楢林鎮山 (med., 1643–1711), 284, 287, 331, 386
Narabayashi Sōken 楢林宗建 (Dutch Learning, fl. 1846), 335
Narcotics merchants, xvi
Narutaki 鳴瀧, Nagasaki, 340
National defense, 292, 315, 337, 341, 345, 373, 395–396, 398, 399
  naval, 345
  and Western science and technology, 278, 291, 294, 296, 302, 334, 394
"National Learning" (*Kokugaku* 國學), 238, 248, 249, 307–309, 395
National unity, 3, 160, 161, 168, 172, 228
  Buddhist threat to, 158, 160
  Christian threat to, 160

476 Index

Natural law, 200
Natural order
  cyclical behavior in, 91–92
  and fate calculation, 55
  irregularities in, 53
  and the political world, 52
Natural phenomena
  Chinese observation of, 52
  Jesuits' attitude to, 159
Natural sciences, 303–304
  in curriculum of Jesuit college, 194
Nautical calendar, 176, 178, 179, 180, 203, 360
Nautical instruments, 170
Naval Academy (Kaigunsho 海軍所), 398
Naval defense. See National defense
Naval techniques, 397
Navigation, xxix, 156, 169, 174–181, 204, 237, 301, 361, 370, 371, 396
  Western. See Western navigation
Navigational charts, 176, 179, 203
Navigational table
  Dutch, 371
Needham, Joseph, 36, 137n.42, 274, 275
Negative numbers. See Numbers
Nei-ching. See Huang-ti nei-ching
Neo-Confucianism, 8, 108, 120, 139, 167, 187, 198, 215, 239–240, 241, 249, 251, 259, 261, 282, 299, 303–304. See also Shushigaku
  attacks on Catholicism, 244
  separated from Buddhist learning, 241
  criticism of, 246, 247, 300, 302
  and medicine, 277, 278
Nervous system, 381, 391
New Buddhism, 103, 106, 113, 120
  priests of, 120, 127, 141–142
New Japanese-English Dictionary (Kenkyusha), xxx
New Year's Day, in Islamic calendar, 70
Newton, Isaac, 173, 304, 314, 352, 353, 355, 371
Nichigetsu kaigō sanpō 日月會合算法, 253
Nichiren 日蓮, 113
Nieuw Nederduitsch en Fransch Woordenboek (Halma), 332, 333
Nigi ryakusetsu 二儀略說 (Outline theory of terrestrial and celestial globes), 260, 261
Nihon kairo sokuryōzu 日本海路測量圖 (Surveyor's coastal map of Japan), 342
Nihon shoki 日本書紀 (Chronicles of Japan, also known as Nihongi 日本紀; compiled A.D. 720), 44, 45, 249
Nihonkoku kenzaisho mokuroku 日本國見在書目錄 (Catalogue of books seen in Japan), 133
Nin'an 仁安 era (1166–1169), 147
Nine chapters on the mathematical art. See Chiu-chang suan-shu
Nineteen-year period (chang 章), 47, 70
Ninshō 忍性 (Buddhist priest, fl. 1259), 141
Nintoku 仁德, Emperor (d. ca. A.D. 399), tomb of, 24
Nirvana, 142
Nishi Amane 西周 (philosopher, 1829–1897), 306
Nishi Genpo 西玄甫 (med., ?–1684), 286
Nishi Kichibei 西吉兵衞 (interpreter, ?–1666), 259, 286
Nishi school, Edo, 317
Nishi Zenzaburō 西善三郎 (interpreter, ?–1768), 332
Nishikawa Joken 西川如見 (astron., 1648–1724), 261–262, 286, 304, 313, 349
Nishikawa Seikyū 西川正休 (astron., 1693–1756), 261, 262, 286, 313, 349, 350, 351
Nisshūdō 日習堂 (Dutch Learning academy), 335
Nitrates, 152

Nobunaga. *See* Oda Nobunaga
Nodical months, 355
Nōgyō zensho 農業全書 (Complete work on agriculture), 238
Noro Genjō 野呂元丈 (Dutch Learning, 1693–1761), 315–316, 350
North Pole, 53
Northern-Southern Dynasties era (420–589), 44, 50
Northern Sung period, 129
  calendar changes, 122
  medicine, 144
Notations, 275
  algebraic, 130, 208, 267, 268, 271
  numerical, 74
Nozawa Sadanaga 野澤定長 (math., fl. 1664), 265
Nu 奴 (northern Kyushu state), 3n.4
Numbers
  decimal units: *to* 十 (10), *momo* 百 (100), *chi* 千 (1,000), *yorozu* 萬 (10,000), *soyorozu* 十萬 (100,000), *momoyorozu* 百萬 (1,000,000), *chiyorozu* 千萬 (10,000,000), 134
  even 135
  negative, 74, 79, 81, 130
  odd, 135
  positive, 79, 130
Numerals
  Arabic, 180, 208
  Indian, 49, 76
Numerical calculations, 54
  on abacus, 205
  on counting board, 266–271, 275
  using Arabic numerals, 180
Numerical coefficients, 81, 82
Numerical notations. *See* Notations
Numerology, 9, 78–79, 206
  Indian, 49
  and predictions, 134–135

Obama 小濱 fief, 380
Ōbaku 黃檗 (Zen) sect, 231n.15
Observation, methods of, 99
  medical, 221

Obstetrics, 217, 386, 388
Ōchōfū 王朝風 (courtly life-style), 114
Octant, 176, 356
Oda Nobunaga 織田信長 (1534–1582), 148, 158, 160, 161, 162, 164, 166, 167, 172, 173, 181, 188, 199
Odawara 小田原, 171, 184
Ōe 大江 household, 115, 116, 118
Ōei jūhachinen ryōreki dankan 應永十八年領曆斷簡, 58
Office for investigating barbarian documents (Bansho shirabesho 蕃書調所), 396
Office for the Translation of Barbarian Texts (Bansho wage goyō 蕃書和解御用). *See* Translation Office of the Tenmongata
Office of Temples and Shrines (*Jisha bugyō* 寺社奉行), 257
Official interpreters, 331
Official Prescriptions manual, 209, 210, 213, 214. *See also Ho-chi-chü fang*
Ogata Kōan 緒方洪庵 (med., 1810–1863), 330, 335
Ogawa Masaoki 小川正興 (cal., fl. 1673), 253
Ogyū Sorai 荻生徂來 (Conf., 1666–1728), 238, 241, 247–248, 249, 251, 274, 297, 298, 300, 304, 306, 307
Ōhashi Totsuan 大橋訥庵 (Conf., 1816–1862), 305
Ōjin 應神, Emperor (d. ca. A.D. 310), tomb of, 24
Oka Kenkai 岡研介 (med., fl. 1831), 330
Okamoto Genya 岡本玄冶 (med., 1587–1645), 279
Okamoto Ippō 岡本一抱 (med., 1686–1754), 280
Okanoi Gentei 岡野井玄貞 (med., fl. 1643), 255
Okayama 岡山 domain, 245
Okumura Kisaburō 奧村喜三郎

(surveyor), 345–346
Omens. *See* Portents
Ōmiya 大宮, calendar issued at, 199
Ōmura 大村 fief, 160
On'isei. *See* "Privileged-rank system"
Onmyōdō 陰陽道 (Chinese-style divination), 35, 114, 124, 125, 127, 128, 198
Onmyōji 陰陽師 (or *on'yōshi*; practitioners of *onmyōdō*), 124, 125
Onmyōryō 陰陽寮 (Institute of Divination), 31, 33, 35–36, 38, 50, 53, 54, 55, 56, 72, 74, 116, 123, 126, 128, 133
  buildings destroyed by fire, 124
  declining prestige of, 199
Onozukara no michi 自然の道 ("natural way" of the Japanese), 307
Ontleedkundige tafelen (Dutch translation of a German anatomical atlas), 317
"Openings" through which disease agents can enter the body, 96
Operations. *See* Surgical operations
Ophthalmology, 147, 214, 217, 289, 343, 347, 393
Opium War (1840–1842), 292, 347
Oracle bones, 7
Ōraimono 往來物 (didactic exchange of letters), 135–136, 203
Oral hygiene, 214, 217, 289, 375
Oranda-banashi 紅毛談 (Tales of Holland), 323
Oranda chikyū zusetsu 阿蘭陀地球圖說 (Dutchmen's illustration of the earth), 351
Oranda honzō wage 阿蘭陀本草和解 (Japanese version of Dutch materia medica), 315
Oranda iwa 和蘭醫話 (Medical tales from Holland), 391
Oranda kaheikō 和蘭貨幣考 (On Dutch currency), 316
Oranda kinjūchūgyozu wage 阿蘭陀禽獸蟲魚圖和解 (Japanese version of Dutch explanations of drawings of birds, beasts, insects, and fish), 315
Oranda monji ryakkō 和蘭文字略考 (Primer of the Dutch language), 316
Oranda naikei ihan teikō 和蘭內景醫範提綱 (Outline of a Dutch anatomical text), 330, 384
Oranda yakusen 和蘭藥選 (Selected drugs of Holland), 330, 385
Oranda zenku naigai bungōzu 和蘭全軀內外分合圖 (Anatomical chart of the human body), 382
Orandajii 和蘭字彙 (Dutch glossary), 333
Organs, of the human body, 378, 379
  internal, 322, 378, 380, 381
Osaka 大坂 (Naniwa 難波), 17, 101, 230, 378
  amateur astronomers in, 351
  calendar, 199
  Ishiyama Honganji temple, 173
  Kaitokudō academy, 305
  Senjikan private academy, 355
  Shitennōji temple, 105
  Sumiyoshi shrine, 105
Osaka uprising (1837), 346
Ōsu 大洲 fief, Shikoku, 245
Ōshio Heihachirō 大鹽平八郎 (shogunate samurai, 1793–1837), 246n.28, 346
Ōta Kenryū 太田見龍 (med., 1725–1812), 309
Ōtomo 大友 fief, 160
Ōtsuki Gentaku 大槻玄澤 (med., 1757–1827), 329, 332, 334, 336, 337, 341, 357, 386
Ōtsuki Shunsai 大槻俊齋 (med., 1806–1862), 377
Ōuchi 大內 household, 152

Outcasts, 380, 382
Overseas trade, xxvii, 14, 19, 104, 105, 153, 156, 163, 165, 168, 204, 334
　Chinese, 107, 148, 150–151, 175, 224–225
　Dutch, 160–161, 225–226, 284, 338
　English, 161, 225
　European, 156
　Japanese prohibition of, 104, 106, 161
　Japanese restrictions on, 225
　Ming prohibition of, 106, 152
　Portuguese, 160, 225
　private transactions, 151
　prohibitions of, 152
Overseas travel, 113, 149, 179
　restrictions on, 19, 161, 163, 165, 225
　restrictions rescinded, 398
Ozeki San'ei 小關三英 (med., 1787–1839), 344, 346, 384
Ozuki 小槻 household, 116, 118, 132, 203

Pacific, crossing of, 177, 178
Padua University, 381
Paekche 百濟 state, 4, 5, 12, 14, 20, 21, 25, 45
　requests from Yamato court for medical specialists, etc., 44
Paper, 109n.7, 112, 169
Papermaking, 169, 185
Paradise Temple. *See* Gokurakuji
Paré, Ambroise (surgeon, 1517?–1590), 220–221, 222, 284, 286, 386
Pascal (1623–1662), 364
Pasteur, Louis (1822–1895), 220
Pastimes, 78, 79
Pathological theory, 99
Pathology, 140, 209, 210, 281, 374, 379, 393
　Buddhist ideas on, 142, 143
　terminology, 306
Peasant omen interpretation, 127
Peasant revolts, 161, 165, 293

Peasant-warriors, 111
Peasants, 6, 12, 163, 293–294
　foreign influence among, 165
　wealthy, 308
Pediatrics, 147, 217, 289, 307, 386, 393
Pendulum clock, 356
Peking 北京, 151, 202, 208
　Islamic observatory in, 122
Pen-ts'ao chi-chu 本草集注 (Collected annotations on materia medica), 86
Percentages, 75
Perry, Matthew C., 292, 293, 294, 296, 323, 334, 337, 338, 370, 371, 372, 394, 395
Persia and the Persians, 104, 122
Phaeton (British frigate), 292
Pharmacology, 9, 385. *See also* Materia medica
"Phase energetics," 86, 139, 209–214, 280, 282, 373
Philippines, 154, 157, 177, 259
"Philosophy," Japanese terms for, 306
Phoenicians, 175
Physical exercises, 98
Physicians, 44, 119, 141, 251, 377, 378
　Buddhism and, 279, 282–283, 288
　and clerical dress, 289
　hondō, 289
　of *Li* school, 279
　in shogunate employment, 302
Physics, xxviii, 305, 371, 396, 397
　terminology, 306
Physiology, 95–96, 99, 140, 209, 210, 305, 384, 393
　Buddhist ideas on, 142
　Chinese, 95–97
　technical terms, 306, 385
Pi, value of, 75, 76, 81–82, 129, 362
Pilots, 178
Pinax microcosmographicus (Remmlin), 382
Pinhole device, to produce sharply defined shadow, 258

Pinto, Fernão Mendes (Portuguese traveler), 171–172
Pirate-traders. *See* Wakō
Place values, 74, 79, 80
Planetary motion, 52, 55, 61, 71, 200
  elliptic orbits of, 201
Planetary tables, 69, 125
Planets, 49, 61, 252
  apparent and true courses of, 352
  conjunction or occultation of, 53
  "invisible," 49
  orbits of, 54
  positions of, 48, 50, 56, 69, 121
Plants, medicinal uses of, 316
Platonic norms, reflected in Jesuit college's curriculum, 194
Plumb line, 176, 341
Pole star, 300
Political power,
  and military cliques, 133
  and the nobility, 133
Political prisoners, vivisection of, 379
Pollution, and disease, 395
Polo, Marco, 109n.7
Population growth
  in China, 107, 149
Porcelain, 169, 226
Porkert, Manfred, 86, 95
Portents, 52, 54, 55, 128
Portugal and the Portuguese, xxv, 157, 159, 160, 171, 178, 189, 201, 202, 219, 292
Portuguese language, 284, 330
Portuguese priests, 189
Portuguese ship, 158
Positive numbers. *See* Numbers
Post-mortem dissections. *See* Dissections
Postal system, 398
"Practical learning" (*jitsugaku* 實學), 311
Prayers, 114
  Buddhist, 21
Prediction. *See* Astronomical predictions, Prognostication
Pregnancy, 89, 388, 391
Prescriptions. *See* Medicinal prescriptions
Priest-practitioners, 142, 213, 217
Priestly ranks, 217, 243
Printing, 108, 109n.7, 112, 169, 184–186, 224, 398. *See also* Reprints
  of Buddhist texts and Chinese classics, 184, 185
  of calendrical books, 252–253
  of Catholic doctrinal works, 185
  copperplate, 189, 384
  metal type, 108n.6, 185
  movable type, 108n.6, 169, 184, 185, 186,
  oldest existing printed book in Japan, 184
  oldest extant printed matter in the world, 184n.36
  romanization in, 185
  of secular material for commercial purposes, 184, 185, 186, 230
  woodblock, 107, 128, 184, 185, 186, 380.
Printing-press, 185, 186, 189
Private academies, 39–40, 41, 118, 224, 230, 235, 236, 263, 295
  for calendrics, 355
  Chinese (Ming), 150
  Confucian, 243, 245, 247–248
  Dutch Learning (*Rangakujuku*), 334–336, 383, 386
  for Kōmō surgery, 285
  medical, 288–289, 295, 302
  for National Learning, 249, 307
  for *wasan*, 276, 301, 366
Private ownership
  of Chinese classics used in the University, 124
  of land, 13, 123
"Privileged-rank system" (*on'isei* 蔭位制), 40
"Problem succession" (*idai keishō* 遺題繼承), 206, 263–266, 271, 290
Professional appointments, 116, 117, 118
Prognosis, 135

Prognostication, 45, 63, 198
  individual fate, 52, 54, 55, 66–67, 74, 78, 90, 114, 120, 124, 125, 127, 128, 199
  numerological, 134–135
Progressions
  arithmetical, 76, 131, 208
  geometric, 208
Prohibition Against Heterodox Teachings (1790), 298
Proof, absence of demand for, in Chinese mathematics, 275
Proportions, 75
Protective associations (*kabu nakama* 株仲間), 230n.12
Protestants, 157
Provincial colleges. *See* Kokugaku
Provincial unrest, 111
Pseudosciences, 9
Psychology
  terminology, 306
Psychophysical principles in human activities, 142–143
Ptolemy and the Ptolemaic system, xv, 201, 258, 351, 352
Pu. See Seventy-six-year cycle
Pu-k'ung. See Amoghavajra
Public finance, 33
Publishing, government directives on, 376–377
Pulguksa 佛國寺 temple, Korea, 184n.36
Pulse, 216n.72
Pumps, 183, 233n.18
"Pure Land" (*Jōdo* 淨土) sect, 113
Purification rituals, 44
Puzzles. *See* Games and puzzles

Quadrant, 176, 179, 180, 203, 312, 356, 358
Quicksilver. *See* Mercury

Racial types, in Japan, 2n.2
Rahu ("invisible" planet), 49
Rangaku. See Dutch Learning
Rangakujuku. See Dutch Learning academies
Rangaku kotohajime 蘭學事始 (Beginning of Dutch studies), 322n.34
Rangefinding, 232
Ranpeki 蘭癖 (mania for things Dutch), 373
Ranpō. See Dutch medicine
Rarande rekisho kanken ラランデ曆書管見 (A private review of Lalande's *Astronomie*), 357
Rationalism, 30, 46, 187
Ratios, 75
Rebellions. *See* Riots and rebellions
"Red-hairs" surgery. *See Kōmō geka*
"Red Seal Ships" (*Shuinsen* 朱印船), 153, 154, 155, 156, 161, 162, 164, 168, 172, 176, 178, 179
Refining techniques for gold and silver, 183
Reformation (European), 156
Reforms, 1, 5–6, 11–14
  conservative, 347
Reign name changes. *See* Era names
Reincarnation, 143
Rekidō. See Calendrical astronomy course
Rekigaku seimō 曆學正蒙, 253
Rekihō shinsho 曆法新書, 254, 354
Rekirin mondōshū 曆林問答集 (Collection of dialogues on the calendar), 54n.72, 198
Rekishō shinsho 曆象新書 (New treatise on calendrical phenomena), 304, 353
Remmlin, Johann, 382
Renaissance (European), 156
Renaissance of learning and science (Japanese). *See* Intellectual renaissance, or outburst
Reprints
  of calendrical texts, 252, 253
  of classical texts, 107–108, 129
  of mathematical texts, 206
  of medical encyclopedia, 214
  by woodblock, 185
Research, 99
  anatomical, 381
  biological, 391

Research (continued)
  calendrical, 251
  chemical, 391
  medical, by Buddhist priests, 121
  on Western military technology, 396, 397
Research and training institutes, 397
Research students, 19
Reservoir construction, 25
Revenue, sources of, 164
Rhijne, Willem ten (Dejima doctor, 1647–1700), 228, 287
Rho, Jacob (Jesuit), 368
Ri. See Li
Rib surgery, 388
Ricci, Matteo (Jesuit scholar, 1552–1610), 202, 208, 259, 361, 370
Rice cultivation, 2, 24
  in China, 107
Rigaku 理學 (term for branch of natural science), 306
Right angle theorem, 75
Rikugunsho. See Military Academy
Rinzai 臨濟 (Buddhist sect), 119, 120, 135, 143
Riots and rebellions
  in northern Korea, 2
  peasant, 161, 165, 293
  at Shimabara, 161, 163, 165
  urban, 294
Riparian construction, 25
Rites and ceremonies, 120, 126
  at the Yamato court, 23–24
  Taoist, 87
Ritsuryō Institute of Medicine. See Ten'yakuryō
Ritsuryō (seido) 律令 (制度), xxvi, xxvii, 11–14, 16, 18, 20, 27, 82, 103, 110–111, 115, 120, 133
  academic system, 31–42, 45, 79, 194
  Buddhism's role in, 21, 41
  breakdown of, 22, 27, 30, 32, 41, 140
  and Chinese learning, 29
  and human dissections, 379
  and medicine, 100, 140–141
  and Shinto, 24
Ritual healing. See Jugon
Rituals to ward off evils, 126
Rōkoku 漏刻 (timekeeping), 35
Roman calendar, 71n.78
Roman Catholicism and Roman Catholics, 157, 158, 160, 165, 220, 231, 244, 310–311
  charitable work of, 217–219
  criticism of, 240
  Japanese sympathizers of, 161
  persecution and suppression of, 161, 162, 165, 167, 186, 189, 195, 227, 235, 242, 299
Romanization
  of Dutch words, 316
  of Japanese and Chinese words, xxi
  of Korean words, xxi
  in printing, 185
Rome, 177
  Jesuit headquarters in, 218
Rongo. See Lun yü
Rōnin 浪人 (lordless samurai), 236, 245, 247, 276
  mathematicians, 364, 366
Root extraction methods, 130, 132
Ruijūhō 類聚方 (Classified prescriptions), 373–374
Rural sector, 293–294
  and urban sector differentiated, 163, 229–230
Russia and the Russians, 292, 377
  Treaty of amity (1854), 395
Russian language, 377–378, 398
Ryōrenji 了蓮寺 temple, Kyoto, 354
Ryūha 流派 (school or sect), 367
Ryukyu 琉球 Islands, 226n.5
Ryōbyōin 療病院 (medical facility), 102, 105
Ryōgi shūsetsu 兩儀集說 (An explanation of collected materials on celestial and terrestrial globes), 262

Sado 佐渡 Island, 182
"Sado kinzan kinbori no zu" 佐渡金山金掘之圖, 182
Saga 佐賀 domain, 397
Saga 嵯峨, Kyoto, 279
Saichō 最澄 (Dengyō Daishi 傳教大師; Buddhist priest, 767–822), 10n.12, 41–42
Saijō 最上 school, 368
Saijō-ryū 最上流 school of mathematics, 363
Saikaidō 西海道 region (Kyushu), 38
Sailing techniques, 175, 176, 178
Sairan igen 采覽異言 (Selected strange accounts), 311
Saitokujin 歲德神 (God of the Year's Virtue), 57
Sakai 堺 (Izumi 和泉 district), 163, 171, 184, 214, 219
Sakai Tadamochi 酒井忠用 (daimyo, fl. 1754), 380
Sakamoto 坂本, 171
Sakuma Shōzan 佐久間象山 (Dutch Learning, 1811–1864), 246n.28
Samurai 侍 (or 武士), xxiii, 79, 105, 111, 112, 119, 120, 126–127, 163, 165, 166, 167, 195, 216, 230, 231, 242, 247, 289, 299, 399
  education of, 166, 188, 224, 295, 334
  in the isolation period, 228–229
  low-ranking, 207, 294
  and mathematics, 207, 274, 276, 302, 366
  and national defense, 396, 397, 398
  scholars, 235–236, 263
  in the *Shōshikai*, 345
  technology and, 170
San-chiao 三焦 (*sanshō* in Japanese), 95, 379
San-t'ung li 三統曆 calendar, 48
Sandō. *See* Mathematics course
Sangaku 算額 (votive plaque), 276–277

Sangaku kōchi 算學鉤致 (Inquiry into mathematics), 264n.41
Sano Yasusada 佐野安貞 (med., fl. 1760), 381
Sanpō chokkai 算法直解, 265
Sanpō hatsumōshū 算法發蒙集, 265
Sanpō ketsugishō 算法闕疑抄 (Selected solutions of mathematical methods), 264, 265
Sanpō kongenki 算法根源記 (Recorded bases of mathematical method), 264, 265
Sanpō meikai 算法明解, 265
Sanpō shigenki 算法至源記, 265
Sanpō shinso 算法新書 (New text on mathematical method), 272, 366
Sanron 產論 (Obstetrics), 388, 391
Sanronyoku 產論翼 (Obstetrics, Enlarged), 392
Sanryōroku 參兩錄, 265
Sansho. *See* San-chiao
Sanso 算俎, 265
Santa Buenaventura (Japanese-built ship), 177
Satō Masaoki 佐藤正興 (math., fl. 1669), 264, 265
Satō Taizen 佐藤泰然 (Dutch Learning, fl. 1842), 335
Satsuma 薩摩 domain, 178, 397, 399
Saturn's rings, 356
Sawaguchi Kazuyuki 澤口一之 (math., fl. 1670), 265, 266, 271
Sawano Chūan 澤野忠庵 (an apostate Jesuit, formerly Christavão Ferreira, 1580–1650), 259–260, 284
Schaedel, Juliaen (mortar technician), 232
Schall von Bell, Johann Adam (Jesuit), 368
Schamberger, Caspar (Dejima doctor, fl. 1649–1651), 286. *See also* Caspar school
Scholasticism, xxv, 195
Schools, xv, 196

Schools (*continued*)
  fief (*hankō* 藩校), 224, 288, 295, 298, 302, 308, 367
  Jesuit, 188
  medical, 288
Science
  Buddhism and, 41, 121
  Chinese. *See* Chinese science
  Chinese-style, xxi, 4n.8, 196
  European. *See* Western science and technology
  as a function of government, 7, 8
  institutionalized, 45
  interaction between traditional and modern, xiv
  prehistory of, 42–43
  revival of, 103, 121
  and Western techniques, 170
Science and Culture in Modern Japan, xxx, 397, 399
Scientific instruments, 183
Scientific Revolution (European), xv
Scientific revolutions, theory of, xv
Scientific studies, 229
Scientific terms, 306
Sea battles, 173
the Seasons, 70, 262
Seibutsugaku 生物學 (biology), 306
Seii genbyō ryaku 西醫原病略 (Outline of Western physiology), 385
Seii tai shōgun 征夷大將軍 ("Barbarian-quelling Generalissimo"), 164
Seijukan 躋壽館 (private medical academy), 302, 323, 375
Seijutsu hongen taiyō kyūri ryōkai shinsei tenchi nikyū yōhōki 星術本源太陽窮理了解新制天地二球用法記 (The ground of astronomy . . .), 352
Seikiron 生機論 (physiology text), 330
Seikyō yōroku 聖敎要錄 (Essentials of revered learning), 247
Seimi kaisō 舎密開宗 (chemistry text), 331

Seireki shinsho 西曆新書 (A new treatise on Western astronomy), 359
Seiri. *See Seirigaku*
Seiri hatsumō 生理發蒙 (Enlightenment on physiology), 385
Seirigaku 生理學 (physiology), 306, 385
Seisetsu naika sen'yō 西說內科撰要 (Essentials of Western internal medicine), 329, 376, 385, 393
Seiyō igakusho. *See* Center for Western Medicine
Seiyō kibun 西洋紀聞 (Memorandum concerning the West), 310–311
Seiyō sanpō 精要算法 (Detailed key to mathematical methods), 363, 364–365, 366
Seki 關 school of *wasan*, 346, 350, 361, 362, 364, 366, 367, 371
Sekii 昔尹 (son of Shibukawa Harumi), 349n.54
Seki "idan" 斥醫斷 (Refutation of "Idan"), 374
Seki Takakazu 關孝和 (math., ?–1708), 238, 253, 264, 265, 266–267, 271, 274, 275, 276, 362, 365–366, 369
Semiseclusion period, xxiv–xxv, xxvii, 19, 42, 74, 83, 84, 103–114, 187, 199, 203, 205, 213, 348
Sendai, 仙臺, 177
Sengoku 戰國 period (1467–1603), 163, 165, 166, 168, 287
Sengoku daimyo 戰國大名 (daimyo of the Warring States era), 161, 166, 168
Senjikan 先事館 (private academy), Osaka, 355
Senmyōreki 宣明曆 (*Hsüan-ming li* in Chinese) calendar, 49, 51, 72, 128, 199, 238, 252, 253, 254, 255, 256
Senpu 先負, 66
Senshō 先勝, 66

Serf labor, 7
Serum therapy, 220
Setchūha. See Eclectic School of Medicine
Seventy-six-year cycle (*pu* 蔀), 47, 70–71
Sexagenary cycle, 57, 58, 61, 63, 66
Sextant, 176, 341, 356
Sexual hygiene, 140
Seyakuin 施藥院 (medical facility), 102
Shaketsu shujutsu zu 瀉血手術圖 (Picture of Siebold performing a blood-letting operation), 339
Shakku 赤口, 66
Shamanism, 90, 109
Shang 商 dynasty (ca. 16th century B.C.), 7
 calendars, 46, 69
Shang-han tsa-ping lun 傷寒雜病論 (sometimes abbreviated *Shang-han lun* 傷寒論; On Cold Damage Disorders), 85, 138, 139, 210, 211, 281, 283, 309, 374, 378
Shang lun p'ien 尚論篇 ("On Cold Damage Disorders" Reaffirmed), 281
Shells (projectiles), 232
Shen Kua 沈括 (Chinese, cal., fl. 1088), 107
Shen-nung pen-ts'ao ching 神農本草經 (The Divine Cultivator's classic on materia medica), 85–86, 91, 97–98
Shi 師 (practitioner), 37
Shiba Kōkan 司馬江漢 (astron., 1738–1818), 297, 314, 352, 354
Shibamura Moriyuki 柴村盛之 (math., fl. 1657), 265
Shibukawa Harumi 澁川春海 (also read Shibukawa Shunkai; cal., 1639–1715), 254, 255–258, 275, 299, 348
Shibukawa Kagesuke 澁川景佑 (cal., 1787–1856), 254, 314, 315, 357, 359, 371
Shih 史 (history, also "one who makes the calendar"), 52
Shih chi 史記 (Records of the Grand Historian), 55
Shih tsung 世宗 (Ming emperor), 214
Shikandō 絲漢堂 (Dutch Learning academy), 335
Shikibushō 式部省 (ministry for ceremonials), 32
Shikoku 四國 island, 245
Shimabara revolt (*Shimabara no ran* 島原の亂), 161, 163, 165
Shimamura Teiho 島村鼎甫 (med., 1830–1881), 385
Shimazu army, 219
Shimazu 島津 fief, 184, 189
Shimogamo 下賀茂 Shrine, Kyoto, 23
Shimonoseki 下關 straits, 341, 343
Shingon 眞言 sect, 41, 42, 141
Shingū Ryōtei 新宮凉庭 (Dutch Learning, fl. 1839), 335
Shinkan Jujirekikyō 新勘授時曆經, 253
Shinkiron 愼機論 (On caution at a vital juncture), 345
Shinkō rekisho 新巧曆書 (Astronomy by the new technique), 314, 359
Shinpen (*newly revised*) *Jinkōki* 新編塵劫記, 206, 264, 265
Shinpō rekisho 新法曆書 (Calendrical treatise by the new method), 254, 359
Shinrigaku 心理學 (psychology), 306
Shinsen shōjiroku 新撰姓氏錄 (New compilation of the register of families), 20
Shinto 神道, 23–24, 28, 44, 114, 124, 158, 187, 245, 249, 301
 heliocentric theory and, 354
 and National Learning, 308
Shinto ceremonial items in tombs, 25
Shinto deities, 276
Shintō kireiden 神道奇靈傳 (On

486   Index

the mystic spirits of Shinto), 309
Shinto priests, 307, 308
Shinto rituals, 23–24
Shipbuilding, 156, 169, 174–181, 232, 396, 397
  restrictions on, 233
Shipping, restrictions on, 225, 345
Ships
  confiscation of, over 500-*koku* capacity, 232
  flat-bottomed, 175, 233
  iron-plated, 173
  Japanese-style, 233
  merchant. *See* Merchant ships
  for troops, horses, and supplies, 177
  war, 179, 396
  Western-style, 177, 233
Ship's fittings, 183
Ship's log, 176
Shirandō 芝蘭堂 (Dutch Learning academy), 334, 336
Shishisai 思思齋 (literary name of Naka Ten'yū), 335
Shishisaijuku 思思齋塾 (academy of Naka Ten'yū). 335
Shitennōji 四天王寺 temple (Osaka), 101, 105
Shizuki Tadao 志筑忠雄 (astron., 1760–1806), 304, 313–315, 330, 352–353, 354, 361, 371
Shōchō 消長 (secular diminution of tropical year-length), 355
Shōchōhō. See Law of variation
Shock, and early surgery, 221
Shōgakuin 奬學院 (University affiliate school), 116
Shogunate Institute of Medicine. *See* Igakkan
Shōheikō 昌平黌 (校)(Shōhei School), 243. *See also* Yushima Seidō
Shōheizaka 昌平坂 slope, Edo, 243
Shōheizaka Gakumonjo 昌平學問所 (Institute of Learning at Shōheizaka), 243n.26. *See also* Yushima Seidō

Shokan bumono 諸勘分物 (Various approaches to area and volume), 205–206
Shōkokuji 相國寺 temple, Kyoto, 242
Shokubutsugaku 植物學 (botany), 306
Shokuden 職田, 123
Shokugaku keigen 植學啓原 (botany text), 330
Shōmu 聖武, Emperor (r. 724–749), 27
Shōsendō 象先堂 (Dutch Learning academy), 335, 336
Shoshidai 所司代 (magistrate of Kyoto), 380
Shōshikai 尚齒會 (Elders-respecting circle), 345, 346, 384
Shōshō 少小 (pediatrics) course, 37
Shōsōin 正倉院 Repository, Nara, 27, 84n.88
Shōtoku 稱德, Emperor (r. 764–770), 184n.36
Shou-shih li. See Jujireki
Shou-shih li-ching 授時曆經 (calendrical system), 253, 254
Shu-li ching-yun 數理精蘊 (Essential treasury of mathematical principles), 370
Shu-shu chi-i 數術記遺 (a lost mathematical text), 79
Shu-shu chiu-chang 數書九章 (Mathematical treatise in nine sections), 129
Shugei shuchiin 綜藝種智院 (Buddhist academy), 41
Shūki sanpō 拾璣算法 (Selected jewels of mathematical method), 365
Shushigaku 朱子學 (philosophy of Chu Hsi), 235, 238, 240, 241, 242, 243–244, 248, 249, 295, 298, 299, 300, 301, 305. *See also* Neo-Confucianism
  attacks on, 245, 246, 247, 248, 297
Shutōjo. *See* Vaccination center
Siam, 154

Siberia, 377
Sickness. *See* Disease
Sidotti, Giovanni Battista (Jesuit missionary, 1668–1715), 300, 310–311
Siebold, Philipp Franz Balthasar von (Dejima doctor, 1796–1866), 228, 288, 333, 334, 338–344, 347, 359, 373, 376, 377, 384, 387, 391, 393
Siege warfare, 173, 232
Sighting calculations, 232
Silk, 225, 226
  weaving, 169
Silla 新羅 state, 4, 5, 12, 14, 25
  Buddhism in, 21
  unification of Korea, 19–20
  worsened relations with, 19
da Silva, Duarte (lay brother, fl. 1560), 194
Silver, 233
  mines, 181–183
  refining of, 183
  trade, 181, 225, 226
Sin, as cause of disease, 97
Sine functions, 49, 76
"Single poison" theory of disease, 374
Sino-Japanese relations, 28, 30, 153
"Six Bowels" (*liu-fu* 六腑), 95, 144, 379
Six Dynasties period (222–589), 4, 6, 56
  mathematics, 77
"Six Energetic Configurations" (*liu-ch'i* 六氣), 86, 211
Skeleton, 381
Skin diseases, 219
Skin irritation, treatment of, 43
Sky. *See* Heavens
Slaves and servants, 12
Smallpox, 377
Smellie, William (1697–1763), 391
Smelting, 169
Smoked glasses, 356
Social change
  during "semiseclusion," 103
  technology and, 170

Social class. *See* Status system
Social reforms, 248
Social stability, 166
Society of Jesus, 157
Sŏkkat'ap 釋迦塔 pagoda, 184n.36
Solar calendars, 69, 179, 360
Solar motion, 48, 49, 52
Solar system, 352
Solar year, 71n.78, 355
Solstices, 38, 47, 57, 122, 258, 259
Somatic functional systems (Chinese medicine), 96
Sōmō no Kokugaku 草莽の國學 ("grass-roots National Learning"), 308
Sōshu 創腫 (surgery) course, 37
Southeast Asia, Japanese trading communities in, 154–155
"Southern Barbarians" (*Nanbanjin* 南蠻人), 159
"Southern Barbarians' smelting" (*Nanbanbuki* 南蠻吹), 183
"Southern Barbarians' surgery" (*Nanban geka* 南蠻外科), 220, 284
Southern Seas, 152
Southern Sung era (1127–1279), 105, 129
  calendar changes, 122
  mathematics, 129, 132
Sōzokuki 草賊記 (essay by Kumazawa Banzan), 246
Spain and the Spanish, 157, 178, 181, 219, 292
Specialization of medical practice. *See* Medical specialists
Sperm, 391
Spine, 381
Spleen, 61, 93, 95
Splenetic functions, 212, 213
"Splenetic" system, 95
Spring and Autumn era (481–403 B.C.), 45
Square roots, 74, 75, 81, 132, 206, 208
Ssu-fen li 四分曆 calendar, 47, 48, 70
Ssu-ma Ch'ien 司馬遷 (135? B.C.–

93? B.C.), 55
Ssu-yüan yü-chien 四元玉鑑 (Precious mirror of the four origins), 131, 204
Star chart, 250
Stars, navigation by, 179, 180, 203
Statics, 173
Status system, 12–13, 229, 294, 357
  and education, 39–40, 41
  hereditary, 164
  mathematics and, 274, 364
  physicians and, 289
Steam engines, 396, 397
Steamship travel, 398
Steel, 172
Steenstra, Pybo (fl. 1711–1722), 359
Sterilization, 221
Stomach, functions of, 212, 213
Stone Age, in Japan, 2
Stones in the bladder, 388
Student-priests, 14, 16, 17–18
Students
  cultural role of, 14, 17, 18–19
  and social class privilege, 40
Su Ching 蘇敬 (Chinese, med., fl. 659), 86
Su-wen 素問 (part of the "Yellow Emperor's inner classic"), 85, 210
Su-wen ju-shih yün-ch'i lun-ao 素問入式運氣論奧, 210
Suan-ching shih-shu 算經十書 ("ten mathematical manuals"), 75
Suan-fa t'ung-tsung 算法統宗 (A systematic treatise on arithmetic), 204, 206, 207, 264
Suan-hsüeh ch'i-meng 算學啟蒙 (Introduction to mathematical studies), 131, 204, 264, 266
Subtraction, 74, 75, 79
  on the abacus, 205
Sugar, 226
Sugawara 菅原 household, 115, 116, 118
Sugawara no Minetsugu 菅原岑嗣 (med., fl. 868), 309
Sugimoto, Masayoshi 杉本正慶, xvii, xviii

Sugita Genpaku 杉田玄白 (med., 1738–1818), 317, 322, 332, 380, 386
Sugita Ryūkei 杉田立卿 (med., 1786–1845), 386
Sugiyama Sadaharu 杉山貞治 (math., fl. 1670), 265
Sui 隋 court, diplomatic missions to. See *Kenzuishi*
Sui 隋 period (589–618), 5, 6, 8, 25, 72, 143
Sui shu 隋書 (History of the Sui dynasty), 76
Sukuyō unmei kanroku 宿曜運命勘錄 (A record of fate prognostication according to the mansions and planets), 125
Sukuyōdō 宿曜道 (Japanese practice of Indian astrology), 57, 69, 74, 125, 127, 128
Sukuyōji 宿曜師 (practitioners of *sukuyōdō*), 125, 126, 128, 133
Sukuyōkyō. See *Hsiu-yao ching*
Sulphur, 152, 181, 226
Sumeru, Mt. (Shumisen 須彌山), 354
Sumiyoshi 住吉 Shrine, Osaka, 105
Sun, 49
  altitude at noon, 176
  motions of, 49, 72
  navigation by, 179, 180, 203
  orbit of, 54, 304
  position of, 50, 121
  revolution of, 356
  shooting the, 178
Sun-declination tables, 203
Sun and moon, conjunction of, 48, 72
Sun spots, 356
Sun-tzu suan-ching 孫子算經 (The mathematical manual of Master Sun), 76, 80, 135
Sundials, 122
Sung 宋 era (960–1279), 6, 8, 66, 103–104, 105, 107, 108, 114, 127, 143, 150, 172, 175, 184, 187, 210, 239–240, 297
  astronomy and astrology, 73,

121, 122
  calendar changes during, 122
  calendrical systems, 355
  culture, 106
  gunpowder, 170
  mathematics, 74, 80, 129, 132, 133
  medicine, 86, 94, 96, 138–139, 209–210, 279, 379
  smallpox vaccine discovered, 377
Sung Ying-hsing 宋應星 (Chinese, technology, fl. 1637), 231n.16
Sunrise and sunset times, 57
Suō 周防 province, 382
Surgery and surgeons, 43, 87, 98, 214, 217, 218, 224, 284, 289, 331, 347, 393
  Chinese-style, 287
  Dutch. *See* Kōmō geka
  low regard for, in traditional medicine, 285
  Western. *See* Western surgery
Surgical instruments, 215, 221, 287
Surgical operations, 339, 387–388, 393
Surveyor's pole, 76
Suzuki Sōden 鈴木宗傳 (med., fl. 1772), 382
Swain, David, xvii, xix
Sweating, 212
Swords and swordsmiths, 112, 168, 172
Syllabic script. *See* Kana
Symptoms of disease, 87, 138, 373, 378
Syndromes (symptomatic), 85, 91
Synodic month, 47, 48, 69, 70, 355
Synodic periods, 55
Syntagma anatomicum (Vesling), 381

Ta-ming li 大明曆 calendar, 76
Ta-t'ung li 大統曆 calendar, 123, 151, 255
Ta-yen li 大衍曆 calendar, 49, 51
Tale of sine, cosine, and inverse sine functions, 368
Tables of definite integrals, 362

Taboos, 57, 114, 125
Tachibana 橘 household, 116
Tacking, 176, 178
Tafel anatomia, 317, 322
Tagaya Tsunesada 多賀谷經貞 (math., fl. 1667), 265
Tahara 田原 fief, 344
T'ai-chi 太極 (Grand Polarity), 240
T'ai-chu li 太初曆 calendar, 48
T'ai-i shu 太醫署 (T'ang medical institute), 101
T'ai-pu shu 太卜署 (T'ang Bureau of Divination), 35, 53
T'ai-shih chu 太史局 (T'ang Institute of Astrology), 35, 52, 53, 73
Taian 大安, 66
Tai-fang 帶方, 3n.5
Taihō 大寶 Code (701), 12, 16, 36
Taiju 大儒 (great [Confucian] scholars), 304
Taika 大化 Reforms (646), 1, 11–12, 15, 21, 45, 228
Taira clan. *See* Heike clan
Taira no Kiyomori 平清盛 (samurai, 1118–1181), 104
Tairyō 體療 (internal medicine) course, 37
Taisei honzō meiso 泰西本草名疏 (Introduction of Linneaus' classifications), 330
Taiseiden 大成殿 hall, Yushima Seidō, 237n.20, 243
Taiseikan 大成館 (Dutch Learning academy), 335
Taiyō kyūri 太陽窮理 ("tracking of the sun's orbit"), 304
Takahashi Kageyasu 高橋景保 (astron.), 341, 343, 357–359
Takahashi Yoshitoki 高橋至時 (cal., 1764–1804), 254, 305 314, 315, 337, 355, 357
Takano Chōei 高野長英 (Dutch Learning, 1804–1850), 333, 344, 345, 346, 384–385
Takatsuki 高槻, 188
Takayama Ukon 高山右近 (vassal of Hideyoshi; 1552–1614),

188–189
Takebe Kataaki 建部賢明 (math.,
　　1661–1716), 276n.57
Takebe Katahiro 建部賢弘 (math.,
　　1664–1739), 253, 268, 271,
　　312, 313, 349, 362, 364, 368,
　　369
Takeda Katsuyori 武田勝頼
　　(daimyo, 1546–1582), 172
Takeda Masayoshi 武田昌慶
　　(med., 1338–?), 213–214
Taki 多紀 household, 288, 295, 302,
　　323, 375
Taki 瀧 (Siebold's mistress), 340
Taki Keizan 多紀桂山 (med.,
　　1755–1810), 375, 376
Taki Motokata 多紀元堅 (or
　　Genken; med., 1796–1857),
　　376
Taki Motonori 多紀元徳 (med.,
　　1732–1801), 375
Taki Mototaka 多紀元孝 (or
　　Genkō; med., 1695–1766),
　　375
Taki Motoyasu 多紀元簡 (or
　　Genkan). See Taki Keizan
Tally Trade (Agreement), 19, 106,
　　148, 150, 151–152, 156, 161,
　　168, 174–175, 198, 203
Tama no gotoku marukimono 玉の如く
　　丸き物 ("ball-like round
　　things"), 205
Tanaka Yoshizane 田中由眞
　　(math., fl. 1679), 265
Tanba 丹波 household, 140, 141
Tanba Motoyasu 丹波元泰 (or
　　Gentai; med.), 375
Tanba no Kaneyasu 丹波兼康
　　(med., fl. Nin'an era [1166–
　　1169]), 147
Tanba no Masatada 丹波雅忠
　　(med., 1021–1088), 140–141
Tanba no Yasuyori 丹波康頼
　　(med., fl. 982), 88, 140, 143,
　　375, 376
Tanegashima 種子島 (island), 157,
　　171, 201
Tanegashima Hisatoki 種子島久時
　　(daimyo, fl. 1607), 171n.16
Tanegashima Tokitaka 種子島時堯
　　(daimyo, 1528–1579),
　　171n.16
T'ang 唐 court, 12, 14
　　envoys to, 13, 14–20
　　medical services, 100–101
T'ang culture, 14, 25, 27, 29
　　decline of enthusiasm for, in
　　　Japan, 16, 19
T'ang 唐 period (618–907), 5, 6, 8,
　　9, 10, 11, 19, 46, 66, 106,
　　107, 210
　　and Buddhism, 21
　　calendars, 49, 72
　　decline of scientific creativity,
　　　45
　　mathematics, 74, 75, 76, 77, 81,
　　　132
　　medicine, 86, 91, 94, 95, 143
T'ang medical institutes, 87, 101
T'ang university (Kuo-tzu-chien
　　國子監), 75
　　course on mathematics, 129
Tanning, 380
Tao 道 (circuit), 101
Tao Hung-ching 陶弘景 (Chinese,
　　med., 456–536), 86, 91
Taoism and Taoists
　　and alchemy, 90–91
　　and astronomy, 50
　　and calendrical science, 57
　　in China, 1, 7, 9, 10, 11, 46, 108,
　　　124
　　in Japan, 30
　　and medicine, 86–87, 90, 91, 94
Tashiro Sanki 田代三喜 (med.,
　　1465–1537), 139, 214, 215,
　　216, 277
Tax collection, 207
Taxation, xxviii, 12, 13, 19, 27,
　　110–111, 164
Tçuzzu, João Rodrigues (Jesuit,
　　1561?–1633), 194–195
Te Lai 德來 (Tokurai in Japanese;
　　med., 459), 44
Tea, medicinal effects of, 143
Tea ceremony, 224, 231, 364

Teaching personnel
　court rank of, 83
　in Jesuit seminaries, 189
　in *ritsuryō* academic program 33,
　　35, 36, 38, 83
Technical diffusion, 21, 24–27
Technical terms, for Western
　sciences, 306, 384–385
Technicians, 18, 20, 27, 232
　Korean immigrants as, 25
　slaves as, 12
Technology, 20, 103, 163, 168,
　231–233
　Chinese. *See* Chinese technology
　Western. *See* Western technology
Teiki 帝紀 (Genealogy of the
　imperial family), 42n.51
Tekitekisaijuku 適適齋塾 (Dutch
　Learning academy), 335
Telescope, 180, 181, 201, 349, 350,
　356
Temple schools. *See Terakoya*
Temples. *See* Buddhist temples
Ten Celestial Stems, 57, 60–61, 62,
　63
Ten mathematical manuals, 129,
　132, 133
Tenant farmers, 294
Tenchi 天智, Emperor (r. 661–
　671), 55
Tenchi nikyū yōhō 天地二球用法
　(The use of celestial and
　terrestrial globes), 351
Tendai 天臺 sect, 42
Tengenjutsu. *See* "Heaven-origin
　algebraic method"
Tenmei 天明 era (1781–1798),
　293n.2
"Tenmon bunya no zu" 天文分野
　之圖 (star chart), 250
Tenmondō 天文道 (astrology), 35
Tenmongata 天文方 (Bureau of
　Astronomy), 257–258, 262,
　298, 300, 301, 303, 312, 314,
　334, 337, 341, 347, 348, 349,
　350, 351, 353, 356, 357, 359,
　377, 396
Tenpō 天保 calendar, 337, 360

Tenpō 天保 era (1830–1844), 347
Tenpōreki 天保暦 (calendrical
　system), 254, 313, 314, 315,
　359, 360
Tenryūji 天龍寺 temple (Kyoto),
　105
Ten'yakuryō 典藥寮 (Institute of
　Medicine [*ritsuryō*]), 31–32,
　36, 38, 84, 87, 88, 94, 101,
　102, 116, 118, 123, 140, 288
Tenzanjutsu 點竄術 (written
　algebra), 365–366
Teppōki 鐵炮記 (Record of
　firearms), 171
Terakoya 寺小屋 ("temple schools"),
　119, 136, 204, 275–276, 295
Terrestrial globes, 54, 122, 195
Tetsugaku 哲學 (philosophy), 306
Textual criticism and research, 223,
　241, 307, 375
Therapy. *See* Medical treatments
Three Kingdoms era (3rd–4th
　centuries), 9, 98, 387
Thunberg, Carl Pieter (Dejima
　doctor, 1796–1866), 228
T'ien-ching huo-wen 天經或問
　(Queries on the classics of
　heaven), 261, 349, 350
T'ien-kuan shu 天官書 (Record of
　heavenly offices), 55
T'ien-kung k'ai-wu 天工開物
　(Exploitation of the works
　of nature), 231n.16
T'ien-tsu 天子 (Heaven's son), 52
T'ien-wen chih 天文志 (Astrological
　treatise), 48, 54–55
T'ien-wen lueh 天文略 (An outline
　of celestial phenomena), 261
T'ien-yüan shu. *See* "Heaven-origin
　method"
Timekeeping, 50, 53, 73
To 弟 (*yin* aspect), 57
Toda Mosui 戸田茂睡 ("Japanese
　Learning," 1629–1706), 249
Tōdaiji 東大寺 temple, Nara, 21,
　26
Tōfukuji 東福寺 temple, Kyoto,
　136, 204

492   Index

Tōhoku 東北 district, 368
Toita Yasusuke 戸坂保佑 (math., 1708–1784), 363, 369
Tokugawa 徳川 era (17th to mid-19th centuries), xxii, xxv, 50, 103, 125, 148, 149, 160–161, 166, 167, 174, 182, 202, 208, 224, 228, 239, 240, 242, 246, 276, 288, 294, 295, 297, 299, 306, 380, 394. *See also* Edo period
Tokugawa hegemony, 228
Tokugawa household, 234, 343
Tokugawa Iemitsu 徳川家光 (shogun, 1604–1651), 165, 243
Tokugawa Ienari 徳川家齊 (shogun, r. 1787–1837), 343
Tokugawa Ienobu 徳川家宣 (shogun, r. 1709–1712), 244, 310
Tokugawa Iesada 徳川家定 (shogun, r. 1853–1858), 377
Tokugawa Ietsugu 徳川家繼 (shogun, r. 1713–1716), 244
Tokugawa Ietsuna 徳川家綱 (shogun, r. 1651–1680), 243, 247
Tokugawa Ieyasu 徳川家康 (shogun, r. 1603–1605, d. 1616), 148, 160, 161, 162, 164–165, 166, 167, 177, 181, 185, 216, 235, 242, 243, 255, 311
Tokugawa Mitsukuni 徳川光圀 (daimyo, 1628–1700), 237n.20, 256
Tokugawa 徳川 shogunate, xxvi, 125, 166, 234
Tokugawa Tsunayoshi 徳川綱吉 (shogun, r. 1680–1709), 235, 243
Tokugawa Yoshimune 徳川吉宗 (shogun, r. 1716–1745), 224, 244, 262, 291, 295, 311, 312, 313, 315, 316, 336, 348, 349, 350, 368, 372
Tokyo University, 359

medical faculty of, 378
Tombs
  ceremonial items in, 25
  of the Emperor Nintoku, 24
  of the Emperor Ōjin, 24
Tomobiki 友引, 66
Ton'ishō 頓醫抄 (Selected medical cures), 143, 144, 145, 146, 379
Torii Yōzō 鳥居耀藏 (shogunate official; 1804–1874), 345, 346
Torikabuto 草烏頭, 388
Tosa 土佐 province (domain), 244, 399
Tōshōdaiji 唐招提寺 (temple), 26
Towns
  castle, 163, 230
  commercial, 163
Townsmen's culture, 166
Toyotomi Hideyoshi 豊臣秀吉 (1536–1598), 148, 153, 160, 161, 162, 164, 165, 166, 167, 177, 181, 185, 189, 216, 219, 242
  seal of, 154
Trade, 204
  foreign. *See* Overseas trade
Trade agreements, 395
Trading ships. *See* Merchant ships
Traditional culture
  court nobility and, 231
  departure from, 166, 168
  preservation of, 165–166, 167, 196
Traditional learning and science, xxii, xxiii, 294, 295, 304, 310, 311, 316
  Japanese scholars' dissatisfaction with, 291, 296
  and Western learning, 347–348, 360
Training of adminstrative personnel, 13, 31, 32
  medical doctors, 32
Trajectory problems, 174, 232
Transit (astronomical instrument), 356

Index   493

Translation Office of the Tenmongata (Bansho wage goyō 蕃書和解御用), 301, 303, 314, 334, 336, 337, 341, 344, 357
   developed into the Center for Western Learning, 396
   separated from the Tenmongata, 359
Translations of Buddhist sutras, 9, 10
Translations into Japanese of Western sources, 337, 344, 346, 351–359, 398
   astronomy, 313–315, 331, 349, 351–352, 357, 359
   Christian literature, 195
   Dutch books, 323, 329, 330
   gynecology, 386
   medicine, 317, 331, 334, 337, 384, 385–386, 393
   obstetrics, 386
   pathology, 385
   pediatrics, 386
   pharmacology, 385
   physics, 371
   physiology, 384
   publication of, 323, 347
   surgery, 386–387, 393
   of the *Tafel anatomia*, 322
Trapezoid, equipartition of, 129
Travel. *See* Overseas travel
Treaties of amity (1854, 1855), 293, 395
Triangles, right-angled, 75, 76
"Tribute-trade," 14, 150
Tribute
   to the Chinese court, 3, 4, 14, 106, 151
   from Paekche, 45
Trigonometrical functions, 350, 368
Trigonometrical tables, 176, 371
Trigonometry, 180, 361, 370
Tropical year, 47, 69. 70, 355
   two day's discord between the official calendar and, 255
Ts'ao Shih-wei 曹士蔿 (Chinese, cal., fl. 780), 125

Ts'e-yüan hai-ching 測圓海鏡 (Sea-mirror of circle measurements), 130
Tsu Ch'ung-chih 祖沖之 (Chinese, math., 430–501), 76, 81
Tsuboi Shindō 坪井信道 (Dutch Learning, 1795–1848), 336
Tsuchimikado 土御門 household, 123, 125, 199, 255, 257–258, 350, 351, 359
Tsuchimikado officials, and calendrical reform, 256, 257, 313
Tsuchimikado Yasukuni 土御門泰邦 (cal., fl. 1750s), 254
Ts'ui Hao 崔浩 (Chinese, cal.), 50
Tsūjutsu 通術 (general methods [of *wasan*]), 363
Tsukamoto Akitaka 塚本明毅 (cal., 1833–1885), 360
Tsuru Gen'itsu 鶴元逸 (med., fl. 1759), 374
Tu-li-yü-ssu ching 都利聿斯經, 125
"Tung-i ch'uan" 東夷傳 (Record of the eastern barbarians [Japanese] of the *Hou Han shu*), 3n.4
T'ung-t'ien li 統天曆 calendar, 355
Tunneling, 183
Twelve major channels (*ching mo* 經脈), 95, 145
Twelve Terrestrial Branches, 57, 60–61, 62, 63
   special Japanese names for: *tatsu* 建, *nozoku* 除, *mitsu* 滿, *taira* 平, *sadamu* 定, *toru* 執, *yaburu* 破, *ayau* 危, *naru* 成, *osamu* 收, *hiraku* 開, *tozu* 閉, 67
Twenty-four Fortnightly Periods, 62, 67, 68, 256
Two-heavens dualism, 261–262

Uchida Itsumi. *See* Uchida Yatarō
Uchida Yatarō 內田彌太郎 (or Itsumi 五觀; math., 1805–1886), 345–346, 360, 369
Udagawa Genshin 宇田川玄眞 (Dutch Learning, 1769–

1834), 332, 336, 384
Udagawa Genzui 宇田川玄随 (Dutch Learning, 1755–1797), 329, 332, 336, 376, 385, 393
Udagawa Yōan 宇田川榕菴 (Dutch Learning, 1798–1846), 336, 341
Udagawajuku [?] 宇田川塾 (Dutch Learning academy), 335
Uesugi 上杉 family, 119
"Underdevelopment," xiii
United States
   Japan's ports forced open, 292
   trade agreements, 395
   Treaty of amity (1854), 293, 395
Universe
   as macrocosm, 91
   shape and dimensions of, 75
University. See Daigakuryō (Japan), T'ang university (China)
"Unsolved problems" (*idai* 遺題), 206, 263, 266, 271
Uraga 浦賀, 292
   surveying of coast at, 345
Uranus (planet), 356
Urban and rural sectors, differentiation of, 163, 229–230, 293–294
Urban riots, 294
Uri 瓜 (small gourd), fetus' size comparable to, 391
Urinary bladder, 380
Ushigome 牛込
   astronomical observatory at, 312

Vaccination, 220, 377–378
Vaccination center (Shutōjo 種痘所), 336, 378, 398
Vatican, envoys to, 189
Velocity, concept of, 173
Vesalius, Andreas (1514–1564), 221, 381
Vesling, Johann (professor of anatomy, 1598–1649), 381
Vessels, of artificial gold, for conferring immortality, 91
Vietnam, 154
Village states, 2, 24
Vital pneuma (*yüan-ch'i* 元氣; *genki* in Japanese), 282
Vivero y Velasco, Rodrigo (governor-general of the Philippines, ?–1636), 177, 181
Vivisection, 379
Volume, measurement of, 75, 82, 205, 206, 361
Vomiting, 212
Votive plaques. See Sangaku
Voyage de 1803 à 1806 (von Krusenstern), 343
Voyages on the high seas, ban on, 233

Wada Yasushi 和田寧 (math., 1787–1840), 361, 362, 369
Wada-Giles system of romanization, xxxi
Waei gorin shūsei 和英語林集成 (first printed Japanese-English dictionary), 333n.40
Wagaku. See "Japanese Learning"
Wagaku kōdansho. See Lecture Hall for Japanese Learning
Wahō 和方 (Japanese Practice) school of medicine, 309–310
Waka 和歌 (traditional Japanese verse form), 248–249
Wakan gōun 和漢合運, 253
Wakan sanpō taizen 和漢算法大全, 265
Wake 和氣 household, 140, 141, 288, 376
Wake no Hiroyo 和氣廣世 (med., fl. 781), 88
Wakō 倭寇 (pirate-traders), 105–106, 150, 152, 153, 156, 161, 178
   "later *wakō*," 168, 175
Wang Ch'ing-jen 王清任 (Chinese, med., 1768–1831), 379
Wang Hsiao-t'ung 王孝通 (Chinese, math., fl. 625), 76

Index   495

Wang P'ing 王萍, xvii
Wang Ping 王冰 (Chinese, med., fl. 762), 86, 94, 210
Wang Shu-ho 王淑和 (Chinese, med., fl. 3rd century), 85, 281
Wang Yang-ming 王陽明 (Chinese philosopher, 1472–1529), 150, 235, 238, 240–241, 242, 244, 245, 246, 297, 299
Warfare, methods of, 172, 174
Warizansho 割算書 (Manual of division), 205, 206, 255, 271
Warning of heaven, 47, 53
Warring States era (China, 403–221 B.C.), 1, 7, 45, 75, 90
Warring States era (Japan, 1467–1603), 111, 128, 133, 136, 147, 161, 171, 181, 184, 199
Wars, Sengoku era, 168
Warship training center (Gunkan kyōjusho 軍艦教授所), 398
Warships, 179, 396
Wasan 和算 (a special kind of Japanese mathematics), xxviii, 78, 84, 207, 208, 238, 239, 263–265, 271, 301–302, 346, 360–362
  problems and problems-solving, 363, 364, 365, 366
  secrecy and, 275, 365, 366
  status groups and, 274–276
  systematization and simplification of, 363, 364
  and Western mathematics, 315, 360–361, 368, 370, 371, 372
Wasan schools, 366–368
Watanabe Kazan 渡邊華山 (Dutch Learning, 1793–1841), 344, 345, 346
Watches
  mechanical, 180–181
  oldest extant, 181
Watchmaking, 181
Water clocks, 50
Waterways, 164
Wātoru yakuseiron 窊篤兒藥性論 (Wātoru's pharmacology), 385
Weapons, 24, 232
Weather, 43
Weaving, 2, 20, 169
Wei 魏, Northern, 50
Weight system of decimals, 80
West, the
  attitudes toward, 237, 245, 299–301, 310–311
  flow of information about Japan to, 238–239
  need for information about, 396
Western aggression, 292, 293
Western anatomical book, first published translation of, 383
Western astrology, 56
Western astronomy, 176, 178, 180, 198, 201, 203, 258–259, 260, 262, 298, 300, 304, 313, 314, 315, 337, 349, 350, 355, 370, 384
Western aurifaction, 90
Western calendar. *See* Gregorian calendar
Western cosmology, 239, 252, 284, 286, 298, 299, 310, 354
Western cultural influx, xviii, xxi–xxii, xxiv, xxv, xxvi, xxix, 149, 157
Western domination in Asia, 347
Western expansion, into East Asia, 157
Western influence, 224
  fears of, 165, 169
Western learning, 159, 188–194, 239, 301, 303, 304, 310, 360, 394
  challenge to Chinese-style learning, 295–297
  and government policy, 336, 347
  resistance against, 305, 382
"Western Learning" (*Yōgaku* 洋學), 278, 371n.85, 395
Western mathematics, 78, 82, 131, 173–174, 176, 178, 180, 208, 263, 302, 315, 348, 361, 362, 364, 370, 371, 372
  and defense system, 396

496   Index

Western mathematics (*continued*)
　pure, 361, 370
　treatises on, by Matteo Ricci, 208
　　and *wasan*, 368, 371, 372
Western medicine, 87, 95, 99, 217–
　　222, 278, 284–288, 329, 334,
　　336, 340, 348, 373, 376, 377,
　　387
　difficulties experienced by
　　Japanese medical scholars,
　　383–384
　discoveries of the early 19th
　　century, 391
　instruction in, 340, 341, 387, 393,
　　394, 396
　modern, 220
　not extensively studied by *Kōmō*
　　surgeons, 287, 288
　organ-centered, 85, 394
　restrictions on, 347
　shift from Japanese traditional
　　medicine to, 372, 391, 393,
　　398
　after Siebold, 393–395
　transmission of, to Japan, 215,
　　288, 291
　Wahō and, 309
"Western medicine" (*Yōhō* 洋方),
　　323
Western military technology, 370,
　　395–396, 397
Western navigation, 175–181, 232,
　　262, 371
Western science and technology,
　　xiii-xiv, xv, xvi-xvii, xxi-
　　xxiii, xxiv, xxv, xxix, xxx,
　　156, 169–170, 180, 278, 291,
　　294, 296, 301, 302, 304, 305–
　　306, 311, 323, 347–348, 370,
　　372, 395, 398–399
　failure to make permanent
　　impact, in preisolation
　　period, 197–198
　and military strength, 293
　modern, 303
　and national defense, 337, 397
　National Learning and, 308
　seen as a growing menace, 305

shogunate program for adopting,
　　334, 397–398
transmitted to Japan, on
　　advanced level, 313
Western ships, 176, 178, 179, 180
Western surgery, 220–222, 239,
　　278, 284, 287, 310, 329, 347,
　　394
　practice of, 394
Western trade, 155, 156, 157
Western weapons, 232
　and strategy, 395
"White Rabbit of Inaba" (folk
　　tale), 43
Winds, and navigation, 175
Wo 倭 (Wa in Japanese; ancient
　　Chinese name for Japan),
　　3n.4
"Wo-jen ch'uan" 倭人傳 (Record
　　of the Japanese) in the
　　Wei-chih 魏志 (History of
　　Wei), 3n.5
Woodblock printing. *See* Printing
Woolen goods, 226
*Woordenboek der Nederduitsche
　　Fransche laalen: Dictionnaire
　　Flamand et Français*, 332
World War II, 395
Wounds, 98, 215
　combat—, 214, 221
Writing, 21, 29, 43, 113
Written calculations, 267, 268, 271
Written language and symbols, 42,
　　43
Wu 武, Empress, 9
Wu 吳 state, 44
Wu-chi li 五紀曆 calendar, 51
Wu-hsing. *See* Five Phases
Wu-hsing ta-i 五行大意
　　(Fundamental principles of
　　the Five Phases), 124

Xavier, Francis (Jesuit, 1506–
　　1552), 200

Yaita Kinbei 八板金兵衛 (or
　　Kiyosada 清定; swordsmith,
　　fl. mid-16th century), 171

Yakkei taiso 藥經太素 (Outline of materia medica), 88
Yakken 譯鍵 (Translation key), 333, 383
Yakuen 藥園 (herb cultivation) course, 37
Yakushima 屋久島 island, 310
Yamada Masashige 山田正重 (math., fl. 1659), 265
Yamaga Sokō 山鹿素行 (Conf., 1627–1685), 238, 241, 245, 247, 282
Yamagata Bantō 山片蟠桃 (Conf., 1748–1821), 297, 305, 314, 352, 354
Yamaguchi 山口, 171, 184
Yamai 病 (or *byō*), 146
Yamaji Nushizumi 山路主住 (math., 1704–1772), 350, 362, 365, 369
Yamaji Tomotaka 山路諧孝 (cal., mid-19th century), 254, 314, 359
Yamatai 邪馬臺 state, 3n.5
Yamato 大和, 3, 4, 12
 emperors, 5, 12
Yamato court, 12, 21
 ceremonies, 23
 Korean emissaries to, 5
 missions to Chin and Southern courts (5th century), 16
 requests to Paekche to send medical specialists, etc., 44
Yamato honzō 大和本草 (Materia medica of Japan), 238
Yamato-e 大和繪, 104n.2
Yamatoreki 大和暦 (revised calendar system), 256
Yamawaki Tōyō 山脇東洋 (med., 1705–1762), 284, 317, 318, 372, 373, 378, 380, 381
Yamazaki Ansai 山崎闇齋 (Conf., 1618–1682), 244, 245, 247
Yanagawa Shunzō 柳河春三 (math., fl. 1857), 331
Yang Hui 楊輝 (Chinese, math., Southern Sung), 129, 131
Yaso no henpō 耶蘇の變法 ("variation of Catholicism"), 246
Yasuoka 安岡 (surname), 336
Year, length of, 47, 71n.78
 subject to cyclic variation, 355
Yellow Emperor's Inner Classic. *See Huang-ti nei-ching*
Yen 燕 state, 2
Yi 李 dynasty (1392–1910), 110, 153, 185
"*Yin*-nourishing school" (*Yang-yin p'ai* 養陰派; *yōinha* in Japanese), 213
Yin-yang 陰陽 active and passive principles, 7, 8, 9, 46, 53, 55, 57, 77, 124, 135, 240, 259, 260
 correspondences, 92
 divination, 125
 lunar eclipses and, 300
 medicine and, 84, 85, 91–93, 94–95, 97, 210–211, 213, 280, 281, 283, 373
 six-sequence cycle, 210–211
Yin-yüan 隱元 (Ingen in Japanese; priest), 231n.15
Yō no yō 用の用 (mathematics "useful for some use"), 365
Yōgaku. *See* "Western Learning"
Yōhō. *See* "Western medicine"
Yōi shinsho 瘍醫新書 (New text on surgery), 329, 386
Yōinha. *See* "*Yin*-nourishing school"
Yōka shinsen 瘍科新選 (New selections on surgery), 386
Yokohama. *See* Kanagawa
Yōmeigaku 陽明學 (philosophy of Wang Yang-ming, *q.v.*)
Yong Na-san 容螺山 (Korean, fl. 1643), 255
Yoriki 與力 (police officer), 246
Yōrō 養老 Code (718), 12, 16
Yoshida Chōshuku 吉田長叔 (med., 1779–1824), 393
Yoshida Ikyū 吉田意休 (med.), 217
Yoshida Mitsuyoshi 吉田光由 (math., 1598–1672), 205, 206, 253, 255, 263–265

Yoshida Sōkei 吉田宗桂 (med., fl. 1539–1547), 214
Yoshimasu Tōdō 吉益東洞 (med., 1702–1773), 284, 372, 373, 374, 375, 381
Yoshimitsu. *See* Ashikaga Yoshimitsu
Yoshio 吉雄 household, 335n.'c'
Yōyō seigi 幼幼精義 (Details of pediatrics), 386, 393
Yōzan yōhō 洋算用法 (mathematics text), 331
Yu 禹 (legendary emperor), 64
Yü Ch'ang 喩昌 (Chinese, med., fl. 1648), 281
Yü I 游藝 (Chinese, astron., fl. 1675), 261
Yüan 元 dynasty (1279–1368), 105, 108, 109, 110, 129, 170, 172, 175, 184, 235
  astronomy, 121, 122, 355
  mathematics, 132, 133, 204
  medicine, 139, 209, 212, 213, 214, 215, 277, 278, 279, 280–281, 282, 283, 388
Yüan-chia li 元嘉曆 calendar, 51
Yume monogatari 夢物語 (Tale of a dream), 345
Yurian kōjōden 由利安攻城傳 (Report on Juliaen's siege warfare), 232
Yūrin 有隣 (med., fl. 1360s), 146
Yushima Seidō 湯島聖堂, 238, 243–244, 288, 295, 298, 299 302, 303

Za 座 (merchant guild), 230n.12
Zen 禪 Buddhism, 10n.12, 113
Zen priests, 58, 120, 127–128, 135, 136, 151, 158, 187, 195, 239, 241
Zero
  "dot" symbol for, 49, 76
  denoted on counting board, 79
  "0" for, 130
Zodiac, 50
Zoology, 305, 340, 341
Zōshi 臟志 (On the viscera), 317, 318, 380, 381, 382
Zōzu 臟圖 (Charts of the viscera), 317, 318